A Concise Geologic
Time Scale

A Concise Geologic Time Scale

2016

James G. Ogg, Gabi M. Ogg,
Felix M. Gradstein

ELSEVIER

AMSTERDAM • BOSTON • HEIDELBERG • LONDON • NEW YORK • OXFORD • PARIS
SAN DIEGO • SAN FRANCISCO • SINGAPORE • SYDNEY • TOKYO

Elsevier
Radarweg 29, PO Box 211, 1000 AE Amsterdam, Netherlands
The Boulevard, Langford Lane, Kidlington, Oxford OX5 1GB, UK
50 Hampshire Street, 5th Floor, Cambridge, MA 02139, USA

Notices
Knowledge and best practice in this field are constantly changing. As new research and experience broaden our understanding, changes in research methods, professional practices, or medical treatment may become necessary.

Practitioners and researchers must always rely on their own experience and knowledge in evaluating and using any information, methods, compounds, or experiments described herein. In using such information or methods they should be mindful of their own safety and the safety of others, including parties for whom they have a professional responsibility.

To the fullest extent of the law, neither the Publisher nor the authors, contributors, or editors, assume any liability for any injury and/or damage to persons or property as a matter of products liability, negligence or otherwise, or from any use or operation of any methods, products, instructions, or ideas contained in the material herein.

British Library Cataloguing-in-Publication Data
A catalogue record for this book is available from the British Library

Library of Congress Cataloging-in-Publication Data
A catalog record for this book is available from the Library of Congress

ISBN: 978-0-444-63771-0

For information on all Elsevier publications
visit our website at https://www.elsevier.com/

  Working together
to grow libraries in
Book Aid International developing countries

www.elsevier.com • www.bookaid.org

Publisher: Candice Janco
Acquisition Editor: Louisa Hutchins
Editorial Project Manager: Marisa LaFleur
Production Project Manager: Mohanapriyan Rajendran
Designer: Greg Harris

Typeset by TNQ Books and Journals

CAPTION for COVER PHOTO:
Late Triassic in Italian Dolomites. The Lagazuoi peak is the resumption of late Carnian and Norian shallow-water carbonates (Dolomia Principale) following a regional termination of prograding carbonate platforms and influx of siliciclastics (the slope-forming Heiligkreuz and Travenanzes formations at its base). The brief carbonate crisis is part of a global mid-Carnian warming and humid episode that appears to coincide with the eruption of the Wrangellia large igneous province. Photo courtesy of Austin McGlannan.

CONTENTS

INTRODUCTION

Geologic time scale and this book

A standardized geologic time scale is the framework for deciphering and understanding the long and complex history of our planet Earth. We are constantly improving our knowledge of that history including the intertwined feedbacks among the evolution of life, the climatic and geochemical trends and oscillations, the sea-level withdrawals and transgressions, the drifting tectonic plates and major volcanic upheavals, and the radioisotopic and astronomical-cycle dating of deposits. In turn, this knowledge of past relationships and feedbacks enable us to make predictions about our own future impacts on our planet.

The challenges and major accomplishments of geoscientists are to integrate these diverse interpretations of the global stratigraphic record, to apply an age model ("linear time") to that geologic record, and to assign a standardized and precise international terminology of subdivisions. The publications of *A Geologic Time Scale 1989* (Harland et al., 1989), of *A Geologic Time Scale 2004* (Gradstein et al., 2004; under the scientific auspices of the International Commission on Stratigraphy, ICS), and of *The Concise Geologic Time Scale* (Ogg et al., 2008) spurred dedicated research and collective activities to bring about improvements in stable-isotope stratigraphy, radioisotopic and cyclostratigraphic dating of stage boundaries, and formal definitions of stage boundaries.

Any synthesis of this geologic time scale is a status report in this grand undertaking. This *Concise Geologic Time Scale 2016* handbook presents a brief summary of the current scale and some of the most common means of global stratigraphic correlation and age calibration as graphics with brief explanatory texts. These rely extensively on the two-volume *Geologic Time Scale 2012* (*GTS2012*) compilation (Gradstein et al., 2012), and readers who desire more background or details should use that reference. This handbook does incorporate some selected important advances in stratigraphic scale calibration, in new ratified or candidate international divisions and in their scaling to numerical time.

Each chapter in this handbook, which generally spans a single geologic system or period, includes:

1. International divisions of geologic time, with graphics for ratified bases of series/epoch definitions (Global Boundary Stratotype Sections and Points (GSSPs)). GSSPs for stage-level divisions are diagrammed in *GTS2012*, at the website of the Geologic TimeScale Foundation, and at the websites of the ICS subcommissions.
2. Major paleontological zonations, geomagnetic polarity reversals, selected geochemical trends (usually isotopic ratios of carbon and of oxygen), interpreted sea-level history, and other events or zones.
3. Explanation of the derivation and uncertainties for the current numerical age model of the stratigraphic boundaries and events, and a summary of any incorporated revised ages assigned to stage boundaries compared to GTS2012.
4. Selected references and websites for additional information.

The stratigraphic scales in the diagrams are a small subset of the compilations and databases in GTS2012 and other syntheses. One can generate custom charts from these databases using the public *TimeScale Creator* visualization system available at www.tscreator.org (which mirrors to https://engineering. purdue.edu/Stratigraphy/tscreator/).

International divisions of geologic time and their global boundaries (GSSPs)

A common and precise language of geologic time is essential to discuss Earth's history. Hence, a chart of international ratified stratigraphic units (e.g., Fig. 1.1) is a vital part of the scientific toolbox carried by each earth scientist to do his or her job. Ideally, each stage boundary is defined at a precise Global Boundary Stratotype Section and Point (GSSP) (e.g., McLaren, 1978; Remane, 2003). This GSSP is a point in the rock record of a specific outcrop at a level selected to coincide with one or more primary markers for global correlation (lowest occurrence of a fossil, onset of a geochemical anomaly, a distinctive geomagnetic polarity reversal, etc.). The majority of ratified GSSP placements and the terminology for the geologic stages of Silurian through Quaternary were selected to correspond closely to traditional European usage (e.g., Emsian, Campanian, Selandian). In contrast, those in the Cambrian and Ordovician were established after an international effort to identify a set of global events that could be reliably correlated, therefore many of the ratified GSSPs have new stage names (e.g., Fortunian, Katian) (Fig. 1.1).

Divisions of the preserved rock record, geologic time, and assigned numerical ages are separate but related concepts which are united through the GSSP concept. Chronostratigraphic ("rock time") units are the rocks formed during a specified interval of geologic time. Therefore, the Jurassic "System" is the body of rocks that formed during the Jurassic "Period." A similar philosophy of clarifying whether one is discussing rocks or time applies to stratigraphic successions in which the terms of "lower, upper, and lowest occurrence" have corresponding geologic time terms of "early, late, and first appearance" when describing the geologic history. (Note: The international geochronologic unit for the chronostratigraphic "stage" is confusingly called an "age"; therefore, those columns are labeled "stage/age" on our diagrams to distinguish from the adjacent "age" column that is measured in millions of years.)

Biologic, chemical, sea-level, geomagnetic, and other events or zones

Geologic stages are recognized, not by their boundaries, but by their content. The rich fossil record remains the main method to distinguish and correlate strata among regions, because the morphology of each taxon is the most unambiguous way to assign a relative age. The evolutionary successions and assemblages of each fossil group are generally grouped into zones. We have included selected zonations and/or event datums (first or last appearances of taxa) for widely used biostratigraphic groups in each system/period. However, as vividly illustrated by many studies, most biological first/last appearance datums are diachronous on the local to regional levels due to migrations or facies dependences of the taxa, to different taxonomic opinions among paleontologists, and other factors. In some cases, GSSPs that had been ratified based on their presumed coincidence with a single primary biostratigraphic marker are now being reevaluated or reassigned when it was discovered that the sole marker was not

PHANEROZOIC and PRECAMBRIAN CHRONOSTRATIGRAPHY

Table 1 (Cenozoic – Mesozoic)

Eon	Era	Period	Series/Epoch	Stage/Age	Age Ma	GSSP
Phanerozoic	Cenozoic	Quaternary	Holocene	*Anthropocene* *		
				Upper		
				Middle	4.2 ka	
				Lower	8.2 ka	
					11.8 ka	
			Pleistocene	Upper	126 ka	
				"Ionian"	773 ka	
				Calabrian	1.80	
				Gelasian	2.58	
		Neogene	Pliocene	Piacenzian	3.60	
				Zanclean	5.33	
			Miocene	Messinian	7.25	
				Tortonian	11.63	
				Serravallian	13.82	
				Langhian	15.97	
				Burdigalian	20.44	
				Aquitanian	23.03	
		Paleogene	Oligocene	Chattian	28.1	
				Rupelian	33.9	
			Eocene	Priabonian	38.0	
				Bartonian	41.0	
				Lutetian	47.8	
				Ypresian	56.0	
			Paleocene	Thanetian	59.2	
				Selandian	61.6	
				Danian	66.0	
	Mesozoic	Cretaceous	Upper	Maastrichtian	72.1	
				Campanian	84.2	
				Santonian	86.5	
				Coniacian	89.8	
				Turonian	93.9	
				Cenomanian	100.5	
			Lower	Albian	113.1	
				Aptian	126.3	
				Barremian	130.8	
				Hauterivian	134.7	
				Valanginian	139.4	
				Berriasian	145.0	
		Jurassic	Upper	Tithonian	152.1	
				Kimmeridgian	157.3	
				Oxfordian	163.1	
			Middle	Callovian	166.1	
				Bathonian	168.3	
				Bajocian	170.3	
				Aalenian	174.2	
			Lower	Toarcian	183.7	
				Pliensbachian	191.4	
				Sinemurian	199.4	
				Hettangian	201.4	
		Triassic	Upper	Rhaetian	~ 209.6	
				Norian	~ 228.5	
				Carnian	237.0	

* Anthropocene under discussion

Table 2 (Mesozoic – Paleozoic)

Eon	Era	Period	Series/Epoch	Stage/Age	Age Ma	GSSP
	Mesozoic	Triassic	Middle	Ladinian	237.0	
				Anisian	241.5	
					246.8	
			Lower	Olenekian	249.8	
				Induan	251.9	
		Permian	Lopingian	Changhsingian	254.2	
				Wuchiapingian	259.8	
			Guadalupian	Capitanian	265.1	
				Wordian	268.8	
				Roadian	272.3	
			Cisuralian	Kungurian	282.0	
				Artinskian	290.1	
				Sakmarian	295.0	
				Asselian	298.9	
Phanerozoic	Paleozoic	Carboniferous	Pennsylvanian Upper	Gzhelian	303.4	
				Kasimovian	306.7	
			Middle	Moscovian	314.6	
			Lower	Bashkirian	323.2	
			Mississippian Upper	Serpukhovian	330.9	
			Middle	Visean	346.7	
			Lower	Tournaisian	358.9	
		Devonian	Upper	Famennian	372.2	
				Frasnian	382.7	
			Middle	Givetian	387.7	
				Eifelian	393.3	
			Lower	Emsian	407.6	
				Pragian	410.8	
				Lochkovian	419.2	
		Silurian	Pridoli		423.0	
			Ludlow	Ludfordian	425.6	
				Gorstian	427.4	
			Wenlock	Homerian	430.5	
				Sheinwoodian	433.4	
			Llandovery	Telychian	438.5	
				Aeronian	440.8	
				Rhuddanian	443.8	
		Ordovician	Upper	Hirnantian	445.2	
				Katian	453.0	
				Sandbian	458.4	
			Middle	Darriwilian	467.3	
				Dapingian	470.0	
			Lower	Floian	477.7	
				Tremadocian	485.4	
		Cambrian	Furongian	Stage 10	489.5	
				Jiangshanian	494	
				Paibian	497	
			Series 3	Guzhangian	500.5	
				Drumian	504.5	
				Stage 5	509	
			Series 2	Stage 4	514	
				Stage 3	~ 520	
			Terreneuvian	Stage 2	~ 530	
				Fortunian	541.0	

Table 3 (Precambrian)

Eon	Erathem/Era	System/Period	Age Ma	GSSP/GSSA
Proterozoic	Neoproterozoic	Ediacaran	541	
		Cryogenian	635	
		Tonian	720	
	Mesoproterozoic	Stenian	1000	
		Ectasian	1200	
		Calymmian	1400	
	Paleoproterozoic	Statherian	1600	
		Orosirian	1800	
		Rhyacian	2050	
		Siderian	2300	
Archean	Neoarchean		2500	
			2800	
	Mesoarchean		3200	
	Paleoarchean		3600	
	Eoarchean		4000	
	Hadean (informal)		~4560	

Units of the international chronostratigraphic scale with estimated numerical ages.

Colors are according to the Commission for the Geological Map of the World.

Subdivisions of the Phanerozoic (~541 Ma to Present) and the base of the Ediacaran are defined by a basal Global Boundary Stratotype Section and Point (GSSP), whereas the Precambrian units are formally subdivided by absolute age (Global Standard Stratigraphic Age, GSSA).

Stratigraphic information and details on international and regional geologic units can be found on the websites of the **Geologic TimeScale Foundation** *https://engineering.purdue.edu/stratigraphy* and the **ICS** *www.stratigraphy.org.*

Figure 1.1 Units of the international chronostratigraphic time scale with estimated numerical ages from the GTS2016 age model.

reliable for precise correlations. These are discussed within the relevant chapters.

Trends and excursions in stable-isotope ratios, especially of carbon 12/13 and strontium 86/87, have become an increasingly reliable method to correlate among regions. Carbon 12/13 stratigraphy, like magnetostratigraphy, can be utilized in both marine and nonmarine basins. Some of the carbon-isotope excursions are associated with widespread deposition of organic-rich sediments and with eruptions of large igneous provinces. The largest magnitude excursions occur during the Proterozoic through Silurian, but the causes of some of these remain speculative.

Ratios of oxygen 16/18 are particularly useful for the glacial–interglacial cycles of Pliocene–Pleistocene, and are important in the interpretation of past temperature trends through the Phanerozoic. However, the conversion of oxygen-isotope ratios to temperature requires knowing the oxygen-isotope composition of seawater through time. The tropical seawater temperatures derived from Paleozoic and Mesozoic data from phosphatic and carbonate fossils that assume an ocean oxygen-isotope composition similar to the Cenozoic tend to be anomalously warm, indeed at levels that would be lethal to modern marine life. Therefore, Veizer and Prokoph (2015) hypothesized that there has been a progressive drift in ocean chemistry and that the derived temperature values should be adjusted. We have shown comparisons of the derived and the adjusted temperatures in some of the diagrams in this book.

Sea-level trends, especially rapid oscillations that caused widespread exposure or drowning of coastal margins, are associated with these isotopic-ratio excursions in time intervals characterized by glacial advances and retreats. The synchronicity and driving cause of such stratigraphic sequences in intervals that lack major glaciations are disputed.

We have included major sequences as interpreted by widely used selected publications, but many of these remain to be documented as global eustatic sea-level oscillations. A discussion of eustasy and sequences is by Simmons (in *GTS2012*).

Geomagnetic polarity chronozones (chrons) are well established for correlation of the magnetostratigraphy of fossiliferous strata to the magnetic anomalies of Late Jurassic through Holocene. Pre-Late Jurassic magnetic polarity chrons have been verified in some intervals, but exact correlation to biostratigraphic zonations remains uncertain for many of these. The geomagnetic scales on the diagrams in this booklet are partly an update of those compiled for *GTS2012*.

Assigned numerical ages

Although the GSSP concept standardizes the units of both chronostratigraphy and geologic time, the numerical age model ("linear time") assigned to those boundaries and events is a more abstract interpretation based on extrapolation from radioisotopic-dated levels, astronomical cycles, relative placement in magnetic polarity zones, or other methods. Those age models are always being refined; but ideally the ratified GSSPs are fixed. *GTS2012* presented a suite of comprehensive age models for each Phanerozoic period and for the Cryogenian and Ediacaran periods of the Proterozoic.

Numerical ages in this book are abbreviated as "a" (for annum), "ka" for thousands, "Ma" for millions, and "Ga" for billions of years before present. The moving "Present" has led many Holocene workers to use a "BP2000," which assigns "Present" to the year AD 2000. For clarity, elapsed time or duration is abbreviated as "yr" (for year), "kyr" (thousands of years) or "myr" (millions of years). Ages are given in years before "Present" (BP).

In the years between the assembly of *GTS2012* in late 2011 and the preparation of this concise handbook in late 2015, many significant enhancements have occurred. These include enhanced astronomical time scales, publication of additional or refined radioisotopic dates, revised definitions for some stage boundaries through ratified GSSPs or new preferred primary markers for candidate GSSPs, and other advances. Even though we preferred to be conservative and retain as many ages from *GTS2012* as possible, some of these significant advances in geochronology were incorporated. Therefore, in addition to rescaling of zonations and events within stages, some of the assigned numerical ages for some geologic stage boundaries required revisions from the age models used in *GTS2012* (Table 1.1). Each chapter includes a brief explanation of uncertainties in such age assignments and possible future improvements in precision and accuracy.

TimeScale Creator database and chart-making package

Onscreen display and production of user-tailored timescale charts is provided by *TimeScale Creator*, a public JAVA package available at www.tscreator.org (which mirrors to https://engineering.purdue.edu/Stratigraphy/tscreator/). The internal database includes over 200 columns of all major biostratigraphic zonations, regional scales, geomagnetic polarity scales, geochemical trends, sea-level interpretations, major large igneous provinces, hydrocarbon occurrences, etc. Additional online data packs can be added that have the lithostratigraphy of regions scaled to the standardized GTS (e.g., map-interfaces to all Australia basins in collaboration with Geoscience Australia, British basins with the British Geological Survey, etc.), microfossil zonations with images of all taxa and links to Nannotax and other external websites for each taxon, human civilization scales, evolutionary charts of life, etc.

In addition to screen views and a scalable-vector graphics (SVG) file for importation into popular graphics programs, the onscreen display has a variety of display options and "hot-curser-points" to open windows providing additional information on definitions and method of assigning ages to zones and events. Cross-plotting routines enable conversion of outcrop/well data to standardized geologic time diagrams. Tutorials provide instruction on making one's own data packs.

The database and visualization package are envisioned as a convenient reference tool, chart-production assistant, and a window into the geologic history of our planet. These are progressively enhanced through the efforts of stratigraphic and regional experts, and contributions are welcome.

Geologic Time Scale 2020

At the time of this writing, a major comprehensive update of the Geologic Time Scale is underway, targeted for publication in 2020 in collaboration with Elsevier Publishing. A majority of international stage boundaries (GSSPs) should be established by that date, including the base of the Berriasian (base of the Cretaceous). The entire Cenozoic and significant portions of the Mesozoic–Paleozoic will have high-resolution scaling based on astronomical tuning or orbital cycles. The book will be a full-color, enhanced, improved, and expanded version of *GTS2012*, with detailed coverage of zonal biostratigraphy, stable and unstable isotope stratigraphy, sequence stratigraphy, global eustasy, and many other integrated aspects of Earth's fascinating history.

**Table 1.1 Modified ages of stage boundaries in this book relative
to *The Geologic Time Scale 2012***

Chronostratigraphic unit	Age in this book	Age in *GTS 2012*	Summary
Middle Pleistocene	0.773	0.781	Enhanced accuracy
Calabrian	1.80	1.806	Enhanced accuracy
Gelasian	2.58	2.59	Enhanced accuracy
Priabonian	37.97	37.7	Changed marker for base
Bartonian	41.03	41.15	Revised cyclostratigraphic dating
Campanian	84.19	83.6	Revised radioisotopic dating
Santonian	86.49	86.3	Changed marker for base
Coniacian	89.75	89.8	Enhanced accuracy
Albian	113.14	113.0	Placement change for boundary
Hauterivian	134.7	133.9	Revised ammonite and cyclostratigraphic dating
Oxfordian	163.1	163.5	Revised boundary definition
Toarcian	183.7	182.7	Revised radioisotopic dating
Pliensbachian	191.36	190.8	Revision of stage boundaries
Sinemurian	199.4	199.3	Revision of stage boundaries
Hettangian	201.36	201.31	Revised radioisotopic dating
Anisian	246.8	247.1	Revision of stage boundaries
Olenekian	249.8	250.0	Revision of stage boundaries
Induan	251.902	252.16	Revised radioisotopic dating
Changhsingian	254.15	254.2	
Kungurian	282.0	279.3	Revised spline fit
Gzhelian	303.4	303.7	Revised cyclostratigraphic dating
Kasimovian	306.7	307.0	Changed marker for base
Moscovian	314.6	315.2	Changed marker for base
Stage 3 (base of Series 2)	ca. 520	521	Implied precision on this estimate is removed
Stage 2	ca. 530	529	Implied precision on this estimate is removed
Cryogenian	720	850	Change of boundary definition

Selected publications and websites

Cited references

Gradstein, F.M., Ogg, J.G., Smith, A.G. (Eds.), 2004. *A Geologic Time Scale 2004*. Cambridge University Press, Cambridge. 589 pp.

Gradstein, F.M., Ogg, J.G., Schmitz, M.D., Ogg, G.M., (coordinators), with, Agterberg, F.P., Anthonissen, D.E., Becker, T.R., Catt, J.A., Cooper, R.A., Davydov, V.I., Gradstein, S.R., Henderson, C.M., Hilgen, F.J., Hinnov, L.A., McArthur, J.M., Melchin, M.J., Narbonne, G.M., Paytan, A., Peng, S., Peucker-Ehrenbrink, B., Pillans, B., Saltzman, M.R., Simmons, M.D., Shields, G.A., Tanaka, K.L., Vandenberghe, N., Van Kranendonk, M.J., Zalasiewicz, J., Altermann, W., Babcock, L.E., Beard, B.L., Beu, A.G., Boyes, A.F., Cramer, B.D., Crutzen, P.J., van Dam, J.A., Gehling, J.G., Gibbard, P.L., Gray, E.T., Hammer, O., Hartmann, W.K., Hill, A.C., Paul, F., Hoffman, P.F., Hollis, C.J., Hooker, J.J., Howarth, R.J., Huang, C., Johnson, C.M., Kasting, J.F., Kerp, H., Korn, D., Krijgsman, W., Lourens, L.J., MacGabhann, B.A., Maslin, M.A., Melezhik, V.A., Nutman, A.P., Papineau, D., Piller, W.E., Pirajno, F., Ravizza, G.E., Sadler, P.M., Speijer, R.P., Steffen, W., Thomas, E., Wardlaw, B.R., Wilson, D.S., Xiao, S., 2012. *The Geologic Time Scale 2012*. Elsevier, Boston, USA. 1174 p. (2-volume book).

Harland, W.B., Armstrong, R.L., Cox, A.V., Craig, L.E., Smith, A.G., Smith, D.G., 1989. *A Geologic Time Scale 1989*. Cambridge University Press. 263 pp. (and their previous *A Geologic Time Scale 1982*).

McLaren, D.J., 1978. Dating and correlation, a review. In: Cohee, G.V., Glaessner, M.F., Hedberg, H.D. (Eds.), *Contributions to the Geologic Time Scale*. Studies in Geology, **vol. 6**. AAPG, Tulsa, pp. 1–7.

Ogg, J.G., Ogg, G., Gradstein, F.M., 2008. *The Concise Geologic Time Scale*. Cambridge University Press. 177 p. (book). Translated in Japanese in 2012.

Remane, J., 2003. Chronostratigraphic correlations: their importance for the definition of geochronologic units. *Palaeogeography, Palaeoclimatology, Palaeoecology* **196**: 7–18.

Veizer, J., Prokoph, A., 2015. Temperatures and oxygen isotopic composition of Phanerozoic oceans. *Earth-Science Reviews* **146**: 92–104.

Websites (selected)

In addition to many excellent books on historical geology, paleontology, individual periods of geologic time and other aspects of stratigraphy, there is now an extensive suite of websites on the history of Earth's surface and its life. These are continuously updated and enhanced. Some selected ones (biased slightly toward North America) are:

Geologic TimeScale Foundation—engineering.purdue. edu/Stratigraphy—diagrams of GSSPs for all stage boundaries, time-scale charts, and supporter of GTS2012/GTS2020 syntheses.

TimeScale Creator—www.tscreator.org—free JAVA program for Earth history visualization, suites of enhanced datasets, online "TSC-Lite," etc. (*mirrors to site at Purdue University*)

International Commission on Stratigraphy—www. stratigraphy.org—for current status of all stage boundaries, the *International Stratigraphic Guide*, links to subcommission websites, etc.

Palaeos—The Trace of Life on Earth (originally compiled by Toby White)—www.palaeos.com—and others it references at end of each period. There is also a WIKI version being compiled at Palaeos.org. The *Palaeos* suite has incredible depth and is written for the general scientist.

Smithsonian Paleobiology—"Geologic Time"—paleobiology.si.edu/geotime—After entering, select the desired Period or Eon by clicking on (*Make a Selection*) in upper right corner of screen.

Museum of Paleontology, **University of California**— www.ucmp.berkeley.edu/exhibits—thousands of pages about history of life on Earth. Main "exhibit" sections include Life through Time, Tour of Geologic Time, and Understanding Evolution.

Paleontology Portal—paleoportal.org—Exploring North American geologic history (with geologic maps of each state), main fossil sites, and fossil gallery.

Palaeocast—www.palaeocast.com—A free web series exploring the fossil record and the evolution of life on earth through an extensive suite of well-presented paleontology podcasts (ca. 60) with accompanying slideshows, news stories, and a future Virtual Natural History Museum. Launched in 2012 with education and outreach grants from the Paleontological Society and the Palaeontological Association.

Fossilworks (*Paleobiology Database* gateway)— fossilworks.org—A suite of search and analytical tools for using the large relational PaleoDB database of global fossil occurrences (ca. 350,000 taxa; 57,000 references; contributed by over 400 scientists in 30 countries) to generate paleomaps, diversity curves, etc. The online tool sets were developed by John Alroy.

PAST (PAlaeontological STatistics)—folk.uio.no/ohammer/past—free software for scientific data analysis, with functions for data manipulation, plotting, univariate and multivariate statistics, ecological analysis, time series and spatial analysis, morphometrics and stratigraphy; developed by Øyvind Hammer, Natural History Museum, University of Oslo.

Fossil Mall—www.fossilmall.com—Even though it is a commercial site, it maintains an extensive educational outreach content with superb photographs and an impressive synthesis column of major events in Earth history and evolution.

Virtual Fossil Museum—www.fossilmuseum.net—"An Education Resource Dedicated to the Diversity of Life" with extensive photographs and details of macrofossils organized by period, by tree-of-life, and by taxa group. Numerous contributors, coordinated by Roger Perkins (bioinformatics with evolutionary biology interest) since 1999, and constantly expanded and enhanced through 2015 (last viewed).

Wikipedia online encyclopedia (a public effort)—en.wikipedia.org/wiki/Geologic_time_scale—directs users to excellent summaries of each geologic period and most stages, plus links at the bottom of each page to other relevant sites.

Plate Reconstructions (images and animations), some selected sites:

Paleomap Project (by Christopher **Scotese**)—www.scotese.com/.

Reconstructing the Ancient Earth (Ron **Blakey**)—http://cpgeosystems.com.

Additional collections of links to stratigraphy of different periods and paleontology of various phyla are at www.geologylinks.com, and other sites. The world-wide web array of posted information grows daily. However, as lamented at the current Virtual Fossil Museum homepage "Back in 1999, there was a nice site maintained by UC Berkeley, and a number of other sites that, like VFM, were built and maintained by passionate amateurs. Most of these amateur sites are long gone, and some can't even be found in Internet archives." Fortunately, some like Palaeos, were resurrected and maintained by the next generation of enthusiasts.

2

PLANETARY TIME SCALE

K.L. Tanaka[1], W.K. Hartmann[2]

[1]U.S. Geological Survey, Flagstaff, AZ, United States; [2]Planetary Science Institute, Tucson, AZ, United States

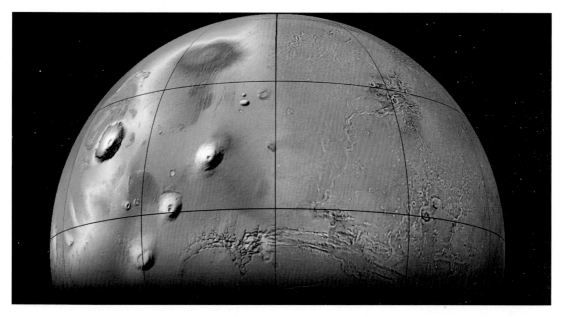

Northern part of the western hemisphere of Mars. Left half shows a color elevation, shaded-relief view highlighting the immense volcanic shields of the Tharsis rise. Right half shows a true-color view of the vast Valles Marineris and Kasei Valles canyon systems, which connect to the dark basin of Chryse Planitia at upper right. From Tanaka et al., 2014; Image data from National Aeronautics and Space Administration (NASA).

Introduction

Formal stratigraphic systems have been developed for the surfaces of Earth's Moon, Mars, and Mercury (Fig. 2.1). The time scales are based on regional and global geologic mapping, which establishes relative ages of surfaces delineated by superposition, transaction, morphology, and other relations and features. Referent map units are used to define the commencement of events and periods for definition of chronologic units. The following summary is mainly reproduced from Tanaka and Hartmann (2008, 2012).

Relative ages of these units in most cases can be confirmed using size–frequency distributions and superposed craters. For the Moon, the chronologic units and cratering record are constrained by radiometric ages measured from samples collected from the lunar surface. This allows a calibration of the areal density of craters versus age, which

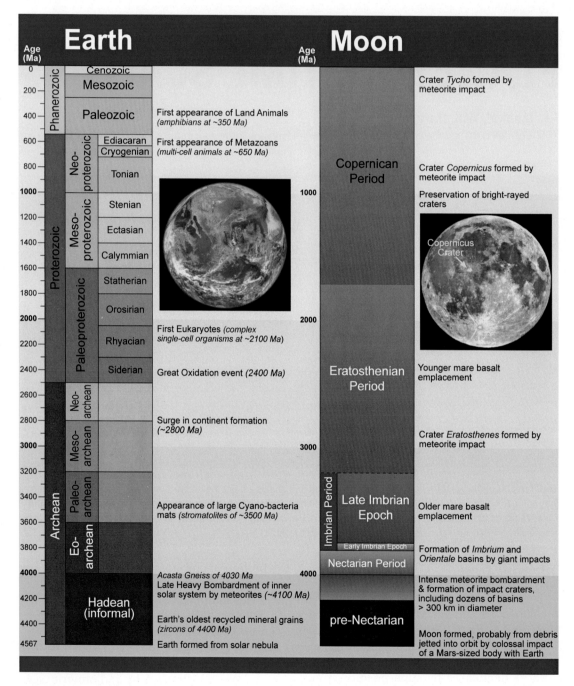

Figure 2.1 *(Continued)*

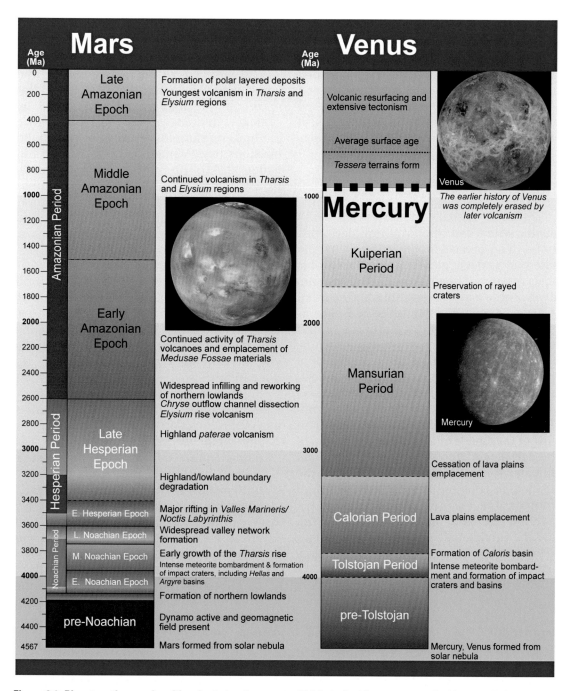

Figure 2.1 Planetary time scale with selected major events. Thick dashed line separates the Venus and Mercury time scales. Diagram revised by G. Ogg from Tanaka and Hartmann (2012).

permits model ages to be measured from crater data for other lunar surface units. Model ages for other cratered planetary surfaces are constructed by two methods: (1) estimating relative cratering rates with Earth's Moon and (2) estimating cratering rates directly based on surveys of the sizes and trajectories of asteroids and comets (e.g., Hartmann, 2005).

The Moon

The first formal extraterrestrial stratigraphic system and chronology was developed for Earth's Moon beginning in the 1960s, first based on geologic mapping using telescopic observations (Shoemaker and Hackman, 1962). These early observations showed that the rugged lunar highlands are densely cratered, whereas the *maria* (Latin for "seas") form relatively dark, smooth plains consisting of younger deposits that cover the floors of impact basins and intercrater plains. Resolving power of the lunar landscape improved greatly with the Lunar Orbiter spacecraft (Fig. 2.2), which permitted also the first mapping of the farside of the Moon. By the end of the decade and into the 1970s, manned and unmanned exploration of lunar sites by the Apollo and Luna missions brought return of samples. The majority of early exploration involved the lunar nearside (facing Earth), and the stratigraphic system and chronology follow geologic features and events primarily expressed on the nearside (see Fig. 2.3).

The cratering rate was initially very high; uncertain is whether the lunar cratering rate records a relatively brief period of catastrophic "Late Heavy Bombardment" in the inner solar system at ~4.0 Ga, possibly spawned by perturbations in the orbits of the giant outer planets (e.g., Strom et al., 2005). Alternatively, the dense population of highland craters records the gradual trailing off of the accretionary period itself. Telescopic surveys of the numbers, sizes, and

orbits of asteroids indicate that they have been the prime contributor to the lunar cratering record.

The materials of the early crust and the emplacement of extensive lava flows that make up the lunar maria were dated by geologic inferences and by radiometric methods on samples returned by the Apollo missions (e.g., Wilhelms, 1987; Stöffler and Ryder, 2001). Attempts were also made to use the samples to date certain lunar basin-forming impacts and the large craters, Copernicus and Tycho. Two processes have mainly accomplished resurfacing: impacts and volcanism. Analogous to volcanism, impact heating can generate flow-like deposits of melted debris that can infill crater floors or terrains near crater rims. As on Earth, the broadest time intervals are designated "Periods" and their subdivisions are "Epochs" (if not meeting formal stratigraphic criteria, these unit categories are not capitalized).

From oldest to youngest, lunar chronologic units and their referent surface materials and events include:

1. *pre-Nectarian* period, earliest materials dating from solidification of the crust (a suite of anorthosite, norite, and troctolite) until just before formation of the Nectaris basin;
2. *Nectarian* Period, mainly impact melt and ejecta associated with Nectaris basin and later impact features;
3. *Early Imbrian* Epoch, consisting mostly of basin-related materials associated at the beginning with Imbrium basin and ending with Orientale basin;
4. *Late Imbrian* Epoch, characterized by mare basalts post-dating Orientale basin;
5. *Eratosthenian* Period, represented by dark, modified ejecta of Eratosthenes crater; and
6. *Copernican* Period, characterized by relatively fresh bright-rayed ejecta of Copernicus crater.

Figure 2.2 Lunar stratigraphy: (A) Photograph of the Moon. Provided by Gregory Terrance (Finger Lakes Instrumentation, Lima, New York; www.fli-cam.com).

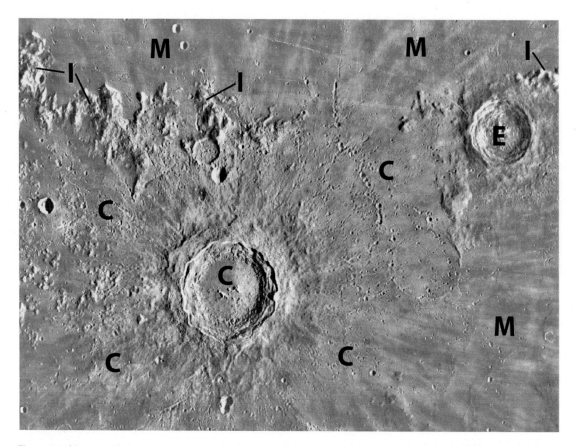

Figure 2.2 (Continued) **(B) Copernicus region of the Moon**. Approximate location of this region is shown on a photograph of the Moon. Copernicus crater (C) is 93 km in diameter and centered at latitude (lat) 9.7°N, longitude (long) 20.1°W. Copernicus is representative of bright-rayed crater material formed during the lunar Copernican Period. Its ejecta and secondary craters overlie Eratosthenes crater (E), which is characteristic of relatively dark crater material of the Eratosthenian Period. In turn, Eratosthenes crater overlies relatively smooth mare materials (M) of the Late Imbrian Epoch. The oldest geologic unit in the scene is the rugged rim ejecta of Imbrium basin (I), which defines the base of the Early Imbrian Epoch (Lunar Orbiter IV image mosaic; north at top; illumination from right; courtesy of US Geological Survey (USGS) Astrogeology Team).

Mars

The Red Planet has a geologic character similar to the Moon, with vast expanses of cratered terrain and lava plains, but with the important addition of features resulting from the activity of wind and water over time. This results in a geologically complex surface history; geologic mapping has assisted in unraveling it, following the approaches developed for studies of the Moon (Fig. 2.4). Beginning in the 1970s with the Mariner 9 and Viking spacecraft, and continuing with a flotilla of additional orbiters and landers beginning in the 1990s, Mars has become a highly investigated planet. Geologic mapping led to characterization of periods and epochs as on the Moon (e.g., reviews in Tanaka, 1986; Kallenbach et al., 2001; Nimmo and Tanaka, 2005; Tanaka et al., 2014) (Fig. 2.1).

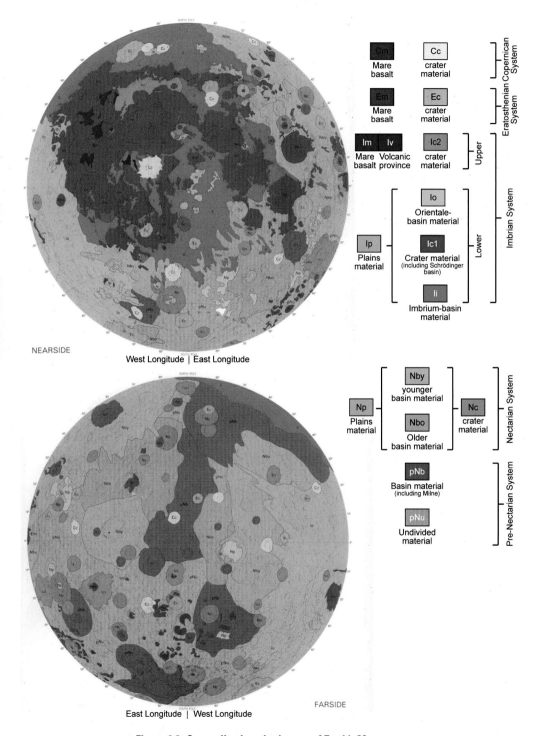

Figure 2.3 Generalized geologic map of Earth's Moon.

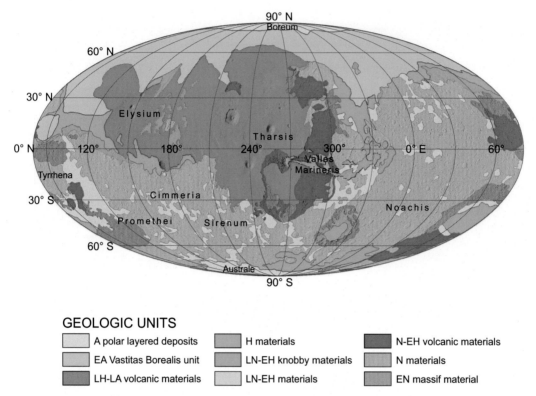

Figure 2.4 Global geologic map of Mars. Generalized geologic map of Mars showing distribution of major material types and their ages. Chronologic unit abbreviations: *N*, Noachian; *H*, Hesperian; *A*, Amazonian; *E*, Early; *L*, Late. *(Adapted from Nimmo and Tanaka (2005).)* Terrain names shown without descriptor terms. Mollweide projection, using east longitudes, centered on 260°E, Mars Orbiter Laser Altimeter (MOLA) shaded-relief base illuminated from the East. On Mars, 1° latitude = 59 km.

The ***pre-Noachian*** period represents the age of the early crust and is not represented in known outcrops, but a Martian meteorite, ALH84001, was crystallized at ~4.5 Ga.

Heavily cratered terrains formed during the ***Noachian*** Period. These include large impact basins of the Early Noachian Epoch, vast cratered plains of the Middle Noachian, and intercrater plains resurfaced by fluvial and possibly volcanic deposition during the Late Noachian when the atmosphere apparently was thicker and perhaps warmer and heat flow was higher.

Hesperian Period rocks are much less cratered and record waning fluvial activity but extensive volcanism, particularly during the Early Hesperian Epoch. Mars Express and Mars Reconnaissance Orbiter data indicate that clay minerals occur in some Noachian strata, whereas hydrated sulfates are mostly in Hesperian rocks. A thick permafrost zone developed as the surface cooled, and much of the fluvial activity during the Late Hesperian Epoch occurred as catastrophic flood outbursts through this frozen zone, perhaps initiated by magmatic activity.

The ***Amazonian*** Period began with expansive resurfacing of the northern lowlands, perhaps by sedimentation within a large body of water. Much lower levels of volcanism and fluvial discharges, coupled with aeolian deposition and erosion continued into the Middle and Late Amazonian Epochs. Continued weathering has led to iron oxidation of surface materials.

The polar plateaus, covered by bright deposits of residual ice as well as seasonally waxing and waning meter-thick CO_2 frost, are among the youngest features on the planet. Ice-rich mantles and glacial-like deposits at middle and equatorial latitudes signal climate oscillations in the relatively recent geologic record.

The NASA rover, Curiosity, is investigating Gale Crater, which formed toward the end of the Noachian Period (Le Deit et al., 2012). This crater was partly filled by fluvial, deltaic, and lacustrine sediments over a few hundred million years during the early part of the Hesperian Period. These deposits were partially exhumed by wind erosion during the middle Hesperian (ca. 3.3 to 3.1 billion years ago) to form the massive Aeolis Mons (Mount Sharp) of cyclic sediment deposits up to 5-km thick and 6000 km² in area within the Gale crater (e.g., Grotzinger et al., 2015). There has been only very slow eolian erosion since the middle Hesperian.

Mercury

The innermost planet was partly imaged by flybys of the Mariner 10 spacecraft in 1974 and 1975, enabling stratigraphic studies that reveal a remarkably similar surface history to that of Earth's Moon (e.g., Spudis and Guest, 1988). Consequently, a Mercurian chronology was developed based on impact basins and craters that may have similar histories to comparable lunar features (Fig. 2.1).

Thus, five major periods have been proposed that correspond to those of the Moon, as follows:

pre-Tolstojan (equivalent to the lunar pre-Nectarian)
Tolstojan (Nectarian)
Calorian (Imbrian)
Mansurian (Eratosthenian)
Kuiperian (Copernican)

Absolute ages for these periods are much more uncertain than for the Moon and Mars.

Venus

The Venusian surface has been investigated extensively with orbiters and landers, most recently by the Magellan orbiter with its mapping radar in the 1990s. Impact crater densities are low. Statistics of nearly 1000 impact craters on its surface indicate that Venus has an average surface age of hundreds of millions of years. Despite its spectacular volcanic surface dotted with thousands of volcanoes and broad fields of lava flows, all of which has been tectonically disrupted to varying degrees, the details of the global geologic evolution of this Earth's twin planet in size are not well constrained. Possibilities range from local to regional events driven by mantle plumes to global volcanic and tectonic evolution driven by atmospheric greenhouse-heating effects on Venusian climate (e.g., Bougher et al., 1997).

Other solar system bodies

The solid surfaces of asteroids and satellites of Jupiter, Saturn, Uranus, and Neptune show varying degrees of cratering that reflect surface ages (e.g., Schenk et al., 2004). Although asteroids are commonly saturated with craters, indicating their primordial origin, some asteroids, comet nuclei, and other bodies demonstrate later resurfacing as their rocky or icy crusts evolved. Dating these surfaces relies on inferences of the populations of projectiles across time and space. Absolute dates are very poorly constrained. Complications in estimates of cratering rates include the relative importance of asteroids in the inner solar system versus that of comets and other icy materials of the Kuiper Belt.

Selected publications and websites

Cited publications

Bougher, S.W., Hunten, D.M., Phillips, R.J., 1997. *Venus II: Geology, Geophysics, Atmosphere, and Solar Wind Environment*. The University of Arizona Press, Tucson. 1362 pp.

Grotzinger, et al., 2015. Deposition, exhumation, and paleoclimate of an ancient lake deposit, Gale crater, Mars (47 authors total) *Science* **350**: 177. http://dx.doi.org/10.1126/science.aac7575 summary; full version (12 pp.) at.

Hartmann, W.K., 2005. Martian cratering 8: isochron refinement and the chronology of Mars. *Icarus* **174**: 294–320.

Kallenbach, R., Geiss, J., Hartmann, W.K., 2001. *Chronology and Evolution of Mars*. Kluwer Academic Publishers, Dordrecht. 498 pp.

Le Deit, L., Hauber, E., Fueten, F., Mangold, N., Pondrelli, M., Rossi, A., Jaumann, R., 2012. Model age of Gale Crater and origin of its layered deposits. In: *Third International Conference on Early Mars: Geologic and Hydrological Evolution, Physical and Chemical Environments, and the Implications for Life (Lake Tahoe, Nevada, 21–25 May 2012): 7045.pdf*. http://www.lpi.usra.edu/meetings/earlymars2012/pdf/7045.pdf.

Nimmo, F., Tanaka, K., 2005. Early crustal evolution of Mars. *Annual Review of Earth and Planetary Sciences* **33**: 133–161.

Schenk, P.M., Chapman, C.R., Zahnle, K., Moore, J.M., 2004. Ages and interiors: the cratering record of the Galilean satellites. In: Bagenal, F., Dowling, T.E., McKinnon, W.B. (Eds.), *Jupiter: The Planet, Satellites and Magnetosphere*. Cambridge University Press, Cambridge, pp. 427–456.

Shoemaker, E.M., Hackman, R.J., 1962. Stratigraphic basis for a lunar time scale. In: Kopal, Z., Mikhailov, Z.K. (Eds.), *The Moon*. Academic Press, London, pp. 289–300.

Spudis, P.D., Guest, J.E., 1988. Stratigraphy and geologic history of Mercury. In: Vilas, F., Chapman, C.R., Matthews, M.S. (Eds.), *Mercury*. The University of Arizona Press, Tucson, pp. 118–164.

Stöffler, D., Ryder, G., 2001. Stratigraphy and isotope ages of lunar geologic units: chronological standards for the inner solar system. *Space Science Reviews* **96**: 9–54.

Strom, R.G., Malhotra, R., Ito, T., Yoshida, F., Kring, D.A., 2005. The origin of planetary impactors in the inner solar system. *Science* **309**: 1847–1850.

Tanaka, K.L., 1986. The stratigraphy of Mars. Proceedings of the Lunar and Planetary Science Conference, 17, Part 1. *Journal of Geophysical Research* **91**: E139–E158.

Tanaka, K.L., Hartmann, W.K., 2008. 2 planetary time scale. In: Ogg, J.G., Ogg, G., Gradstein, F.M. (Eds.), *The Concise Geologic Time Scale*. Cambridge University Press, pp. 13–22.

Tanaka, K.L., Hartmann, W.K., 2012. The planetary time scale. In: Gradstein, F.M., Ogg, J.G., Schmitz, M., Ogg, G., (Coordinators). *The Geologic Time Scale 2012*. Elsevier Publisher, pp. 275–298. (An overview on the geologic history of all inner planets, Earth's Moon, and briefly on the moons of Mars and Jupiter.)

Tanaka, K.L., Skinner Jr., J.A., Dohm, J.M., Irwin III, R.P., Kolb, E.J., Fortezzo, C.M., Platz, T., Michael, G.G., Hare, T.M., 2014. *Geologic Map of Mars: U.S. Geological Survey Scientific Investigations Map 3292, Scale 1:20,000,000, Pamphlet 43*. http://dx.doi.org/10.3133/sim3292. http://pubs.usgs.gov/sim/3292/.

Wilhelms, D.E., 1987. The geologic history of the Moon. *U.S. Geological Survey Professional Paper* **1348** 302 pp., 12 plates.

Selected further reading

Basaltic Volcanism Study Project, 1981. *Basaltic Volcanism on the Terrestrial Planets*. Houston: Lunar and Planetary Institute, Houston. 1286 pp.

Melosh, H.J., 2011. *Planetary Surface Processes*. Cambridge University Press. 500 pp.

Websites (selected)

US Geological Survey Astrogeology Research Program—astrogeology.usgs.gov/, especially: Astropedia: astrogeology.usgs.gov/search/.

Solar System Exploration (NASA)—solarsystem.nasa.gov.

Welcome to the Planets (JPL, NASA)—pds.jpl.nasa.gov/planets/.

Mars Exploration Program (NASA)—marsprogram.jpl.nasa.gov/.

Wikipedia—Lunar Geologic Timescale—en.wikipedia.org/wiki/Lunar_geologic_time_scale; and Geologic history or Mars: https://en.wikipedia.org/wiki/Geological_history_of_Mars.

3

PRECAMBRIAN

The Archean World. Courtesy of the Smithsonian Institution. Painting by Peter Sawyer. [http://ocean.si.edu/slideshow/ocean-throughout-geologic-time-image-gallery]

Status of international subdivisions

The first 4 billion years of Earth's history consist of the Hadean, Archean, and Proterozoic eons. The *Precambrian* simply refers to the time interval and all rocks that formed prior to the beginning of the Cambrian Period (base of Phanerozoic Eon) at 541 Ma.

The interval with no preserved rock record from the formation of the Earth at 4.567 to ca. 4 Ga is named the "*Hadean*" Eon, a term derived from Greek for "*unseen place*" and also referring to the mythical Hades land of the dead (Subcommission on Precambrian Stratigraphy, 2014). The Hadean is followed at ca. 4 Ga by the *Archean* (from the Greek word meaning "*beginning/origin*") and at ca. 2.5 Ga by the *Proterozoic* (from Greek for "*earlier life*") (Fig. 3.1).

A Concise Geologic Time Scale.

Current subdivision of the Precambrian Time Scale

Age (Ma)

Age (Ma)	Eon	Era	Period	Description
0	Phanerozoic	Cenozoic		Age of Mammals
200	Phanerozoic	Mesozoic		Age of Dinosaurs
400	Phanerozoic	Paleozoic		Coals, amphibians and insects
	Phanerozoic	Paleozoic		Mainly only ocean life
600	Proterozoic	Neo-proterozoic	Ediacaran	Ediacaran Fauna
	Proterozoic	Neo-proterozoic	Cryogenian	Glacial deposits
800	Proterozoic	Neo-proterozoic	Tonian	
1000	Proterozoic	Meso-proterozoic	Stenian	Long period of stable one-celled-life ecosystems in apparently constant environments Supercontinent Rodinia (~1300 to 900 Ma)
1200	Proterozoic	Meso-proterozoic	Ectasian	
1400	Proterozoic	Meso-proterozoic	Calymmian	
1600	Proterozoic	Paleoproterozoic	Statherian	
1800	Proterozoic	Paleoproterozoic	Orosirian	Supercontinent Columbia/Nuna formation, then break-up
2000	Proterozoic	Paleoproterozoic	Rhyacian	Increased burial of organic carbon ("L-J" ^{13}C positive excursion)
2200	Proterozoic	Paleoproterozoic	Siderian	Oxygen begins to accumulate in atmosphere; major glaciations
2400	Archean	Neo-archean		Oxygen levels rise in oceans causing banded-iron formations
2600	Archean	Neo-archean		Sedimentary basins on stable or growing continents
2800	Archean	Meso-archean		
3000	Archean	Meso-archean		Growth of nuclei of continents
3200	Archean	Paleo-archean		
3400	Archean	Paleo-archean		
3600	Archean	Eo-archean		Earliest preserved sedimentary rocks and chemical traces of life
3800	Archean	Eo-archean		Oldest preserved pieces of continental crust
4000	Hadean (informal)			Rapid crust formation & recycling; heavy meteorite bombardment. Earliest Life (Prokaryotes, simple-celled) evolved?
4200	Hadean (informal)			
4400	Hadean (informal)			Accretion of Earth; then giant Moon-forming impact event
4567				

Figure 3.1 The current Precambrian time scale. The current Precambrian eons, eras, and periods, from the International Commission on Stratigraphy, based on Plumb and James (1986) and Plumb (1991). Note that Precambrian is not a formal time scale unit and that all divisions of the Precambrian are chronometric (fixed dates at base). Exceptions are the Cryogenian and the Ediacaran. The base of the Cryogenian Period was initially set at 850 Ma (Plumb, 1991), but was revised in 2014/2015 to the ca. 720 Ma date of the onset of the first global glaciation—the criteria for placement of a future GSSP. The base of the Ediacaran is a chronostratigraphic GSSP at the termination of the last Cryogenian glaciation dated as 635 Ma (see next chapter). Only era divisions are shown for the Phanerozoic Eon. In the years since these Precambrian divisions were standardized in 1990, our dating of major events and cycles in Precambrian geologic history have indicated that the current Global Standard Stratigraphic Ages (GSSAs) do not adequately convey this history.

Although microbial life existed throughout the Archean and Proterozoic, the lack of a diverse and well-preserved fossil record prior to the late Ediacaran, coupled with uncertainties in geochemical or other stratigraphic means of correlations, is a challenge to establish a formal chronostratigraphic scale. Radioisotopic dating was the main method for correlating the Precambrian geologic records; therefore, the Subcommission on Precambrian Stratigraphy adopted the use of chronometric GSSAs for the international subdivisions and standardization of interregional geological maps (Plumb and James, 1986; Plumb, 1991). The Archean Eon is subdivided into four eras (rounded to the nearest 100-myr boundaries), and the Proterozoic into three eras and 10 periods (the first eight of which are rounded to the nearest 50-myr boundaries). The two youngest periods, Cryogenian (ca. 720 Ma to 635 Ma) with its major glaciations and the Ediacaran (635–541 Ma) with metazoan life forms, are summarized in the next chapter. The dates for these GSSA boundaries (and the poetic names for the Proterozoic periods) were selected to delimit major events in tectonics, surface conditions, and sedimentation as known in 1990 (Table 3.1).

Summary of Precambrian trends and events, and a potential revised time scale

Since 1990, our knowledge and dating of the development of Earth's tectonic cycles, crustal features, atmosphere and ocean composition, geochemical trends and excursions, major volcanic and impact events, and stages in evolution of life through the Precambrian has vastly increased. Some major trends are displayed in Fig. 3.2.

The shortcomings of the current rounded dates for the chronometric subdivisions of Precambrian time are: (1) a lack of ties of stratigraphic boundaries to the actual rock record, (2) the current divisions do not adequately convey the major events in the fascinating history of our planet, and (3) severe diachroneity of global tectonic events. Hence, major research efforts are underway by the Subcommission on Precambrian Stratigraphy to replace the current GSSA chronometric scheme to one that is more naturalistic with GSSPs. In GTS2012, members of the Subcommission on Precambrian Stratigraphy under the leadership of Martin van Kranendonk, suggested a possible stratigraphic scheme (revised from Bleeker, 2004) that is principally based on sedimentological, geochemical, geotectonic, and biological events recorded in the rock record with potential "golden spikes" (Van Kranendonk et al., 2012) (Figs. 3.2 and 3.3).

The following summary is largely based on the extensive Precambrian synthesis by Van Kranendonk et al. (2012) and Van Kranendonk (2014).

Hadean

The oldest solid materials in the solar system, therefore the oldest rocks that would have been incorporated in the accretion of planet Earth, are considered calcium–aluminum-rich aggregates in chondritic meteorites that are dated as 4.567 Ga; and that date is assigned as the beginning of the Hadean Eon. After the giant Moon-forming impact at ca. 4.5 Ga, the sphere of molten silicate material cooled and differentiated into the core and mantle. The oldest preserved mineral crystals from cooling of magma on Earth are zircons dated 4.4 Ga that were later recycled into weakly metamorphosed sandstone in the Jack Hills of the Yilgarn Craton of Western Australia. One of these zircons has been reanalyzed by high-resolution mapping of radiogenic isotopes to yield a precise 4.374 ± 0.006 Ga date (Valley et al., 2014; reviewed by Bowring, 2014). This early crust was largely destroyed during the Late Heavy

Table 3.1 Nomenclature for periods of Proterozoic Eon in the current International Commission on Stratigraphy (ICS) geologic time-scale with their intended characteristics

Period name	Base (Ma)	Derivation and geological process
Ediacaran	~635	*Ediacara* = from Australian Aboriginal term for place near water "Earliest metazoan life" GSSP in Australia coincides with termination of glaciations and a pronounced carbon-isotope excursion
Cryogenian	~720	*Cryos* = ice; *Genesis* = birth "Global glaciation" Glacial deposits, which typify the late Proterozoic, are most abundant during this interval. Base, formerly at 850 Ma, was re-defined in 2014/2015 as onset of the first global glaciation.
Tonian	1000	*Tonas* = stretch Further major platform cover expansion (e.g., Upper Riphean, Russia.; Qingbaikou, China; basins of northwest Africa), following final cratonization of polymetamorphic mobile belts.
Stenian	1200	*Stenos* = narrow "Narrow belts of intense metamorphism & deformation" Narrow polymetamorphic belts, characteristic of the mid-Proterozoic, separated the abundant platforms and were orogenically active at about this time (e.g., Grenville, Central Australia).
Ectasian	1400	*Ectasis* = extension "Continued expansion of platform covers" Platforms continue to be prominent components of most shields.
Calymmian	1600	*Calymma* = cover "Platform covers" Characterized by expansion of existing platform covers, or by new platforms on recently cratonized basement (e.g., Riphean of Russia).
Statherian	1800	*Statheros* = stable, firm "Stabilization of cratons; Cratonization" This period is characterized on most continents by either new platforms (e.g., North China, north Australia) or final cratonization of fold belts (e.g., Baltic Shield, north America).
Orosirian	2050	*Orosira* = mountain range "Global orogenic period" The interval between about 1900 and 1850 Ma was an episode of orogeny on virtually all continents.
Rhyacian	2300	*Rhyax* = stream of lava "Injection of layered complexes" The Bushveld Complex (and similar layered intrusions) is an outstanding event of this time.
Siderian	2500	*Sideros* = iron "Banded iron formations"(BIFs) The earliest Proterozoic is widely recognized for an abundance of BIFs, which peaked just after the Archean–Proterozoic boundary.

Modified from Plumb (1991).

Bombardment resurfacing of the inner solar system planets and Moon (ca. 4.1 to 3.85 Ga).

The accretion of planet Earth, partial differentiation of its core–mantle, and the formation of the Moon from the ejected residual from a massive impact with Earth all occurred during the "*Chaotian*" interval between these two dates (Van Kranendonk et al., 2012).

Archean

The oldest surviving rocks that have been dated, the Acasta Gneiss Complex of the Slave Craton in Canada, at 4.03 Ga (Bowring and Williams, 1999), form the base of the Archean. The oldest sedimentary rocks with preserved primary features are in the Isua supracrustal belt of the North Atlantic Craton, western Greenland with an age of 3.81 Ga.

The oldest well-preserved structures formed by life are stromatolites from ancient microbial mats in the Dresser Formation of the Warrawoona Group from the humorously named "North Pole" dome region of the Pilbara Craton of Western Australia, dated at ca. 3.481±0.002 Ga (e.g., Van Kranendonk et al., 2008). The oldest known intertidal shoreline deposit, the Strelley Pool Formation of Western Australia, dated at ca. 3.43 Ga, contains stromatolites and candidates for organic microfossils preserved in episodic silica cementation (Brasier et al., 2015). The origins of life itself are not known and remain a major challenge facing science.

Van Kranendonk et al. (2012) suggest using this suite of the oldest rock, the oldest well-preserved sediment, and the oldest biostructure as chronostratigraphic boundaries to delimit the *Acastan* and the *Isuan* periods within a Paleoarchean Era.

Basins formed within the growing cratons during the Mesoarchean Era, and this Era could be subdivided with a GSSP at the base of ca. 3 Ga quartz-rich sandstone in a platform setting. Dating of crustal rocks indicate that there was another widespread growth period of continental crust beginning at ca. 2.78 until 2.63 Ga (e.g., O'Neill et al., 2015) (Fig. 3.2).

The expansion of photosynthetic life in these basins removed carbon dioxide in the form of stromatolite carbonates. However, carbon preserved in kerogen in these stromatolites during the interval from ca. 2.7 to 2.5 Ga has highly negative $\delta^{13}C_{org}$ values (down to −61 per mille), indicative of a dominance of ^{12}C-enriched products from methane-producing organisms or other methanogenesis process. The photosynthesis activity and carbon burial also increased the influx and concentration of oxygen waste products in the atmosphere and oceans. The oxygen dissolving into the marine waters caused precipitation of iron oxides, which resulted in a unique episode of extensive banded iron formations (BIF) beginning at ca. 2.6 Ga. The onsets of these relatively rapid and easily correlated global changes are options for redefining and subdividing the Neoarchean Era into an earlier "Methanian Period" before the methane-producing microbes were inhibited by the rising oxygen levels, followed by a "Siderian Period" for the main episode of BIF deposition as characterized by those in the Hamersley Basin of Western Australia (Van Kranendonk et al., 2012) (Figs. 3.2 and 3.3).

Proterozoic

The rising oxygen levels, increased weathering rates, and burial of carbon led to major changes in the Earth system beginning at ca. 2.42 Ga—just after the traditional placement for the Archean/Proterozoic boundary at 2.5 Ga. Extensive removal of atmospheric carbon dioxide contributed to the near-global "Huronian" glaciations during ca. 2.4–2.25 Ga (e.g., review by Tang and Chen, 2013). When this "Snowball Earth" episode ended, it was a different world. In the oxygenated oceans, the complex-celled eukaryotic life forms with aerobic metabolism appeared and thrived, later evolving into Phanerozoic animals.

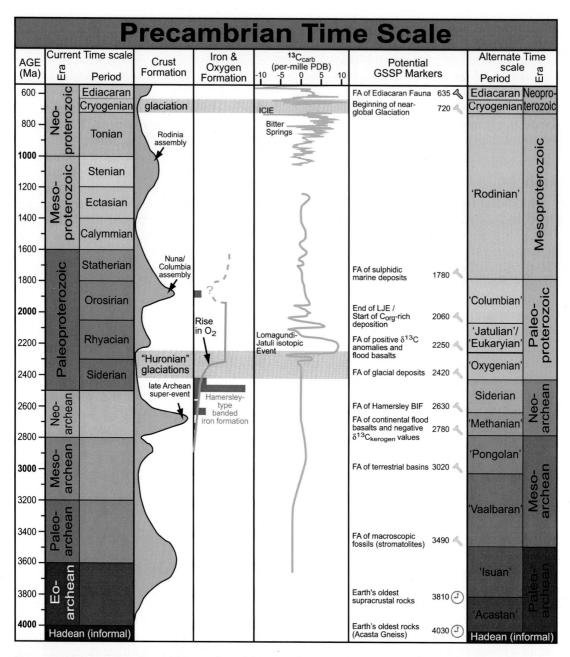

Figure 3.2 Major trends in Precambrian geologic history. (Modified from synthesis diagrams in Van Kranendonk et al. (2012; figs 16.15 and 16.32 in that paper), Van Kranendonk (2014), and O'Neill et al. (2015)). Relative rates of crustal accumulation and possible relationship to supercontinent accretion and breakup are based on the compilation by McCulloch and Bennett (1994; see discussions in O'Neill et al., 2015). Carbon-isotope curves are smoothed versions from the syntheses for the Archean through middle Proterozoic by Halverson et al. (2005), and for the late Proterozoic by Cohen and Macdonald (2015) calibrated by them to the Cryogenian–Ediacaran time scale of Rooney et al. (2015). "ICIE" is the Islay carbon-isotope excursion, and "FA" indicates a first-appearance level or the onset of an episode. The age model is from Van Kranendonk et al. (2012).

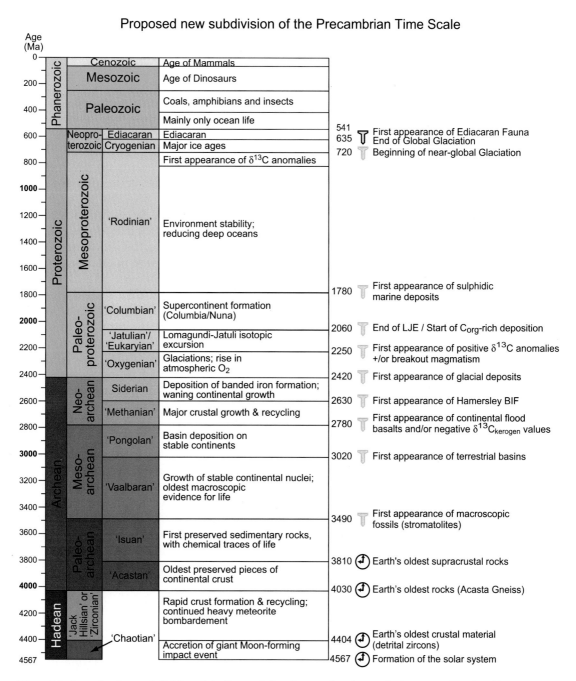

Proposed new subdivision of the Precambrian Time Scale

Figure 3.3 An option for a subdivision of the Precambrian time scale using geologic events. The definitions, age estimates, and nomenclature for these subdivisions are by Van Kranendonk et al. (2012).

Extensive flood basalts erupted onto several continental plates. The isotopic composition of the global carbon cycle, which had been remarkably stable through the late Archean, suddenly underwent the largest positive excursions in $\delta^{13}C_{carb}$ in the entire geologic record. This Lomagundi–Jatuli Excursion (LJE) was named after its initial recognition in the Lomagundi province in Zimbabwe and the Jatuli complex in Russian Karelia.

The LJE event ended suddenly at ca. 2.06 Ma, nearly synchronous with (1) the eruption of one of the world's largest igneous provinces, the Bushveld Complex in southern Africa (e.g., Cawthorn et al., 2006); (2) the largest impact structure preserved on Earth, the Vredefort impact in southern Africa at ca. 2.02 Ga, with ca. 250-km diameter crater, which is larger than the 180 km Chicxulub impact crater that terminated the Cretaceous (e.g., Reimold and Koeberl, 2014); (3) the formation of the earliest major phosphorite deposits; and (4) the beginning of a previously unprecedented accumulation of organic-rich "oil shale" sedimentation in various parts of the world, named the Shunga Event after the Shunga village in northwest Russia where a single deposit alone buried 250 billion tons of organic carbon (e.g., Melezhik et al., 1999). Reviews by Van Kranendonk et al. (2012), Young (2013), and Van Kranendonk (2014) postulate causal relationships among all of these trends and events, including possible influences upon the early evolution of eukaryote life. The global record of these remarkable geologic features may be used to correlate and subdivide the early part of the Paleoproterozoic (Figs. 3.2 and 3.3).

Between about 1.8 and 1.4 Ga, the majority of the continental plates were merged into the Nuna/Columbia supercontinent of uncertain configuration, and again were united between about 1.0 and 0.7 Ga into the Rodinia supercontinent (e.g., Li et al., 2008; Meert, 2012, 2014; Evans, 2013). This Nuna–Rodinia interval is a unique "quiet" time in Earth's history,

which has been termed the "boring billion" (e.g., Young, 2013; Cawood and Hawkesworth, 2014). For the majority of this quiet time, evidence is relatively lacking for the evolution of new life forms, major climatic changes, strontium- or carbon-isotope excursions, new passive margins, and the formation of important ore deposits. Therefore, this interval is difficult to subdivide (Figs. 3.2 and 3.3).

There were major events on the regional scale. At 1.85 Ga, the enormous Sudbury bolide impact left a 200–250-km crater in southern Canada. The North American plate was also affected by the giant Mackenzie volcanic dike swarm in north Canada at 1.27 Ga, by the major Keweenawan flood basalts in the Midcontinent Rift System at 1.12 to 1.09 Ga, and the Franklin giant dike swarm in north Canada and northwest Greenland at 0.72 Ga (Ernst et al., 2008). Other continental blocks experienced similar large igneous provinces (LIPs); but, unlike the common coincidence of LIPs and environmental disruptions through the Phanerozoic, there has not yet been a direct correlation of any of these LIPs with other geochemical excursions that can be used for global correlation. However, toward the end of this interval there are two significant negative excursions in $\delta^{13}C_{carb}$ (Fig. 3.2)—the Bitter Springs event at ca. 810 Ma and the Islay anomaly at 735–740 Ma (e.g., Halverson and Shields-Zhou, 2011; Strauss et al., 2014).

The onset of the Cryogenian "Snowball Earth" glaciations at ca. 720 Ma was preceded by regional glaciations indicated by the Gucheng and Bayisi diamictites near base of Nanhuan System of China at ca. 760 or 740 Ga (e.g., Stratigraphic Chart of China (2015)) and perhaps by the Kaigas Formation of Africa at ca. 740 Ma (reviewed in Shields-Zhou et al. (2012)), although dating of this Kaigas event is uncertain (e.g., Rooney et al., 2015). The Cryogenian and the postglacial Ediacaran periods of the Neoproterozoic are summarized in the next chapter.

Acknowledgments

This brief Precambrian summary of selected highlights and current stratigraphic issues relied heavily on the detailed overview and synthesis by Van Kranendonk et al. (2012), and on discussions with Martin Van Kranendonk and Jim Gehling. Martin Van Kranendonk reviewed an early version of the graphics, database, and text.

Selected publications and websites

Cited publications

Only select publications were cited in this review with an emphasis on aspects of post-2011 updates. Pre-2011 literature is well summarized in the synthesis by Van Kranendonk et al. (2012) and some of the cited publications below.

Bleeker, W., 2004. Towards a "natural" Precambrian time scale. In: Gradstein, F.M., Ogg, J.G., Smith, A.G. (Eds.), *A Geologic Time Scale 2004*. Cambridge University Press, Cambridge, pp. 141–146.

Bowring, S.A., Williams, I.S., 1999. Prisocan (4.00–4.03 Ga) orthogneisses from northwestern Canada. *Contributions Mineralogy and Petrology* **134**: 3–16.

Bowring, S.A., 2014. Closing the gap. *Nature Geoscience* **7**: 169–170.

Brasier, M.D., Antcliffe, J., Saunders, M., Wacey, D., 2015. Changing the picture of Earth's earliest fossils (3.5–1.9 Ga) with new approaches and new discoveries. *Proceedings of the National Academy of Sciences of the United States of America (PNAS)* **112**: 4859–4864. www.pnas.org/cgi/doi/10.1073/pnas.1405338111.

Cawood, P.A., Hawkesworth, C.J., 2014. Earth's middle age. *Geology* **42**: 503–506.

Cawthorn, R.G., Eales, H.V., Walraven, F., Uken, R., Watkeys, M.K., 2006. The Bushveld Complex. In: Johnson, M.R., Anhaeusser, C.R., Thomas, R.J. (Eds.), *The Geology of South Africa*. Geological Society of South Africa/Council for Geoscience, Johannesburg/Pretoria, pp. 261–281.

Cohen, P.A., Macdonald, F.A., 2015. The Proterozoic record of eukaryotes. *Paleobiology* **41**: 610–632. http://dx.doi.org/10.1017/pab.2015.25.

Ernst, R.E., Wingate, M.T.D., Buchan, K.T., Li, Z.X., 2008. Global record of 1600–700 Ma Large Igneous Provinces (LIPs): implications for the reconstruction of the proposed Nuna (Columbia) and Rodinia supercontinents. *Precambrian Research* **160**: 159–178.

Evans, D.A.D., 2013. Reconstructing pre-Pangean supercontinents. *Geological Society of America Bulletin* **125**: 1735–1751. http://dx.doi.org/10.1130/B30950.1.

Halverson, G.P., Hoffman, P.F., Schrag, D.P., Maloof, A.C., 2005. Towards a Neoproterozoic composite carbon-isotope record. *Geological Society of America Bulletin* **117**: 1181–1207.

Halverson, G.P., Shields-Zhou, G., 2011. Chemostratigraphy and the Neoproterozoic glaciations. In: Arnaud, E., Halverson, G.P., Shields-Zhou, G. (Eds.), *The Geological Record of Neoproterozoic Glaciations*, **36**. Geological Society, London, Memoirs, pp. 51–66. http://dx.doi.org/10.1144/M36.4 (Chapter 4).

Li, Z.X., Bogdanova, S.V., Collins, A.S., Davidson, A., DeWaele, B., Ernst, R.E., Fitzsimons, C.W., Fuck, R.A., Gladkochub, D.P., Jacons, J., Karlstrom, K.E., Lu, S., Natapov, L.M., Pease, V., Pisarevsky, S.A., Thrane, K., Vernikovsky, V., 2008. Assembly, configuration and break-up history of Rodinia: a synthesis. *Precambrian Research* **160**: 179–210.

McCulloch, M.T., Bennett, V.C., 1994. Progressive growth of the Earth's continental crust and depleted Mantle – geochemical constraints. *Geochimica et Cosmochimica Acta* **58**: 4717–4738.

Melezhik, V.A., Fallick, A.E., Filippov, M.M., Larsen, O., 1999. Karelian shungite – an indication of 2.0 Ga-old metamorphosed oil-shale and generation of petroleum: Geology, lithology and geochemistry. *Earth-Science Reviews* **47**: 1–40.

Meert, J.G., 2012. What's in a name? The Columbia (Paleopangaea/Nuna) supercontinent. *Gondwana Research* **21**: 987–993.

Meert, J.G., 2014. Strange attractors, spiritual interlopers and lonely wanderers: the search for pre-Pangean supercontinents. *Geoscience Frontiers* **5**: 155–166.

O'Neill, C., Lenardic, A., Condie, K.C., 2015. Earth's punctuated tectonic evolution: cause and effect. In: Roberts, N.M.W., Van Kranendonk, M.J., Parman, S., Shirey, S., Clift, P.D. (Eds.), *Continent Formation Through Time* **389**. Geological Society, London, Special Publications, pp. 17–40. http://dx.doi.org/10.1144/SP389.4.

Plumb, K.A., James, H.L., 1986. Subdivision of Precambrian time: Recommendations and suggestions by the Subcommission on Precambrian Stratigraphy. *Precambrian Research* **32**: 65–92.

Plumb, K.A., 1991. New Precambrian time scale. *Episodes* **14**: 139–140.

Reimold, W.U., Koeberl, C., 2014. Impact structures in Africa: a review. *Journal of African Earth Sciences* **93**: 57–175. http://dx.doi.org/10.1016/j.jafrearsci.2014.01.008.

Rooney, A.D., Strauss, J.V., Brandon, A.D., Macdonald, F.A., 2015. A Cryogenian chronology: two long-lasting synchronous Neoproterozoic glaciations. *Geology* **43**: 459–462. http://dx.doi.org/10.1130/G36511.1.

Shields-Zhou, G.A., Hill, A.C., Macgabhann, B.A., 2012. The Cryogenian Period. In: Gradstein, F.M., Ogg, J.G., Schmitz, M., Ogg, G., (Coordinators). *The Geologic Time Scale 2012*. Elsevier Publ., pp. 393–411. http://dx.doi.org/10.1016/B978-0-444-59425-9.00018-4.

Stratigraphic Chart of China (explanatory notes), in press 2015. *Cryoginian of China (Nanhuan System)*. (preprint provided to J.Ogg by Yin Hongfu, May 2015).

Strauss, J.V., Rooney, A.D., Macdonald, F.A., Brandon, A.D., Knoll, A.H., 2014. 740 Ma vase-shaped microfossils from Yukon, Canada: implications for Neoproterozoic chronology and biostratigraphy. *Geology* **42**: 659–662.

Subcommission on Precambrian Stratigraphy, 2014. Annual report 2014. In: *International Commission on Stratigraphy (ICS) Annual Report 2014. Submitted to International Union of Geological Sciences (IUGS)* at: http://iugs.org/uploads/ICS%202014.pdf.

Tang, H., Chen, Y., 2013. Global glaciations and atmospheric change at ca. 2.3 Ga. *Geoscience Frontiers* **4**: 583–596.

Valley, J.W., Cavosie, A.J., Ushikubo, T., Reinhard, D.A., Lawrence, D.F., Larson, D.J., Clifton, P.H., Kelly, T.F., Wilde, S.A., Moser, D.E., Spicuzza, M.J., 2014. Hadean age for a post-magma-ocean zircon confirmed by atom-probe tomography. *Nature Geoscience* **7**: 219–223.

Van Kranendonk, M.J., Philippot, P., Lepot, K., Bodorkos, S., Pirajno, F., 2008. Geological setting of Earth's oldest fossils in the ca. 3.5 Ga Dresser Formation, Pilbara Craton, Western Australia. *Precambrian Research* **167**: 93–124.

Van Kranendonk, M.J., Altermann, W., Beard, B.L., Hoffman, P.F., Johnson, C.J., Kasting, J.F., Melezhik, V.A., Nutman, A.P., Papineau, D., Pirajno, F., 2012. A chronostratigraphic division of the Precambrian: possibilities and challenges. In: Gradstein, F.M., Ogg, J.G., Schmitz, M., Ogg, G., (Coordinators). *The Geologic Time Scale 2012*. Elsevier Publ., pp. 299–392. http://dx.doi.org/10.1016/B978-0-444-59425-9.00023-8 (An overview on all aspects, including summaries of tectonic cycles, atmosphere-ocean history, climatic episodes and evolution of life; plus age models and a set of suggested chronostratigraphic divisions.).

Van Kranendonk, M.J., 2014. Earth's early atmosphere and surface environments: a review. In: Shaw, G.H. (Ed.), *Earth's Early Atmos. Surf. Environ.*, **504**. Geological Society of America Special Paper, pp. 105–130. http://dx.doi.org/10.1130/2014.2504(12).

Young, G.M., 2013. Precambrian supercontinents, glaciations, atmospheric oxygenation, metazoan evolution and an impact that may have changed the second half of Earth history. *Geoscience Frontiers* **4**: 247–261.

Websites (selected)

Subcommission on Precambrian Stratigraphy (ICS)—*http://precambrian.stratigraphy.org*—partial Website that currently briefly summarizes the official divisions of the Precambrian.

Precambrian Research—*http://www.journals.elsevier.com/precambrian-research/*—Elsevier journal with articles on all aspects of the early stages of Earth's history and nearby planets.

Palaeos: Precambrian—*http://palaeos.com/precambrian/precambrian.htm*—A well-presented suite of diverse topics for a general science audience that was originally compiled by M. Alan Kazlev in 1998–2002.

4

CRYOGENIAN AND EDIACARAN

650 Ma Cryogenian

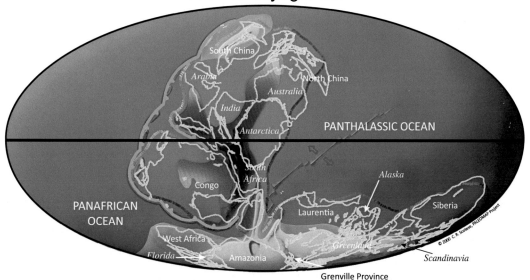

End-Cryogenian paleogeographic reconstruction (ca. 640 Ma). The paleogeographic map of breakup of Rodinia supercontinent was provided by Chris Scotese, although geographic distribution of the continents is uncertain.

Basal definitions and status of international subdivisions

The Cryogenian Period consists of two near-global glacial episodes between ca. 720 Ma and 635 Ma. The Ediacaran Period/System begins with the return of warmer marine conditions that created a distinctive "cap carbonate"; and the Ediacaran Global boundary Stratotype Section and Point (GSSP) is at the base of this cap carbonate in South Australia (Fig. 4.1). Metazoans first appear on Earth during the latter half of the Ediacaran in association with microbial mats.

Cryogenian

The Cryogenian Period was named and defined in 1990 to encompass the major near-global glacial intervals within the chronometric span of 850 to 650 Ma (Plumb, 1991). Two of these glacial episodes deposited extensive diamictite and glaciomarine sediments throughout the globe, including onto the marine shelves in tropical latitudes, thereby giving rise to the "Snowball Earth" model (e.g., Kirschvink, 1992; Hoffman et al., 1998; Arnaud et al., 2011; Smith, 2009). These two glacial episodes and the intervening warm interglacial interval divide the Cryogenian into three globally synchronous units.

The earlier **Sturtian** glaciation, named from the Sturt Formation of the Adelaide Rift Complex of South Australia, has been correlated to diamictite deposits on other continents, which yielded maximum ages of slightly less than ca. 720 Ma (e.g., 717.4 ± 0.1 Ma from underlying volcanics; Macdonald et al., 2010) (reviewed in Shields-Zhou et al., 2012; Rooney et al., 2015). The Subcommission on Cryogenian Stratigraphy (2014) redefined the Cryogenian Period to begin at the onset of that first near-global Sturtian glaciation and assigned a rounded "ca. 720 Ma" chronometric definition until a chronostratigraphic GSSP is selected.

This redefinition of the Cryogenian has placed some precursor cooling episodes into the Tonian Period (reviewed in Shields-Zhou et al., 2012). These include the cold event at base of the Changan Formation (assigned as ca. 780 Ma) that defines the base of the Nanhuan System of China (Wang et al., 2014), the diamictite of the Bayisi Formation of China (either 755 or 768 Ma), and the diamictite of the Kalgas Formation of Namibia (ca. 740 Ma; although dating and correlation of this Kalgas event has been questioned by Rooney et al. (2015)).

The Sturtian glacial episode lasted 60 myr (e.g., Rooney et al., 2015). Whether the Sturtian episode was a single long-lasting Snowball Earth condition or consisted of fluctuating glacial conditions is uncertain; but there do not appear to have been any significant interglacial intervals. The Sturtian suddenly ended at ca. 660 Ma as bracketed by rhenium–osmium (Re-Os) dates of 662.4 ± 4.6 Ma and 659.0 ± 4.5 Ma (Rooney et al., 2015) and a uranium–lead (U-Pb) date of 663 ± 4 Ma from an ash bed above the Sturtian-age diamictite in South China (Zhou et al., 2004). The interglacial interval spanned approximately 20–25 myr; and the initial marine flooding on most of the shelves deposited a distinctive cap carbonate.

The second near-global glaciation, named **Marinoan** after glacial deposits in the Adelaide suburb of Marino in South Australia, began by 637 Ma; based on a U-Pb Isotope Dilution–Thermal Ionization Mass Spectrometry (ID-TIMS) date of 636.4 ± 0.5 Ma above the basal Cottons Breccia in Tasmania, Australia (Calver et al., 2013). The Marinoan glacial episode spanned only 2 myr without any significant interglacial fluctuations, until it ended suddenly at 635 Ma.

Ediacaran

The Ediacaran begins at the sharp contact of the cap carbonate on the Marinoan glaciomarine deposits. At the GSSP in the central Flinders Ranges of South Australia is the contact of the 6-m thick Nuccaleena Dolomite onto the glaciomarine Elatina Formation (Fig. 4.1). Carbon isotopes become progressively negative through the Nuccaleena Dolomite, which is a trend seen in all basal Ediacaran cap carbonates (reviewed in Narbonne et al., 2012).

The Ediacaran will be potentially subdivided into two or three series (Subcommission on Ediacaran Stratigraphy, 2014). One candidate level for a series boundary might be the onset or termination of a relative brief (ca. 2 myr) regional episode named **Gaskiers** Glaciation which is accompanied by a small negative carbon-isotope anomaly (Narbonne et al., 2012). The 250-m thick glacial Gaskiers Formation on the Avalon Peninsula of southeast Newfoundland, Canada, is constrained

Base of the Ediacaran System at Enorama Creek, Flinders Ranges, South Australia

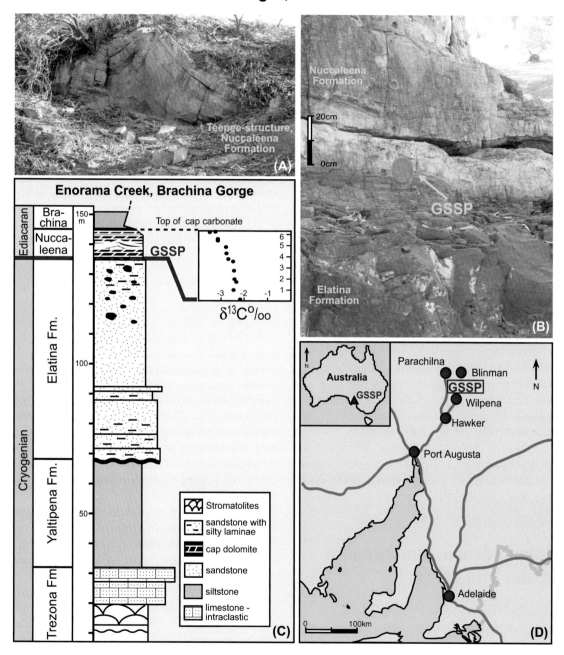

Figure 4.1 GSSP for base of the Ediacaran at Enorama Creek section, central Flinders Ranges, Adelaide Rift Complex, South Australia. The GSSP level is defined as the sharp base of the cap carbonate (Nuccaleena Formation) on the Marinoan glacial and glaciomarine diamictite deposits (Elatina Formation). The Nuccaleena dolomite has cm-scale event beds and enigmatic teepee-like structures that are up to 1 m in amplitude. Carbon-isotope values in the basal cap carbonate are anomalously low and decrease upward. This facies succession with the onset of a negative excursion in $\delta^{13}C_{carb}$ in the cap carbonate is typical of the rapid decay of the global Marinoan ice sheets throughout the world. Stratigraphic diagrams modified from Knoll et al. (2006). Photos by Gabi Ogg.

by U-Pb dating to span 584 to 582 Ma, and similar-aged "Gaskiers" glacial deposits are reported from Massachusetts, Norway, and the Tarim Basin in China (e.g., Condon and Bowring, 2011).

Another candidate for subdividing the Ediacaran is the onset or nadir of the Shuram carbon-isotope excursion, the "*largest negative carbon isotope excursion on Earth*" (e.g., Guerroué, 2010). The Shuram (or Shuram/Wonoka) excursion was named after its recognition in the Shuram Formation of Oman, and is sometimes called the Shuram/Wonoka excursion in reference to its discovery in the Wonoka Formation of South Australia. The exact timing of the full episode is uncertain; but it appears that the Shuram excursion began at ca. 560 Ma, rapidly reached a minimum in about 0.8 myr according to cycle stratigraphy (Minguez et al., 2015; Kodama, 2015), then slowly returned to preexcursion carbon-isotope levels by ca. 550 Ma (Condon and Bowring, 2011) (e.g., Fig. 4.2 has the approximate signature and age placement according to Cohen and Macdonald (2015)). Macroscopic metazoan animals first appear during this Shuram excursion.

Potential biostratigraphic subdivisions of the Ediacaran using biozones of distinctive acritarchs are being evaluated for interregional correlation (reviewed in Narbonne et al., 2012).

Selected main stratigraphic scales and events

(1) Stable-isotope stratigraphy, magnetostratigraphy, and selected events

There are several major negative excursions in carbon isotopes ($\delta^{13}C_{carb}$) during the late Tonian, Cryogenian, and Ediacaran that are important for global correlations (e.g., Halverson et al., 2005). Prior to the relatively low-amplitude Bitter Springs excursion at ca.

810 Ma (Macdonald et al., 2010), the carbon-isotope trend through the Mesoproterozoic through middle Tonian had been very stable (see Precambrian chapter Fig. 3.2). At 735–740 Ma, approximately 15 myr before the Sturtian glaciation at the base of the Cryogenian, the Islay excursion is a sharp, high-amplitude (ca. 10 per-mille excursion) negative anomaly at 735–740 Ma (e.g., Strauss et al., 2014).

Each of the main glacial events (Sturtian, Marinoan, Gaskiers) is preceded by a sharp negative excursion in carbon isotopes that peaks just before the onset of the glacial episode; and the thin cap carbonate that follows the sudden termination of the glacial interval is also in a negative excursion (e.g., Halverson and Shields-Zhou, 2011; Shields-Zhou et al., 2012). This direct association with glaciations led to speculation that aspects of these carbon-isotope excursions may be caused by near-extinctions of marine life, buildup in volcanogenic carbon-dioxide levels in the atmosphere during Snowball Earth conditions until a greenhouse threshold was reached, or other processes (e.g., Hoffman et al., 1998). However, between the glacial episodes there were also pronounced negative excursions, such as the reported Taishir excursion during the mid-Cryogenian, and the major Shuram excursion after the Gaskiers glaciation (Fig. 4.2).

Strontium and sulfur isotopes also have distinctive trends through the Cryogenian and Ediacaran (e.g., reviews in Halverson et al., 2011; Shields-Zhou et al., 2012; Narbonne et al., 2012).

Magnetostratigraphy has been underutilized in the Cryogenian–Ediacaran, although the few studies indicate potential for high-resolution correlation. For example, the nadir of the Shuram carbon-isotope anomaly is near the base of a normal-polarity zone that can be used for interregional correlation (Minguez et al., 2015; Kodama, 2015), and the late Ediacaran may be predominantly reversed polarity with frequent normal-polarity zones (Bazhenov et al., 2016).

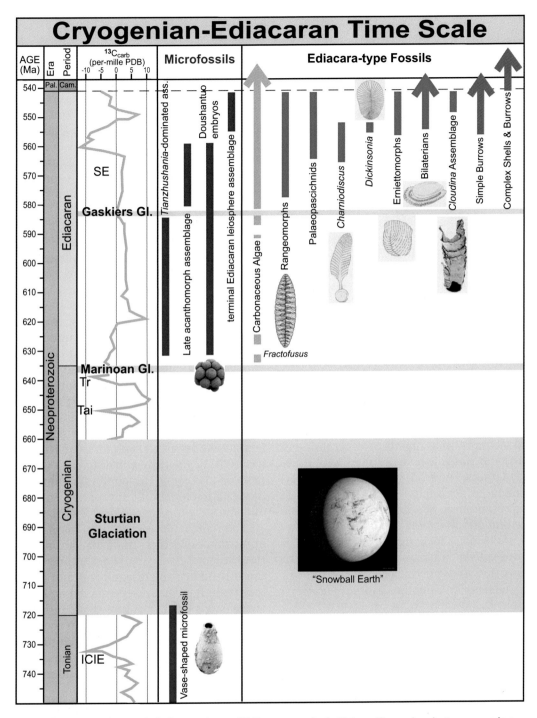

Figure 4.2 Selected major trends in Cryogenian and Ediacaran geologic history. The carbon-isotope curve is a smoothed version modified from the synthesis for the late Proterozoic by Cohen and Macdonald (2015) calibrated by them to the Cryogenian time scale of Rooney et al. (2015)— SE, Tr, Tai, and ICIE are the Shuram, Trezona, Taishir, and Islay carbon-isotope excursions, respectively. Ranges and images of organic-walled microfossils, Ediacaran metazoans, and bioturbation styles are from Narbonne et al. (2012). Additional geochemical trends, biostratigraphic ranges, regional stages, and details on calibrations are compiled in Shields-Zhou et al. (2012) and Narbonne et al. (2012).

The supercontinent of Rodinia was undergoing progressive rifting immediately prior to and through the Cryogenian. The Franklin giant dike swarm in north Canada and northwest Greenland at ca. 720 Ma (Ernst et al., 2008) coincides approximately with the onset of the Sturtian glaciation at the base of the Cryogenian.

The rapid transgressions from the melting Cryogenian and Gaskiers glaciations were accompanied by deposition of organic-rich units on the continental shelves. These are the earliest major source rocks for commercially produced petroleum and natural gas, including fields in Oman and Australia (e.g., reviews in Craig et al., 2009; Ghori et al., 2009).

(2) Biostratigraphy and major trends

The late Proterozoic includes the evolution and radiation of eukaryotes and metazoans. Prior to the advent of the diverse communities of metazoans in the late Ediacaran, the fossil record consists mainly of a succession of microscopic organic-walled spherical or vase-shaped forms. These are grouped under a general name of "acritarchs," but probably consist of representatives of several phyla, including algae and possible metazoan egg cases. However, molecular-clock analyses of the DNA of modern phyla indicate that all major stem groups (red and green algae, amoeba protozoa, ciliates, foraminifera, and metazoans) originated between about ca. 800 to 700 Ma (reviewed in Cohen and Macdonald, 2015). These molecular-clock studies imply that there was a major explosion in eukaryote evolution preceding and during the Sturtian glaciation, although verifying these predictions in the preserved fossil record is a challenge.

A distinctive acritarch, *Cerebrosphaera buickii* (Fig. 4.3B), appears globally at the approximate time of the Bitter Springs carbon-isotope anomaly at 800 Ma, and became extinct before the beginning of the Sturtian glaciation (Grey et al., 2011; Shields-Zhou et al., 2012).

Vase-shaped microfossils (Fig. 4.3A), which have been interpreted as the preserved tests of Amoebozoa or Rhizaria (e.g., Porter et al., 2003) and have potential for biostratigraphic correlation of different facies, appear in abundance only after the Bitter Springs excursion, and disappear from the marine record at the beginning of the Cryogenian (Strauss et al., 2014; Cohen and Macdonald, 2015).

After the Cryogenian glaciations, the diversity of eukaryotes rapidly expanded. Relatively large spiny acanthomorphic acritarchs thrived during the early Ediacaran, and a succession of their biozones has been independently established in Australia and in South China (e.g., Narbonne et al., 2012; Xiao et al., 2014a). One class of large acanthomorphic acritarch called *Tianzhushania* that is present as phosphatized or silicified microfossils in the Doushantuo Formation of South China, has a controversial interpretation of the preserved embryos of early animals (e.g., Xiao and Knoll (2000) and Yin et al. (2013) versus nonembryo interpretation of Huldtgren et al. (2011); see review in Xiao et al. (2014b)). Smooth leiosphaerid acritarch forms are more characteristic of the upper Ediacaran, and vanish at the base of the Cambrian.

The most famous Ediacaran fossils are the appearance of diverse metazoan animals after the Gaskiers glacial episode. Impressions of the soft-bodied to stiffened (but not biomineralized) organisms are preserved on bedding surfaces, especially when a clastic turbidite or storm bed suddenly entombed an ecosystem. Some types are bilateral forms that may have been related to the later Cambrian animals, but most cannot be placed confidently into any post-Ediacaran group.

The earliest "Avalon Assemblage" is preserved in relatively deep water facies in Newfoundland and Britain and dominated by frond-like rangeomorphs, such as *Charnia* (Fig. 4.4). These have fractal-architecture branches from a central stalk, which was attached to the seafloor in some types, and

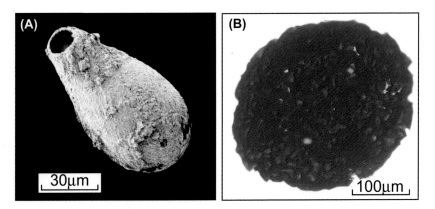

Figure 4.3 Examples of advanced microfossils that became extinct at the beginning of the first Cryogenian global glaciation. (A) A typical vase-shaped microfossil, *Bonniea dacruchares*, from the Chuar Group, western United States *(Photo courtesy of S. Porter; for details, see Porter et al. (2003))*. These and other vase-shaped microfossils are interpreted as tests of Amoebozoa or Rhizaria. (B) The distinctive acritarch, *C. buickii*, from the Hussar Formation, Officer Basin, Australia *(Photo courtesy of K. Grey; see Grey et al., 2011)*. These and images of other typical microfossils are in Shields-Zhou et al. (2012).

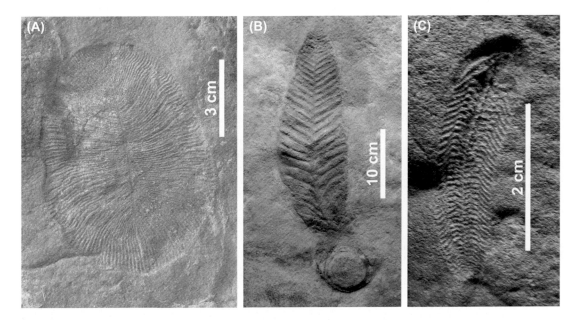

Figure 4.4 Examples of Ediacaran metazoans. (A) *Dickinsonia*, a flat-segmented animal that moved over the seafloor; (B) *Charniodiscus arboreus*, a frond attached to the seafloor by a disk; and (C) the segmented *Spriggina*, which is the first animal with a head; this animal has similarities with early arthropods. Photos by Gabi Ogg taken in the South Australia Museum in Adelaide and in the Flinders Ranges of South Australia (2012).

some reached lengths of over 1 m (e.g., Narbonne et al., 2009; Liu et al., 2015).

The younger and more diverse shallow-water ecosystems of the White Sea Assemblage and Nama Assemblage include rangeomorphs, and the appearances of bilaterians, crawling or gliding animals, shallow-burrowing animals, and evidence of sexual reproduction (e.g., Droser and Gehling, 2015). Excavation of bedding planes below sand beds in the Ediacara Member of the Rawnsley Quartzite in the Flinders Range reveal a range of lifestyles in these complex ecosystems that were developed on microbial mats. For example, the oval-bodied *Dickinsonia* that grew up to 50-cm remained stationary for periods of time while decomposing the microbial mat before moving to the next feeding site (Droser and Gehling, 2015).

The latest Ediacaran has a different ecosystem with an abundance of calcified megafossils of *Cloudina* and *Namacalathus*, which, along with the soft-bodied erniettomorphs (biserially quilted tubes alternately arranged from a central midline) became extinct at the Ediacaran–Cambrian boundary. Indeed, none of the main Ediacaran macrofossils types are preserved in the earliest Cambrian; and there are many hypotheses ranging from ecosystem disruption to predation that explore this mysterious mass extinction (e.g., review in Laflamme et al., 2013). The basal Cambrian has only small shelly fossils (e.g., *Anabarites trisulcatus*; Rogov et al., 2015), acritarch microfossils, and the nonpreserved burrowing animals.

Numerical age model

GTS2012 age model and potential future enhancements

The ages for Cryogenian and Ediacaran glacial episodes, carbon-isotope excursions, and evolutionary stages are constrained by suites of U-Pb and Re-Os dates (e.g., Condon and Bowring, 2011; Noble et al., 2015). Other than relative meter-level positions within stratigraphic sections, there have been only rare applications of cycle stratigraphy to more accurately scale the duration and placement of events and excursions. Indeed, the uncertainties on the placement of events in the schematic summary of Fig. 4.2 are probably greater than 5 myr in many cases.

Revised ages compared to GTS2012

Base of Cambrian (retained **541 Ma** in GTS2012 with qualifier): Temporarily set as nadir of BAsal Cambrian carbon-isotope negative Excursion (BACE) dated as 541 Ma—see discussion on base-Cambrian GSSP. Landing et al. (2013) suggest that 543 Ma may be best estimate for the oldest appearance of *Trichophycus pedum* trace-fossil assemblage.

Base of Cryogenian (720 vs 850 Ma in GTS2012): The base of the Cryogenian Period was initially set at 850 Ma (Plumb, 1991), but was revised in 2014–15 to the ca. 720 Ma date of the onset of the first global glaciation—the criterion for placement of a future GSSP.

Acknowledgments

This brief summary of selected highlights and current stratigraphic issues relied heavily on the detailed overview and synthesis by Shields-Zhou et al. (2012), by Narbonne et al. (2012), and by Van Kranendonk et al. (2012), and on an extensive field trip through the Cryogenian and Ediacaran of South Australia with Jim Gehling. Shuhai Xiao reviewed an early version of the graphics, database and text.

Selected publications and websites

Cited publications

Only select publications were cited in this review with an emphasis on aspects of post-2011 updates. Pre-2011 literature is well summarized in the syntheses by Shields-Zhou et al. (2012) and by Narbonne et al. (2012) and in some of the publications cited below.

Arnaud, E., Halverson, G.P., Shields-Zhou, G., 2011. Chapter 1: The geologic record of Neoproterozoic glaciations. In: Arnaud, E., Halverson, G.P., Shields-Zhou, G. (Eds.), *The Geological Record of Neoproterozoic Glaciations*. Geological Society, **vol. 36**. Memoirs, London, pp. 1–16. http://dx.doi.org/10.1144/M36.1.

Bazhenov, M.I., Levashova, N.M., Meert, J.G., Golovanova, I.V., Danukalov, K.N., Fedorova, N.M., 2016. Late Ediacaran magnetostratigraphy of Baltica: evidence for magnetic field hyperactivity? *Earth and Planetary Science Letters* **435**: 124–135. http://dx.doi.org/10.1016/j.epsl.2015.12.015.

Calver, C.R., Crowley, J.L., Wingate, M.T.D., Evans, D.A.D., Raub, T.D., Schmitz, M.D., 2013. Globally synchronous Marinoan deglaciation indicated by U-Pb geochronology of the cottons Breccia, Tasmania, Australia. *Geology* **41**: 1127–1130. http://dx.doi.org/10.1130/G34568.1.

Cohen, P.A., Macdonald, F.A., 2015. The Proterozoic record of eukaryotes. *Paleobiology* **41**: 610–632. http://dx.doi.org/10.1017/pab.2015.25.

Condon, D.J., Bowring, S.A., 2011. Chapter 9: a user's guide to Neoproterozoic geochronology. In: Arnaud, E., Halverson, G.P., Shields-Zhou, G. (Eds.), *The Geological Record of Neoproterozoic Glaciations*. Geological Society, Memoirs, **vol. 36.** London, pp. 135–149. http://dx.doi.org/10.1144/M36.9.

Craig, J., Thurow, J., Thusu, B., Whitham, A., Abutarruma, Y., 2009. Global Neoproterozoic petroleum systems: The emerging potential in North Africa. In: Craig, J., Thurow, J., Thusu, B., Whitham, A., Abutarruma, Y. (Eds.), *Global Neoproterozoic Petroleum Systems: The Emerging Potential in North Africa*. Geological Society, **vol. 326**. Special Publications, London, pp. 1–25. http://dx.doi.org/10.1144/SP326.1.

Droser, M.L., Gehling, J.G., 2015. The advent of animals: the view from the Ediacaran. *Proceedings of the National Academy of Sciences of the United States of America (PNAS)*, **vol. 112**, pp. 4865–4870. www.pnas.org/cgi/doi/10.1073/pnas.1405338112.

Ernst, R.E., Wingate, M.T.D., Buchan, K.T., Li, Z.X., 2008. Global record of 1600-700 Ma Large Igneous Provinces (LIPs): Implications for the reconstruction of the proposed Nuna (Columbia) and Rodinia supercontinents. *Precambrian Research* **160**: 159–178.

Ghori, K.A.R., Craig, J., Thusu, B., Lüning, S., Geiger, M., 2009. Global Infracambrian petroleum systems. In: Craig, J., Thurow, J., Thusu, B., Whitham, A., Abutarruma, Y. (Eds.), *Global Neoproterozoic Petroleum Systems: The Emerging Potential in North Africa*. Geological Society, **vol. 326**. Special Publications, London, pp. 109–136. http://dx.doi.org/10.1144/SP326.6.

Grey, K., Hill, A.C., Calver, C., 2011. Biostratigraphy and stratigraphic subdivision of Cryogenian successions of Australia in a global context In: (Arnaud, E., Halverson, G.P., Shields-Zhou, G. (Eds.). *The Geological Record of Neoproterozoic Glaciations*. Geological Society, Memoirs, **vol. 36**, London, pp. 51–66.

Guerroué, E.L., 2010. Duration and synchroneity of the largest negative carbon isotope excursion on Earth: the Shuram/Wonoka anomaly. *Comptes Rendus Geoscience* **342**: 204–214. http://dx.doi.org/10.1016/j.crte.2009.12.008.

Halverson, G.P., Hoffman, P.F., Schrag, D.P., Maloof, A.C., 2005. Towards a Neoproterozoic composite carbon-isotope record. *Geological Society of America Bulletin* **117**: 1181–1207.

Halverson, G.P., Shields-Zhou, G., 2011. Chapter 4: chemostratigraphy and the Neoproterozoic glaciations. In: Arnaud, E., Halverson, G.P., Shields-Zhou, G. (Eds.), *The Geological Record of Neoproterozoic Glaciations*. Geological Society, Memoirs, **vol. 36**, London, pp. 51–66. http://dx.doi.org/10.1144/M36.4.

Hoffman, P.F., Kaufman, A.J., Halverson, G.P., Schrag, D.P., 1998. A Neoproterozoic Snowball Earth. *Science* **281**: 1342–1346.

Huldtgren, T., Cunningham, J.A., Yin, C., Stampanoni, M., Marone, F., Donoghue, P.C.J., Bengtson, S., 2011. Fossilized nuclei and germination structures identify Ediacaran "animal embryos" as encysting protists. *Science* **334**: 1696–1699.

Kirschvink, J.L., 1992. Late Proterozoic low-latitude global glaciation: the Snowball Earth Section 2.3. In: Schopf, J.W., Klein, C., Des Maris, D. (Eds.), *The Proterozoic Biosphere: A Multidisciplinary Study*. Cambridge University Press, pp. 51–52.

Knoll, A.H., Walter, M.R., Narbonne, G.M., Christie-Blick, N., 2006. The Ediacaran Period: a new addition to the geologic time scale. *Lethaia* **39**: 13–30.

Kodama, K.P., 2015. The synchroneity and duration of the Shuram carbon-isotope excursion. *Geological Society of America Abstracts with Programs*. **47**(7): 79. https://gsa.confex.com/gsa/2015AM/webprogram/Paper259731.html.

Laflamme, M., Darroch, S.A.F., Tweedt, S.M., Peterson, K.J., Erwin, D.H., 2013. The end of Ediacara biota: extinction, biotic replacement, or Cheshire Cat? *Gondwana Research* **23**: 558–573.

Landing, E., Geyer, G., Brasier, M.D., Bowring, S.A., 2013. Cambrian evolutionary radiation: context, correlation, and chronostratigraphy – overcoming deficiencies of the first appearance datum (FAD) concept. *Earth-Science Reviews* **123**: 133–172.

Liu, A.G., Kenchington, C.G., Mitchell, E.G., 2015. Remarkable insights into the paleoecology of the Avalonian Ediacaran macrobiota. *Gondwana Research* **27**: 1355–1380. http://dx.doi.org/10.1016/j.gr.2014.11.002.

Macdonald, F.A., Schmitz, M.D., Crowley, J.L., Roots, C.F., Jones, D.S., Maloof, A.C., Strauss, J.V., Cohen, P.A., Johnston, D.T., Schrag, D.P., 2010. Calibrating the Cryogenian. *Science* **327**: 1241–1243. http://dx.doi.org/10.1126/science.1183325.

Minguez, D., Kodama, K.P., Hillouse, J.W., 2015. Paleomagnetic and cyclostratigraphic constraints on the synchroneity and duration of the Shuram carbon isotope excursion, Johnnie Formation, Death Valley Region, CA. *Precambrian Research* **266**: 395–408.

Narbonne, G.M., Laflamme, M., Grfeentree, C., Trusler, P., 2009. Reconstructing a lost world: Ediacaran rangeomorphs from Spaniard's Bay, Newfoundland. *Journal of Paleontology* **83**(4): 503–523.

Narbonne, G.M., Xiao, S., Shields, G.H., Gehling, J.G., 2012. The Ediacaran Period. In: Gradstein, F.M., Ogg, J.G., Schmitz, M., Ogg, G., (Coordinators). *The Geologic Time Scale 2012*. Elsevier Publisher, pp. 413–435. http://dx.doi.org/10.1016/B978-0-444-59425-9.00018-4 (An overview on all aspects, including graphics on the ratified GSSP, climatic and geochemical trends, diagrams of biostratigraphic scales, and discussion on the age models.).

Noble, S.R., Condon, D.J., Carney, J.N., Wilby, J.N., Wilby, P.R., Pharaoh, T.C., Ford, T.D., 2015. U-Pb geochronology and global context of the Charnian Supergroup, UK: constraints on the age of key Ediacaran fossil assemblages. *Geological Society of America Bulletin* **127**: 250–265. http://dx.doi.org/10.1130/B31013.1.

Plumb, K.A., 1991. New Precambrian time scale. *Episodes* **14**: 139–140.

Porter, S.M., Meisterfeld, R., Knoll, A.H., 2003. Vase-shaped microfossils from the Neoproterozoic Chuar Group, Grand Canyon: a classification guided by modern testate amoebae. *Journal of Paleontology* **77**: 409–429.

Rogov, V.I., Karlova, G.A., Marusin, V.V., Kochnev, B.B., Nagovitsin, K.E., Grazhdankin, D.V., 2015. Duration of the first biozones in the Siberian hypostratotype of the Vendian. *Russian Geology and Geophysics* **56**: 573–585. http://dx.doi.org/10.1016/j.rgg.2015.03.016.

Rooney, A.D., Strauss, J.V., Brandon, A.D., Macdonald, F.A., 2015. A Cryogenian chronology: two long-lasting synchronous Neoproterozoic glaciations. *Geology* **43**: 459–462. http://dx.doi.org/10.1130/G36511.1.

Shields-Zhou, G.A., Hill, A.C., Macgabhann, B.A., 2012. The Cryogenian Period. In: Gradstein, F.M., Ogg, J.G., Schmitz, M., Ogg, G., (Coordinators). *The Geologic Time Scale 2012*. Elsevier Publisher, pp. 393–411. http://dx.doi.org/10.1016/B978-0-444-59425-9.00017-2 (An overview on all aspects, including summaries of climatic episodes, evolution of life and age models.).

Smith, A.G., 2009. Neoproterozoic timescales and stratigraphy. In: Craig, J., Thurow, J., Thusu, B., Whitham, A., Abutarruma, Y. (Eds.), *Global Neoproterozoic Petroleum Systems: The Emerging Potential in North Africa*. Geological Society, **vol. 326**. Special Publications, London, pp. 27–54. http://dx.doi.org/10.1144/SP326.2.

Strauss, J.V., Rooney, A.D., Macdonald, F.A., Brandon, A.D., Knoll, A.H., 2014. 740 Ma vase-shaped microfossils from Yukon, Canada: implications for Neoproterozoic chronology and biostratigraphy. *Geology* **42**: 659–662.

Subcommission on Cryogenian Stratigraphy, 2014. Annual report 2014. In: *International Commission on Stratigraphy (ICS) Annual Report 2014. Submitted to International Union of Geological Sciences (IUGS)* Available at: http://iugs.org/uploads/ICS%202014.pdf.

Subcommission on Ediacaran Stratigraphy, 2014. Annual report 2014. In: *International Commission on Stratigraphy (ICS) Annual Report 2014. Submitted to International Union of Geological Sciences (IUGS)* Available at: http://iugs.org/uploads/ICS%202014.pdf.

Van Kranendonk, M.J., Altermann, W., Beard, B.L., Hoffman, P.F., Johnson, C.J., Kasting, J.F., Melezhik, V.A., Nutman, A.P., Papineau, D., Pirajno, F., 2012. A chronostratigraphic division of the Precambrian: possibilities and challenges. In: Gradstein, F.M., Ogg, J.G., Schmitz, M., Ogg, G., (Coordinators). *The Geologic Time Scale 2012*. Elsevier Publisher, pp. 299–392. http://dx.doi.org/10.1016/B978-0-444-59425-9.00016-0 (An overview on all aspects, including summaries of tectonic cycles, atmosphere-ocean history, climatic episodes and evolution of life; plus age models and a set of suggested chronostratigraphic divisions.).

Wang, Z.J., Huang, Z.G., Yao, J.X., Ma, X.L., and National Committee of Stratigraphy of China, 2014. Characteristics and main progress of *The Stratigraphic Chart of China and Directions*. *Acta Geoscientifica Sinica*, **35**: 271–276 (with chart attachment).

Xiao, S., Knoll, A., 2000. Phosphatized animal embryos from the Neoproterozoic Doushantuo formation at Weng'an, Guizhou, South China. *Journal of Paleontology* **74**: 767–788.

Xiao, S., Zhou, C., Liu, P., Wang, D., Yuan, X., 2014a. Phosphatized acanthomorphic acritarchs and related microfossils from the Ediacaran Doushantuo Formation at Weng'an (South China) and their implications for biostratigraphic correlation. *Journal of Paleontology* **88**: 1–67.

Xiao, S., Muscente, A.D., Chen, L., Zhou, C., Schiffbauer, J.D., Wood, A.D., Polys, N.F., Yuan, X., 2014b. The Weng'an biota and the Ediacaran radiation of multicellular eukaryotes. *National Science Review* **1**: 498–520.

Yin, Z., Zhu, M., Tafforeau, P., Chen, J., Liu, P., Li, G., 2013. Early embryogenesis of potential bilaterian animals with polar lobe formation from the Ediacaran Weng'an Biota, South China. *Precambrian Research* **225**: 44–57.

Zhou, C., Tucker, R., Xiao, S., Peng, Z., Yuan, X., Chen, Z., 2004. New constraints on the ages of Neoproterozoic glaciations in South China. *Geology* **32**: 437–440.

Websites (selected)

Subcommission on Ediacaran Stratigraphy (ICS)—http://www.paleo.geos.vt.edu/Ediacaran/—Includes "Edies" newsletter.

Snowball Earth – www.snowballearth.org—Website originally developed by Paul Hoffman with National Science Foundation funding to provide online explanations, teaching slides, and extensive bibliography (through 2009).

First Life (by David Attenborough, 2010)—http://firstlifeseries.com, with streaming video at https://www.dailymotion.com/embed/video/xylzdw_first-life-arrival_tech among other sources— "Arrival" is first episode (60 min) of a two-part BBC documentary with extensive coverage of Ediacaran biota, including computer simulations and onsite field work/interviews.

Palaeos: Cryogenian and Ediacaran—http://palaeos.com/proterozoic/neoproterozoic/cryogenian/cryogenian.html and http://palaeos.com/proterozoic/neoproterozoic/ediacaran/ediacaran.htm—A well-presented suite of diverse topics for a general science audience that was originally compiled by M. Alan Kazlev in 1998–2002.

Mistaken Point and Rangeomorph Reproduction—streaming videos at http://www.palaeocast.com/episode-5-mistaken-point/ (60 min, 2012) and http://www.palaeocast.com/episode-50-rangeomorph-reproduction/ (40 min, 2015)—One of the best known and most important Ediacaran localities is at Mistaken Point, Newfoundland, Canada. These podcasts examine aspects of the nature of its biota.

5

CAMBRIAN

510 Ma Cambrian

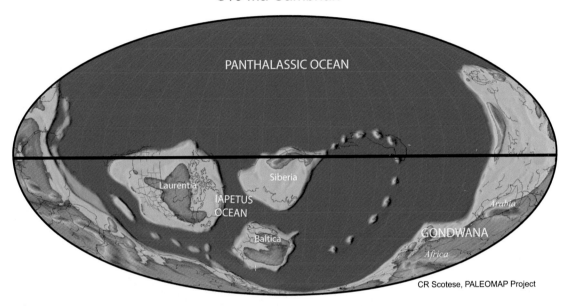

PANTHALASSIC OCEAN

Laurentia

Siberia

IAPETUS OCEAN

Baltica

Arabia

GONDWANA

Africa

CR Scotese, PALEOMAP Project

Mid-Cambrian paleogeographic reconstruction (Sea level+40) from Scotese (2014). Some other authors (e.g., Landing et al., 2013) suggest from facies and biota patterns that most of the continental blocks were in more temperate to tropical paleolatitudes.

Basal definition and status of international subdivisions

The Cambrian is characterized by the appearance of mineralized skeletons of animals. The initial three stages (ca. 25 myr) are major revolutions in Earth's life—(1) the advent of deep complex burrowing of sediments at ca. 540 Ma; (2) the appearance of diverse multicellular animals with "small shelly" mineralized skeletons at ca. 530 Ma; and (3) the appearance of larger trilobites, pelagic agnostoid arthropods, and brachiopods at ca. 520 Ma (e.g.,

synthesis in Landing et al., 2013). The bases of the next seven Cambrian stages, each spanning ca. 5 myr, correspond to widespread appearances of distinctive trilobites, pelagic agnostoid arthropods, or conodonts (Fig. 5.1). These Cambrian biological events are often associated with major oscillations in the carbon cycle.

Terreneuvian series

Fortunian: The **Ediacaran/Cambrian** boundary (base of Terreneuvian Series and Fortunian Stage) was placed at one of the

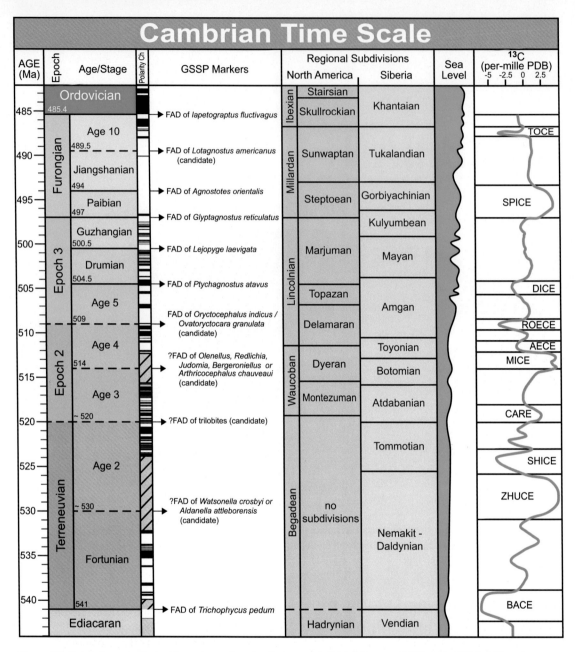

Figure 5.1 Cambrian overview. The main markers for the currently (as of January 2016) ratified Global Boundary Stratotype Sections and Points (GSSPs) of Cambrian stages are the trace fossil *Tr. pedum* for the base of the Cambrian and first-appearance datums (FAD) of cosmopolitan agnostoid arthropod taxa in late Cambrian, as discussed in the text and summarized in Fig. 5.5. ("Age" is the term for the time equivalent of the rock-record "stage.") Magnetic polarity scale is a composite by Peng et al. (2012), which included a Furongian pattern modified from Kouchinsky et al. (2008) and an early Cambrian modified from a Siberian compilation by Varlamov et al. (2008), but most of the polarity pattern awaits verification. Regional subdivisions are a selected subset of the extensive regional correlation chart by Peng et al. (2012). Schematic sea-level curve is modified from Haq and Schutter (2008) following advice of Bilal Haq (pers. comm., 2008); although Babcock et al. (2015) have a slightly different sea-level version that emphasizes that the FADs of the GSSP-marker agnostoid arthropods coincide with rapid regional coastal onlaps. The $\delta^{13}C_{carb}$ curve with major widespread events is modified from Zhu et al. (2006) [*see their text for explanations of their acronyms*]. The vertical scale of this diagram is standardized to match the vertical scales of the first stratigraphic summary figure in all other Phanerozoic chapters.

"greatest enigmas of the fossil record; i.e., the relatively abrupt appearance of skeletal fossils and complex, deep burrows in sedimentary successions around the world" (Brasier et al., 1994). The GSSP level in Newfoundland, Canada, was placed at the beginning of a rapidly diversifying assemblage of trace fossils of burrowers and complex feeding tracks, of which the relatively large burrows called *Phycodes* (now classified as *Treptichnus* or *Trichophycus*) *pedum* is the most distinctive (Fig. 5.2). Underlying deposits have an assemblage ("*Harlaniella podolica*" Ichnozone) of only shallow burrows and surface trails. This *Tr. pedum* deep burrowing appears relatively suddenly in the majority of preserved shelf facies, and is just after the disappearance of *Cloudina* and other typical Ediacaran fossils.

However, the lowest *Tr. pedum* burrows were later found about 4.4 m below the GSSP level (Gehling et al., 2001). Although this offset does not change the main philosophy of the GSSP as representing a major change in Earth's marine ecosystems (e.g., Landing et al., 2013), it has generated discussions on whether to redefine the Ediacaran/Cambrian boundary to coincide with a more precise geochemical or other marker that can be recognized in more settings (e.g., Babcock et al., 2014). One option is to use the beginning or the peak of the "**BA**sal **C**ambrian carbon-isotope negative **E**xcursion" (BACE in Fig. 5.1) (Babcock et al., 2014). Radio-isotopic dating of ash beds in Oman yielded 541.00 ± 0.13 Ma near the BACE peak (Bowring et al., 2007), and this date was used in GTS2012 for the estimated age of the Precambrian/Cambrian boundary (Peng et al., 2012; Narbonne et al., 2012). Landing et al. (2013) suggest that the base of the *Tr. pedum* Assemblage Zone is below the peak of the BACE and suggest an age of ca. 543 Ma. However, the relative appearance of *Tr. pedum* burrowing ecosystems within the BACE is poorly known (Babcock et al., 2014), and carbon-isotope stratigraphy is not possible in noncalcareous sections such as the present GSSP in Newfoundland.

For simplicity, pending future high-resolution correlations and dating, the diagrams of Figs. 5.1 and 5.4 equate the Ediacaran/Cambrian boundary with the BACE peak, the age of 541.0 Ma, and the base of the *Tr. pedum* trace-fossil assemblage zone.

Stage 2: The next major Cambrian evolutionary event was a diversification of animal skeletons of micromollusks and many types of "small shelly fossil" taxa of uncertain affinity with phosphatic or calcareous minerals. Provisional Stage 2 has been proposed to begin with the widespread appearance of these types of small shelly fossils, especially the *Watsonella crosbyi* (a possible micromollusk bivalve rostroconch) and *Aldanella attleborensis* (a possible microgastropod). This biological event is near the onset of a major **ZHU**jianqing **C**arbon-isotope positive **E**xcursion (*ZHUCE*), named after the lower Cambrian Zhujianqing Formation of eastern Yunnan (China). The mollusk *Watsonella crosbyi* had been described under other names, such as *Heraultia (Heraultipegma) sibirica* and *Watsonella yunnanensis*, and after the synonymies were established it was proposed to be the primary marker for the base of Stage 2 (e.g., Li et al., 2011). Landing et al. (2013) examined diachroneity problems with the FAD of *Wat. crosbyi* and other taxa in this boundary interval. They proposed placing the GSSP within the lower range of *Wat. crosbyi* at the peak of ZHUCE, 9.4 m below the top of the Dahai Member in the Laolin section in Yunnan province, South China. This level may have an age close to 531 Ma (e.g., Maloof et al., 2010a,b; Landing et al., 2013).

Series 2

Stage 3: The appearance of the earliest trilobite skeletal remains has been the preferred marker for the base of provisional Series 2 and Stage 3. However, the oldest trilobites in each region are endemic and include *Profallotaspis* species in Siberia, *Fritzaspis generalis* in

Base of the Fortunian Stage of the Cambrian System at Fortune Head, southeastern Newfoundland, Canada

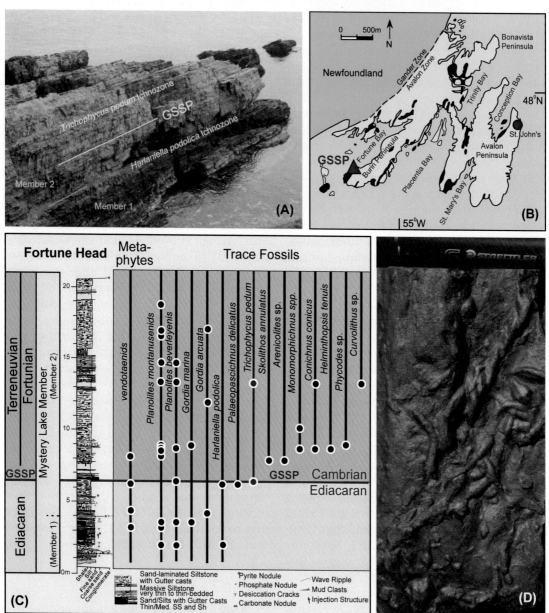

Figure 5.2 GSSP for base of the Cambrian (base of Terreneuvian Series, base of Fortunian Stage) at the Green Point section, Newfoundland, eastern Canada. The GSSP level was selected to coincide with the first-appearance datum (FAD) of the distinctive trace fossil *Trichophycus pedum* (shown in image (D)). Photographs and the ranges of trace fossils (as published when the boundary was ratified in 1992) are from Peng et al. (2012). Occurrences of the *T. pedum* trace fossil were later found 4.4 m lower in this section (Gehling et al., 2001).

Nevada–California, and *Hupetina antiqua* in Morocco; therefore, it is difficult to determine if these are synchronous (e.g., Bushuev et al., 2014; summarized in Landing et al., 2015a; and in Clausen et al., 2015). Other alternatives for interregional correlation and primary markers for a base of Stage 3 include FADs of a micromollusk *Pelagiella subangulata*, a lobopod sclerite *Microdyction effusum*, or an archaeocyathid (Clausen et al., 2015). Another suggested level for the Stage 3 GSSP is the peak of the positive carbon-isotope excursion (***Cambrian Arthropod Radiation isotope Excursion*** or *CARE* of Zhu et al., 2006) that occurs shortly after the appearance of trilobites (Landing et al., 2015a).

Stage 4: The primary marker for provisional **Stage 4** is intended to be the FAD of a widespread trilobite taxon. Current possibilities include a species of the genus *Olenellus* (sensu lato), *Redlichia* (sensu lato), *Judomia*, or *Bergeroniellus*; or the species *Arthricocephalus chauveaui* (e.g., Peng et al., 2012).

Series 3

Stage 5: The *Olenellus* and *Redlichia* genera became extinct at the end of Stage 4 at the onset of a major negative "***Redlichid and Olenellid trilobites Extinction Carbon-isotope Excursion***" (*ROECE*) (Zhu et al., 2006). The peak of this sharp isotopic event, which is near the traditional Lower/Middle Cambrian boundary, would be an important correlation marker for the base of provisional Series 3 and Stage 5. The emergence of new trilobite taxa that appeared in shallow warm waters was highly endemic, therefore possible biostratigraphic markers suggested for a future Stage 5 GSSP are cooler-water polymerid trilobites, such as the FAD of *Ovatoryctocara granulata* or the FAD of *Oryctocephalus indicus* at the base of the next higher biozone, although interregional correlations are uncertain (Clausen et al., 2015). The famous Burgess Shale fauna is in the middle of Stage 5.

Agnostoids, a type of pelagic arthropod often included within the trilobite group, are the most useful for interregional correlation in the middle and late Cambrian. They first appear slightly older than the beginning of Stage 5, and are the primary correlation criteria for the GSSPs of the Drumian through Jiangshanian stages:

Drumian: The GSSP of the second stage of provisional Series 3 coincides with the FAD of agnostoid arthropod *Ptychagnostus atavus* in the Drum Mountains of Utah, United States.

Guzhangian: The GSSP of the third stage of provisional Series 3 coincides with the FAD of agnostoid arthropod *Lejopyge laevigata* in the Louyixi section of northwestern Hunan, China.

Furongian series

Paibian: The GSSP for the base of **Furongian** Series in the Paibi section of northwestern Hunan coincides with the FAD of agnostoid arthropod *Glyptagnostus reticulatus* (Fig. 5.3).

Jiangshanian: The GSSP for the second stage of Furongian Series coincides with the FAD of agnostoid arthropod *Agnostotes orientalis* near Duibian village, Jiangshan County, western Zhejiang Province, China. The Auxiliary boundary Stratotype Section and Point is in the Kyrshabakty section of southern Kazakhstan (Ergaliev et al., 2014).

Stage 10: The final provisional Stage 10 of the Cambrian currently has two main proposed levels for the GSSP. The first is the FAD of agnostoid arthropod *Lotagnostus americanus*. Three candidate GSSP sections using this level are at Wa'ergang, Hunan, China, at Khos-Nelege, western Yakutia Russia, and at Kyrshabakty, southern Kazakhstan (Lararenko et al., 2011; Ergaliev et al., 2014; Peng et al., 2014, 2015). The FAD of agnostoid arthropod *Lotagnostus americanus* in the Wa'ergang section is close to the lowest-known occurrences of the intercontinentally distributed polymerid trilobites

Base of the Paibian Stage of the Cambrian System in the Paibi Section in the Wuling Mountains, NW Hunan Province, China

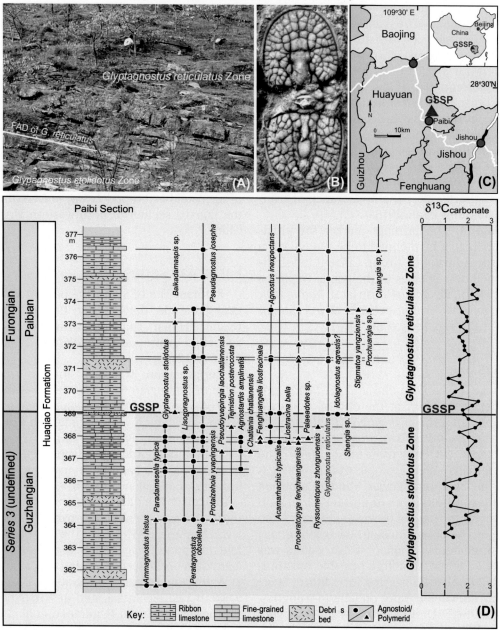

Figure 5.3 GSSP for base of the Furongian Series (base of Paibian Stage) at the Paibi section, Paibi, northwest Hunan, South China. The GSSP level coincides with the lowest occurrence of agnostoid arthropod *Glyptagnostus reticulatus*. Photographs and the stratigraphic section showing ranges of agnostoids and trilobites and the relatively elevated $\delta^{13}C_{carb}$ values of the SPICE interval (see Fig. 5.1) are from Peng et al. (2012).

Hedinaspis regalis and *Charchaqia norini*, and is just above a transgressive surface that is correlative with the onset of a minor positive excursion in carbon isotopes. The second and higher level is the FAD of conodont *Eoconodontus notchpeakensis*, which is near the onset of a distinctive negative carbon-isotope excursion ("**HE**llnmaria–**R**ed Tops **B**oundary," or HERB excursion) and has a candidate GSSP in the House Range of Utah, United States (Landing et al., 2011; Miller et al., 2015).

Selected main stratigraphic scales and events

(1) Biostratigraphy and major trends

The strong faunal provincialism throughout the Cambrian has limited the usefulness of biostratigraphy using benthic and nearshore fauna (e.g., syntheses in Harper and Servais, editors, 2011; especially Álvaro et al., 2011, for trilobites). Acritarchs, a suite of organic-walled microfossils that probably represent different groups of organisms, are more widespread, but their zonation remains rather broad (e.g., review in Landing et al., 2013).

Pelagic organisms that had both a more widespread distribution and left mineralized remains did not appear until the advent of agnostoid arthropods in the Drumian. Agnostoids are often included within the trilobite group, although agnostoids are eyeless with only two thoracic segments, whereas polymerid trilobites have eyes and a minimum of four segments (and eodiscoid trilobites have three segments). The earliest true conodonts (phosphatic jaw elements of a small eel-like animal) appeared in the late Cambrian and are used for biostratigraphy within the Furongian Series.

The early Cambrian pre-trilobite Terreneuvian Series has been a particular challenge for biostratigraphy. The "Small Shelly Fossils" of micromollusks and mineralized skeletons of taxa of uncertain affinities and larger colonial archaeocyathids, an extinct relative of sponges, are very useful for regional zonations (e.g., examples in Fig. 5.4), but global correlations are uncertain. Phosphatic "pseudo-conodont" elements (sometimes inappropriately called "protoconodonts") are present beginning with *Protohertzina* in the middle Terreneuvian.

Series 2 traditionally begins with the appearance of the famous trilobites, which have well-developed zonations that are applied in their separate biogeographic provinces through the rest of the Cambrian (examples in Peng et al., 2012; with selected examples in Fig. 5.4). The widespread pelagic agnostoid arthropods enable the subdivision of Series 3 and the Furongian Series into about a dozen main interregional zones—in ascending order, these are the *Ptychagnostus gibbus, Ptych. atavus, Ptych. punctuosus, Goniagnostus nathorsti, Lejopyge armata, Lejopyge laevigata, Proagnostus bulbus, Linguagnostus reconditus, Glyptagnostus stolidotus, Glypt. reticulatus, Agnostotes orientalis,* and *Lotagnostus americanus* zones; of which the *G. nathorsti* zone is less cosmopolitan. In Fig. 5.4, these agnostoid arthropod zones are partly combined with nonpelagic polymerid trilobites, because hybrid zones are commonly used for regional subdivisions (e.g., correlation diagrams in Peng et al., 2012).

(2) Stable-isotope stratigraphy, magnetostratigraphy, and selected events

The Cambrian had a series of major changes in biota evolution, extinction episodes, carbon-cycle (excursions in $\delta^{13}C_{carb}$), and eustatic sea level; and these events often coincide (e.g., synthesis by Babcock et al., 2015).

The most important tool for global correlation within the Cambrian, especially for the Terreneuvian and Series 2, are major excursions, both negative and positive, in $\delta^{13}C_{carb}$ (Fig. 5.1). These carbon-isotope excursions often coincide with major Cambrian

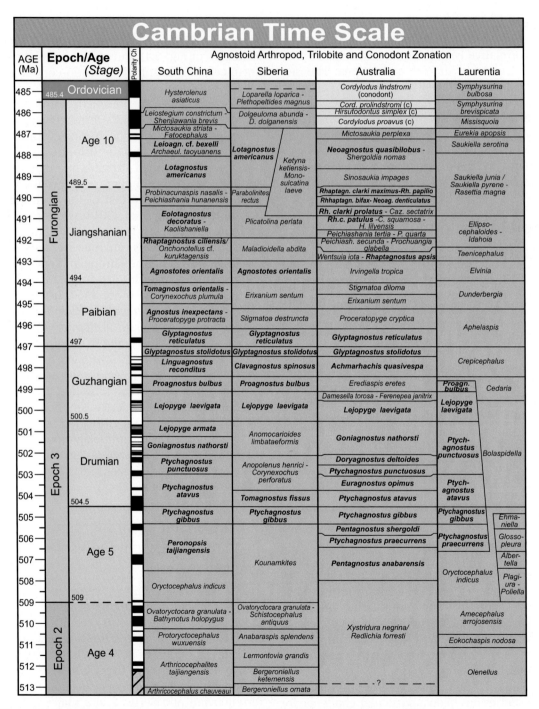

Figure 5.4 (A, B) Selected principal regional biostratigraphic zonal schemes of the Cambrian. ("*Age*" is the term for the time equivalent of the rock-record "*stage.*") Compilation modified from Peng et al. (2012). In the combined zonation columns, agnostoid arthropods are in **bold** and "(c)" denotes a conodont zone.

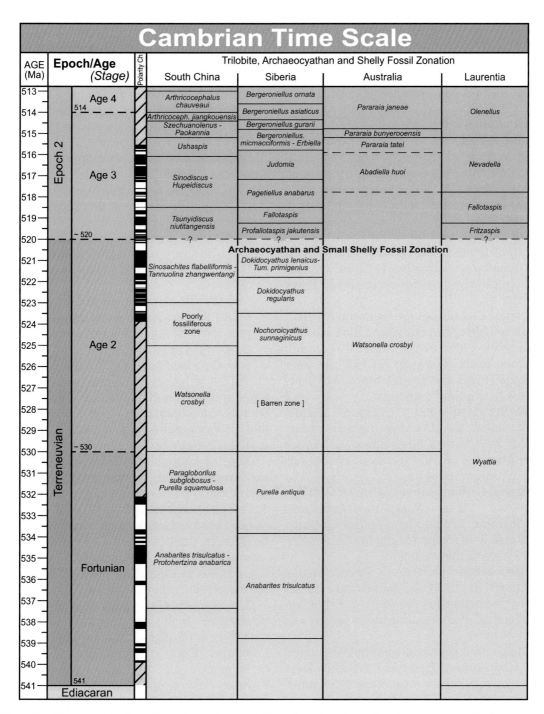

Figure 5.4 (Continued) In the columns for archaeocyathids and small shelly fossils, genera names ending in "-cyathus" are archaeocyathids. Additional zonations, biostratigraphic markers, regional stages, geochemical trends, sea-level curves, and details on calibrations are compiled in Peng et al. (2012) and in the internal data sets within the *TimeScale Creator* visualization system (free at *www.tscreator.org*).

extinction events or faunal turnovers. A set of convenient acronyms for these major carbon-cycle episodes was suggested by Zhu et al. (2006) that also incorporates the SPICE acronym of Saltzman et al. (2000):

BAsal Cambrian Carbon-isotope Excursion (BACE) = large negative excursion. Extinctions of acritarchs and most Ediacaran-type organisms are associated with a strong shift from positive C-13 toward negative values in uppermost Ediacaran.

ZHUjianqing Carbon-isotope Excursion (ZHUCE), named after the basal Cambrian Zhujianqing Formation (eastern Yunnan, China) = large positive excursion in lower Stage 2.

SHIyantou Carbon-isotope Excursion (SHICE), named after Shiyantou Formation (eastern Yunnan, China) = large negative excursion immediately following ZHU-CE in mid-Stage 2 that is coeval with extinction of many small shelly fossil organisms.

Cambrian Arthropod Radiation isotope Excursion (CARE) = large positive excursion near base of Stage 3 associated with appearance of wide variety of arthropods (e.g., earliest trilobites).

MIngxinsi Carbon-isotope Excursion (MICE), named after Mingxinsi Formation (Guizhou, China) = large positive excursion in lower Stage 4 at the time of a major archaeocyathid radiation.

Archaeocyathid Extinction Carbon-isotope Excursion (AECE) = significant negative excursion in mid-Stage 4. Extinction of archaeocyathids occurred in Siberia and elsewhere.

Redlichiid and Olenellid trilobites Extinction Carbon-isotope Excursion (ROECE) = large negative excursion near Stage 4/5 boundary that seems coeval with extinction of redlichiid trilobites in Gondwana and olenellid trilobites in Laurentia, and may be coeval with the Kalkarindji–Antrim large igneous province (LIP) of Australia.

Drumian Isotope-Carbon Excursion (DICE) = negative maximum that nearly coincides with beginning of Drumian stage (although it is mainly at the end of Stage 5). FAD of *Ptychagnostus atavus*, the agnostoid arthropod that is the marker for basal Drumian, occurs in lower part of eustatic rise associated with maximum of DICE.

Steptoean Positive Isotope-Carbon Excursion (SPICE) begins with mass extinction at top of Marjumiid biomere (Marjuman/Steptoean stage boundary; base of Pterocephalid biomere) in the United States. Mass extinction (top of Pterocephalid Biomere; base of Ptychaspid Biomere) is at top of SPICE.

Top of Cambrian carbon-isotope Excursion (TOCE) = large negative excursion accompanying a trilobite mass extinction (end of Ptychaspid Biomere)

Strontium isotopes ($^{87}Sr/^{86}Sr$) slowly decline during the Terreneuvian Series to reach a minimum in Stage 2, and then rapidly rise through Series 2 to a quasiplateau through Series 3 and the Furongian Series (e.g., syntheses by Peng et al., 2012; and McArthur et al., 2012).

Magnetostratigraphy has been underutilized in the Cambrian, even though the limited studies have identified potential polarity scales (e.g., Kirschvink et al., 1991; Kouchinsky et al., 2008; Varlamov et al., 2008; although the Russian publications are sparse on documentation). The schematic polarity pattern in Fig. 5.1 was compiled by Peng et al. (2012) from these and other sources.

The Cambrian is mainly a warm interval of relatively high sea levels. Superimposed on the general high stand are some major lowstands, some of which are interpreted by Babcock et al. (2015) as cooling episodes causing extinctions of marine fauna and with possible extensive glacial ice on the continents. Major rapid rises of sea level that flooded the Cambrian shelves often coincide with the FADs of the agnostoid arthropods that are used as primary markers of Series 3 and Furongian Series stage boundaries (Babcock et al., 2015).

A major LIP, the Antrim and Kalkarindji basalts, erupted over more than 2 million km^2 of west-central and northern Australia at about 511 Ma (minimum age from U-Pb on zircons of 510.7±0.6 Ma), which is nearly synchronous with the estimated age for the extinction of redlichiid and olenellid trilobites that mark the Cambrian Series 2/3 boundary and coincides with the sharp negative carbon-isotope excursion of ROECE (e.g., Glass and Phillips, 2006; Jourdan et al., 2014). This is the only major LIP eruption currently documented within the Cambrian; and, if the proposed causal association is correct, is one of the few well-dated Cambrian stage boundaries.

Numerical age model

GTS2012 age model and potential future enhancements

There are very few high-precision radioisotopic ages from Cambrian levels that can be reliably correlated to cosmopolitan biozones or stage boundaries. The two exceptions are (1) the base-Cambrian (albeit, depending on whether the dating of nadir of the negative excursion in carbon isotopes is preferred over the less-well constrained *Tr. pedum* trace fossil appearance) and (2) the Series 2/3 boundary at 510 Ma, when the mass extinction that preceded the appearances of new trilobite genera may have been partly triggered by the eruption of the Kalkarindji/Antrim large igneous province at 511 Ma. The other stage boundaries were estimated from constraints of selected radioisotopic dates and the relative numbers of biozones (see extensive discussions and chart in Peng et al., 2012). This unsatisfactory situation in the Cambrian for deriving an age model is in contrast to the ability to use cyclostratigraphy, oceanic-spreading rates, extensive U-Pb or Ar–Ar dating suites, or statistical composites of global

biozonations for the age models of the other Phanerozoic periods.

Landing et al. (2013, 2015a,b) have critiqued the age models of GTS2012 for the lower Cambrian and upper Cambrian, respectively. They recommend emphasizing that the interpolated ages of lower Cambrian stage boundaries are highly approximate by rounding these values– e.g., use 530 Ma instead of 529 Ma. They suggest that the bases of the Guzhangian Stage and the Furongian Series (base of Paibian Stage) might be significantly younger, which would imply a shorter duration to the upper Cambrian interval. Although, as emphasized by Landing et al. (2015b), "*However, ultimate resolution of this problem will require re-dating of many rocks dated before widespread use of EARTHTIME protocols, including the use of precise calibrated tracers*" and "*Until high precision dates are determined on the base of the traditional Upper Cambrian and base of the Furongian Series, the rates of biotic replacements and geological developments and the durations of biotic zones in the Middle/Series 3 and Upper Cambrian/Furongian Series remain as 'best guesses'.*" In Figs. 5.1 and 5.4, the GTS2012 age model has been retained for the middle and late Cambrian, but the recommendation by Landing et al. (2015a) to round the estimates in the lower Cambrian has been adopted:

Revised ages compared to GTS2012

Fortunian (base of Cambrian; retained **541 Ma** in GTS2012 with qualifier): Temporarily set as nadir of negative-excursion BACE dated as 541 Ma as—see discussion on base-Cambrian GSSP. Landing et al. (2013) suggest that 543 Ma may be a better age estimate for the oldest appearance of *Tr. pedum* trace fossil assemblage.

Stage 2 ("**ca. 530 Ma**" vs. 529 Ma in GTS2012): Following the recommendation of Landing et al. (2015a) the implied precision on this estimate is removed.

GSSPs of the Cambrian Stages, with location and primary correlation criteria

Stage	GSSP Location	Latitude, Longitude	Boundary Level	Correlation Events	Reference
Stage 10				Agnostoid arthropod, FAD of Lotagnostus americanus *or* conodont FAD of Eoconodontus notchpeakensis	
Jiangshanian	Duibian B Section, Jiangshan County, Zhejing Prov., SE China	28°48.958′N 108°36.896′E	108.12m above base of the Huayansi Formation	Agnostoid arthropod, FAD of *Agnostotes orientalis*	Episodes **35**/4, 2012
Paibian	Paibi, Huayuan County, NW Hunan Province, S. China	28°23.37′N 109°31.54′E	at 396 m above the base of the Huaqiao Formation	Agnostoid arthropod, FAD of *Glyptagnostus reticulatus*	Lethaia **37**, 2004
Guzhangian	Luoyixi, Guzhang County, NW Hunan Province, S. China	28°43.20′ N 109°57.88′ E	121.3 m above the base of the Huaqiao Formation	Agnostoid arthropod, FAD of *Lejopyge laevigata*	Episodes **32**/1, 2009
Drumian	Drum Mountains, Millard County, Utah, USA	39°30.705′N 112°59.489′W	at the base of a dark-gray thinly laminated calcisiltite layer, 62 m above the base of the Wheeler Formation	Agnostoid arthropod, FAD of *Ptychagnostus atavus*	Episodes **30**/3, 2007
Stage 5	*candidate sections are Wuliu-Zengjiayan (east Guizhou, China) and Split Mountain (Nevada, USA)*			Trilobite, FAD of Oryctocephalus indicus / Ovatorycto-cara granulata	
Stage 4				Trilobite, FAD of Olenellus, Redlichia, Judomia, Bergeroniellus *or* Arthricocephalus chauveaui	
Stage 3				*FAD of trilobites*	
Stage 2				Small shelly fossils, FAD of Watsonella crosbyi *or* Aldanella attleborensis	
Fortunian	Fortune Head, Burin Peninsula, E Newfoundland, Canada	47°4'34.47"N 55°49'51.71"W*	2.4m above the base of Member 2 in the Chapel Island Formation	Trace fossil, FAD of *Trichophycus pedum*	Episodes **17**/1&2, 1994; Episodes **30**/3, 2007

* according to Google Earth

Figure 5.5 Ratified GSSPs and potential primary markers under consideration for defining the Cambrian stages (*status as of early 2016*). (Details of each GSSP are available at http://www.stratigraphy.org, https://engineering. purdue.edu/Stratigraphy/gssp/, and in the *Episodes* publications.)

Stage 3 (base of Series 2; "**ca. 520 Ma**" vs. 521 Ma in GTS2012): Following the recommendation of Landing et al. (2015a) the implied precision on this estimate is removed.

Stage 4 (**514 Ma** of GTS2012 retained).

Stage 5 (base of Series 3; **509 Ma** of GTS2012 is retained)—The GTS2012 age estimate is supported by the theorized coincidence of the eruption of the Kalkarindji/Antrim LIP with a minimum age of 510.7 ± 0.6 Ma caused the mass extinction that precedes this stage boundary (e.g., Jourdan et al., 2014).

Estimated uncertainties on assigned ages on stage boundaries

The GTS2012 age model had estimated 2-myr uncertainties on all stage boundaries, except for 1-myr uncertainties for the bases of the Cambrian and Stage 5 (Table 1.2 in Gradstein et al., 2012). As discussed previously, some boundaries, especially for Stage 2 and Stage 3, may have larger uncertainty.

Acknowledgments

This brief Cambrian summary of selected highlights and current stratigraphic issues relied heavily on the detailed overview and synthesis by Peng et al. (2012); and Shanchi Peng reviewed an early version of the graphics, database, and text.

Selected publications and websites

Cited publications

Only select publications were cited in this review with an emphasis on aspects of post-2011 updates. Pre-2011 literature is well summarized in the synthesis by Peng et al. (2012) and in some of the publications cited in the following.

Álvaro, J.J., Ahlberg, P., Babcock, L.E., Bordonaro, O.L., Choi, D.K., Cooper, R.A., Ergaliev, G.Kh., Gapp, I.W., Pour, M.G., Hughes, N.C., Jago, J.B., Korovnikov, I., Laurie, J.R., Lieberman, B.S., Paterson, J.R., Pegel, T.V., Popov, L.E., Rushton, A.W.A., Sukhov, S.S.,

Tortello, M.F., Zhou, Z.Y., Zylinska, A., 2011. Global Cambrian trilobite palaeobiogeography assessed using parsimony analysis of endemicity. In: Harper, D.A.T., Servais, T. (Eds.), *Early Palaeozoic Palaeobiogeography and Palaeogeography* Geological Society London, Memoirs, **vol. 38**, pp. 273–296. http://dx.doi.org/10.1144/M38.19.

Babcock, L.E., Peng, S.-C., Zhu, M., Xiao, S., Ahlberg, 2014. Proposed reassessment of the Cambrian GSSP. *Journal of African Earth Sciences* **98**: 3–10.

Babcock, L.E., Peng, S.-C., Brett, C.E., Zhu, M.,-Y., Ahlberg, P., Bevis, M., Robison, R.A., 2015. Global climate, sea level cycles, and biotic events in the Cambrian Period. *Paleoworld* **24**(1–2): 5–15. http://dx.doi.org/10.1016/j.palwor.2015.03.005.

Bowring, S.A., Grotzinger, J.P., Condon, D.J., Ramezani, J., Newall, M.J., Allen, P.A., 2007. Geochronologic constraints on the chronostratigraphic framework of the Neoproterozoic Huqf Supergroup, Sultanate of Oman. *American Journal of Science* **307**: 1097–1145.

Brasier, M.D., Cowrie, J., Taylor, M., 1994. Decision on the Precambrian-Cambrian boundary stratotype. *Episodes* **17**(1,2): 3–8.

Bushuev, E., Goryaeva, I., Pereladov, V., 2014. New discoveries of the oldest trilobites *Profallotaspis* and *Nevadella* in the northeastern Siberian platform, Russia. *Bulletin of Geosciences* **89**: 347–364.

Clausen, S., Álvaro, J.J., Devaere, L., Ahlberg, P., Babcock, L.E., 2015. The Cambrian explosion: its timing and stratigraphic setting. *Annales de Paléontologie* **101**: 153–160. http://dx.doi.org/10.1016/j.annpal.2015.07.001.

Ergaliev, G.Kh., Zhemchuzhnikov, V.G., Popov, L.E., Bassett, M.G., Ergaliev, F.G., 2014. The Auxiliary boundary Stratotype Section and Point (ASSP) of the Jiangshanian Stage (Cambrian: Furongian Series) in the Kyrshabakty section, Kazakhstan. *Episodes* **37**: 41–47.

Gehling, J.G., Jensen, S., Droser, M.L., Myrow, P.M., Narbonne, G.M., 2001. Burrowing below the basal Cambrian GSSP, Fortune Head, Newfoundland. *Geological Magazine* **138**: 213–218.

Glass, L.M., Phillips, D., 2006. The Kalkarindji continental flood basalt province: a new Cambrian large igneous province in Australia with possible links to faunal extinctions. *Geology* **34**: 461–464. http://dx.doi.org/10.1130/G22122.1.

Gradstein, F.M., Ogg, J.G., Schmitz, M.D., Ogg, G.M., (Coordinators), 2012. *The Geologic Time Scale* 2012. Elsevier, Boston, USA. 1174 p. (2-volume book).

Harper, D.A.T., Servais, T., 2011. In: *Early Palaeozoic Palaeobiogeography and Palaeogeography Geological Society London, Memoirs*, **vol. 38**.

Haq, B.U., Schutter, S.R., 2008. A chronology of Paleozoic sea-level changes. *Science* **322**: 64–68. http://dx.doi.org/10.1126/science.116164.

Jourdan, F., Hodges, K., Sell, B., Schaltegger, U., Wingate, M.T.D., Evins, L.Z., Söderlund, U., Haines, P.W., Phillips, D., Blenkinsop, T., 2014. High-precision dating of the Kalkarindji large igneous province, Australia, and synchrony with the Early–Middle Cambrian (Stage 4–5) extinction. *Geology* **42**: 543–546. http://dx.doi.org/10.1130/G35434.1.

Kirschvink, J.L., Margaritz, M., Ripperdan, R.L., Zhuravlev, A. Yu, Rozanov, A. Yu, 1991. The Precambrian-Cambrian boundary: magneto-stratigraphy and carbon isotopes resolve correlation problems between Siberia, Morocco and South China. *GSA Today* **3**(4): 61–91.

Kouchinsky, A., Bengtson, S., Gallet, Y., Korovnikov, I., Pavlov, V., Runnegar, B., Shields, G., Veizer, J., Young, E., Ziegler, K., 2008. The SPICE carbon isotope excursion in Siberia: a combined study of the upper Middle Cambrian-lowermost Ordovician Kulyumbe River section, northwestern Siberian Platform. *Geological Magazine* **145**: 609–622.

Landing, E., Westrop, S.R., Adrain, J.M., 2011. The Lawsonian Stage—the *Eoconodontus notchpeakensis* (Miller, 1969) FAD and HERB carbon isotope excursion define a globally correlatable terminal Cambrian stage. *Bulletin of Geosciences (Czech Geological Survey)* **86**: 621–640.

Landing, E., Geyer, G., Brasier, M.D., Bowring, S.A., 2013. Cambrian evolutionary radiation: context, correlation, and chronostratigraphy – overcoming deficiencies of the first appearance datum (FAD) concept. *Earth-Science Reviews* **123**: 133–172.

Landing, E., Rushton, A.W.A., Fortey, R.A., Bowring, S.A., 2015a. Improved geochronologic accuracy and precision for the ICS Chronostratigraphic charts: examples from the late Cambrian-Early Ordovician. *Episodes* **38**(3): 154–161.

Landing, E., Geyer, G., Buchwaldt, R., Bowring, S.A., 2015b. Geochronology of the Cambrian: a precise Middle Cambrian U-Pb zircon date from the German margin of West Gondwana. *Geological Magazine* **152**: 28–40. http://dx.doi.org/10.1017/S0016756814000119.

Lazarenko, N.P., Gogin, I.Y., Pegel, T.V., Sukhov, S.S., Abaimova, G.P., 2011. The Khos-Nelege section of the Ogon'or Formation: a potential candidate for the GSSP of Stage 10, Cambrian System. *Bulletin of Geosciences* **86**: 555–568.

Li, G.X., Zhao, D., Gubanov, A., Zhu, M.Y., Na, L., 2011. Early Cambrian mollusk *Watsonella crosbyi*: a potential GSSP index fossil for the base of Cambrian Stage 2. *Acta Geologica Sinica* **85**: 309–319.

Maloof, A.C., Ramezani, J., Bowring, S.A., Fike, D.A., Porter, S.M., Mazouad, M., 2010a. Constraints on early carbon cycling from the duration of the Nemakit-Daldynian–Tommotian boundary $\delta^{13}C$ shift, Morocco. *Geology* **38**: 623–626.

Maloof, A.C., Porter, S.H., More, J.L., Dudás, F.Ö., Bowring, S.A., Higgins, J.A., Fike, D.A., Eddy, M.P., 2010b. The earliest Cambrian record of animals and ocean geochemical change. *Geological Society of America Bulletin* **122**: 1731–1774.

McArthur, J.M., Howarth, R.J., Shields, G.A., 2012. Strontium isotope stratigraphy. In: Gradstein, F.M., Ogg, J.G., Schmitz, M., Ogg, G., (Coordinators), *The Geologic Time Scale 2012*. Elsevier Publ., pp. 127–144.

Miller, J.F., Evans, K.R., Freeman, R.L., Ripperdan, R.L., Taylor, J.F., 2015. The proposed GSSP for the base of Cambrian Stage 10 at the first appearance datum of the conodont *Eoconodontus notchpeakensis* (Miller, 1969) in the house range, Utah, USA. *GFF* **136**: 189–192. http://dx.doi.org/10.1080/11035897.2013.862853.

Narbonne, G.M., Xiao, S., Shields, G.H., Gehling, J.G., 2012. The Ediacaran Period. In: Gradstein, F.M., Ogg, J.G., Schmitz, M., Ogg, G., (Coordinators), *The Geologic Time Scale 2012*. Elsevier Publ., pp. 413–435.

Peng, S.-C., Babcock, L.E., Cooper, R.A., 2012. The Cambrian Period. In: Gradstein, F.M., Ogg, J.G., Schmitz, M., Ogg, G., (Coordinators), *The Geologic Time Scale 2012*. Elsevier Publ., pp. 437–488 (An overview on all aspects, including graphics on the ratified GSSPs of the stages, diagrams and tables for the biostratigraphic scales, and discussion on the age models.).

Peng, S.-C., Babcock, L.E., Zhu, X., Zuo, J., Dai, T., 2014. A potential GSSP for the base of the uppermost Cambrian Stage, coinciding with the first appearance of *Lotagnostus americanus* at Wa'ergang, Hunan, China. *GFF* **136**: 208–213. http://dx.doi.org/10.1080/11035897.2013.865666.

Peng, S.-C., Babcock, L.E., Zhu, X., Ahlberg, P., Terfelt, F., Dai, T., 2015. Intraspecific variation and taphonomic alteration in the Cambrian (Furongian) agnostoid *Lotagnostus americanus*: new information from China. *Bulletin of Geosciences (Czech Geological Survey)* **90**(2): 281–306.

Saltzman, M.R., Ripperdan, R.L., Brasier, M.D., Lohmann, K.C., Robison, R.A., Chang, W.T., Peng, S.C., Ergaliev, G.Kh., Runnegar, B., 2000. Global carbon isotope excursion (SPICE) during the Late Cambrian: relation to trilobite extinctions, organic-matter burial and sea level. *Palaeogeography, Palaeoclimatology, Palaeoecology* **162**: 211–223.

Scotese, C.R., 2014. *Atlas of Cambrian and Early Ordovician Paleogeographic Maps (Mollweide Projection), Maps 81–88, vol. 5, the Early Paleozoic, PALEOMAP PaleoAtlas for ArcGIS, PALEOMAP Project, Evanston, IL.* https://www.academia.edu/16785571/Atlas_of_Cambrian_and_Early_Ordovician_Paleogeographic_Maps.

Varlamov, A.I., Rozova, A.V., Khamentovsky, Yu. Ya., et al., 2008. Introduction. In: Rozanov, A.Yu, Varlamov, A.I. (Eds.), *The Cambrian System of the Siberian Platform. Part 1: The Aldan-Lena Region. XIII International Field Conference of the Cambrian Stage Subdivision Working Group*. Paleontological Institute, Russian Academy of Sciences, Moscow and Novosibirsk, pp. 6–11.

Zhu, M.-Y., Babcock, L.E., Peng, S.-C., 2006. Advances in Cambrian stratigraphy and paleontology: integrating correlation techniques, paleobiology, taphonomy and paleoenvironmental reconstruction. *Palaeoworld* **15**: 217–222.

Websites (selected)

Subcommission on Cambrian Stratigraphy (ICS)—*http://www.palaeontology.geo.uu.se/ISCS/ISCS_home.html*—brief summary of GSSPs, extensive bibliography on chronostratigraphy and other information.

Guide to the Orders of Trilobites—*http://www.trilobites.info*—award-winning site (e.g., Scientific American, GSA Geoscience Information Society, etc.) paleobiology, images, and evolutionary trees for all trilobites (plus agnostoid arthropods) and more. Developed and maintained by Sam Gon III.

Palaeos: Cambrian—*http://palaeos.com/paleozoic/cambrian/cambrian.htm*—A well-presented suite of diverse topics for a general science audience that was originally compiled by M. Alan Kazlev in 1998–2002.

Trilobites Family Album—*http://www.trilobites.us*—taxonomy, images (including agnostoid arthropods) for each order/family.

Burgess Shale—*http://burgess-shale.rom.on.ca* (for the World Heritage Site), and *http://www.palaeocast.com/episode-48-the-burgess-shale/* (for 44-min Podcast with slides by Prof. Simon Conway Morris)—the world's more famous fossil locality in Canada for the spectacular record of earliest animal life during the early stages of the "Cambrian Explosion".

Palaeocast podcasts on Cambrian—streaming videos and slide sets at *http://www.palaeocast.com/category/paleozoic/cambrian/*—topical episodes (typically 45 min, produced during 2013–2015) include Anomalocaridids, Emu Bay Shale locality, Lobopodians, and early animal forms.

6

ORDOVICIAN

456 Ma Ordovician

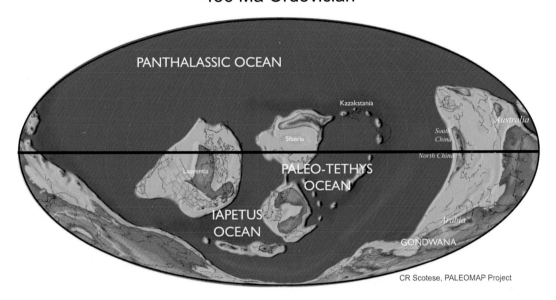

Sandbian (early Late Ordovician) paleogeographic reconstruction (Sea level+120 m maximum flooding surface (MFS)) from Scotese (2014).

Basal definition and international subdivisions

The Ordovician, named after the Welsh Ordovices tribe of Britain, has been subdivided by the Subcommission on Ordovician Stratigraphy into seven stages grouped in three series (Fig. 6.1). Most of these stages do not correspond to any preexisting regional stages, but were developed based on identification of graptolite or conodont first-appearance datums (FADs) that are useful for global correlation. A detailed correlation table of these international stages to other regional units is in Bergström et al. (2008), and descriptions

of GSSPs and primary correlation criteria for these stages are in Cooper et al. (2012) and on the Subcommission website (*ordovician.stratigraphy.org*).

The base of the Ordovician was placed at a wave-cut platform in Newfoundland at a level intended to be just below the FAD of pelagic graptolites and to coincide with the lowest occurrence of conodont *Iapetognathus fluctivagus* in the section. However, some later studies concluded that the FAD of true *Iapeto. fluctivagus* is in Bed 26 above the FAD of the graptolites, and that the current GSSP is in the middle of the range of a similar-looking *Iapeto. preaengensis* (Terfelt et al., 2011, 2012),

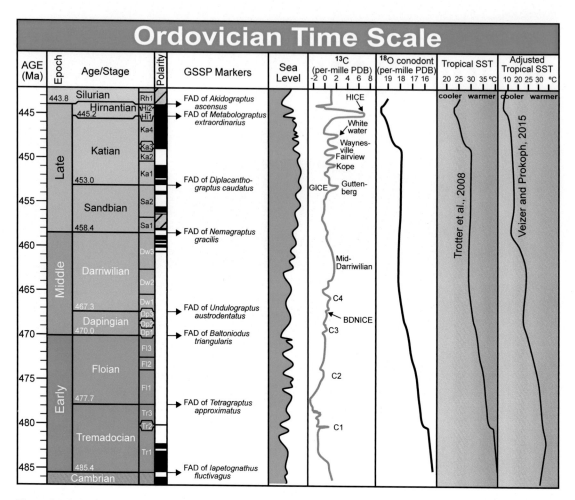

Figure 6.1 Ordovician overview. Main markers for GSSPs of Ordovician stages are first-appearance datums (FADs) of graptolite taxa (except conodont FADs for the GSSPs of the Tremadocian and the Dapingian stages) as detailed in Fig. 6.6. ("Age" is the term for the time equivalent of the rock-record "stage.") Magnetostratigraphy is from Pavlov and Gallet (2005), but many intervals have not been studied or verified. Schematic sea-level curve is modified from Haq and Schutter (2008) following advice of Bilal Haq (pers. comm., 2008). The δ13C curve with major widespread events is modified from Bergström et al. (2008) with enhancements of the Early and Middle Ordovician from Edwards and Saltzman (2014) and of the Late Ordovician from Melchin et al. (2013) and Bergström et al. (2014). (*HICE*, Hirnantian positive Isotope Carbon Excursion (ICE); *GICE*, Guttenberg positive ICE; *BDNICE*, Basal Dapingian negative ICE; and the C-set is from Edwards and Saltzman (2014).) The generalized δ18O curve and estimates of tropical sea-surface temperatures from conodont apatite is averaged from Trotter et al. (2008) and shown for comparison is the adjusted tropical sea-surface temperature of Veizer and Prokoph (2015) that they derived from a synthesis of oxygen-18 values from carbonate fossils. The vertical scale of this diagram is standardized to match the vertical scales of the first stratigraphic summary figure in all other Phanerozoic chapters. *PDB*, PeeDee Belemnite 13C standard; *SST*, sea-surface temperature.

although others have defended the original taxonomic assignments and FAD at the GSSP (Miller et al., 2014). The decision has not yet been made whether to retain the current GSSP level or to shift it slightly to again coincide with a less ambiguous and more widespread biological datum (e.g., FAD of conodont *Cordylodus intermedius* or other taxa; Terfelt et al., 2012) that is accompanied by a geochemical datum (e.g., Azmy et al., 2014; Ordovician News, 2015, p. 9; Fig. 6.2).

The GSSP for the base of **Middle Ordovician** (Dapingian Stage) in Hubei, South China, coincides with the lowest occurrence of conodont *Baltoniodus triangularis* (Fig. 6.3). This level is followed closely by the FAD of conodont *Microzarkodina flabellum* and is considered close to the FAD of graptolite *Isograptus victoriae victoriae*, although that taxon is not found at the GSSP section (Wang et al., 2009). The base of the **Late Ordovician** (Sandbian Stage) in southwestern Sweden coincides with the lowest occurrence of graptolite *Nemagraptus gracilis* (Fig. 6.4).

Selected main stratigraphic scales and events

(1) Biostratigraphy and major trends

Graptolites and conodonts are the main fossil groups for correlation of Ordovician marine strata, with chitinozoans playing a secondary role. Graptolites, the floating colonies of microscopic animals that are preserved in shales as flattened traces, appeared at the base of the Ordovician, underwent a major surge in diversity during the late Floian with other peaks in early Darriwilian, mid-Sandbian, and latest Katian, then dramatically collapsed during the Hirnantian (e.g., Cooper et al., 2014). Their relatively short stratigraphic ranges and widespread distribution are ideal for standardized Ordovician graptolite zonations and as primary markers

in placement of stage boundaries in relatively clay-rich and low-oxygen depositional settings (e.g., Cooper et al., 2012; Fig. 6.5).

The microscopic phosphatic conodonts, thought to be the teeth and jaw elements of an eel-like vertebrate, are mainly used for correlation of calcareous strata. They, like graptolites, lived in major biographic provinces—warm water (e.g., North American midcontinent) and cooler water (e.g., North Atlantic realm) (Fig. 6.5). Organic-walled chitinozoan microfossils, which may have been the floating egg cases of a marine animal, appeared at the beginning of the Ordovician nearly at the same time as graptolites and peaked during the Darriwilian. Other fossil groups used in Ordovician biostratigraphy include organic-walled acritarchs, the calcareous shells of brachiopods and the phosphatic skeletons and molds of trilobites.

During the Middle Ordovician, especially through the Darriwilian Stage, the global ecosystems underwent the Great Ordovician Biodiversification Event as marine diversity nearly tripled. The drivers for this "*most significant and sustained increase of marine biodiversity in Earth history*" are complex, including increases in provincialism and elevated sea levels flooding the continents (e.g., reviews in Webby et al., 2004; Harper et al., 2015).

The final, and relatively brief, Hirnantian Stage of the Ordovician had two major glacial episodes that caused a double wave of extinctions—the first phase during the initial cooling, and the second phase during the rising sea levels and widespread anoxic conditions as the second glaciation retreated (e.g., reviews by Melchin et al., 2013; Armstrong and Harper, 2014; Harper et al., 2014). The combined effect of these two Hirnantian mass extinctions rank with the end-Permian and the end-Cretaceous events as the largest three mass extinctions during the Phanerozoic. The melting of the main mid-Hirnantian glaciation was accompanied by a major episode

Base of the Tremadocian Stage of the Ordovician System at Green Point, Western Newfoundland

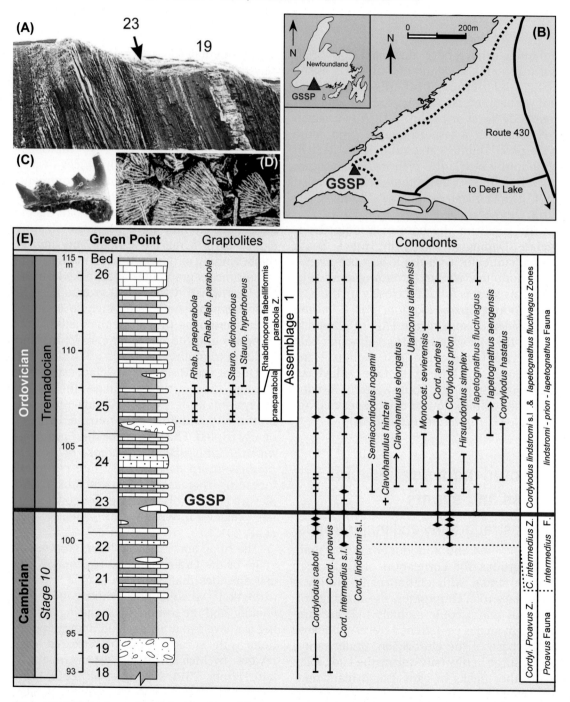

Figure 6.2 GSSP for base of the Ordovician (base of Tremadocian Stage) at the Green Point section, Newfoundland, eastern Canada. Strata in outcrop are overturned. The GSSP level at Bed 23 in the section was intended to coincide with the lowest occurrence of conodont *Iapetognathus fluctivagus*; but later studies concluded that its true lowest occurrence is in Bed 26. Specimen that had been interpreted as *Iapeto. fluctivagus* in image (C) is 0.5-mm long. The lowest appearance of graptolites is 4.8 m above the GSSP; and the specimen of graptolite *Rhabdinopora flabelliformis parabola* in image (D) is 17-mm long. (Stratigraphic column and ranges of taxa are modified from the original GSSP publication (Cooper et al., 2001), images of fossils are from Cooper et al. (2012), and the outcrop photograph is by S.H. Williams.)

Base of the Dapingian Stage of the Ordovician System at Huanghuachang, near Yichang City, Hubei Province, China

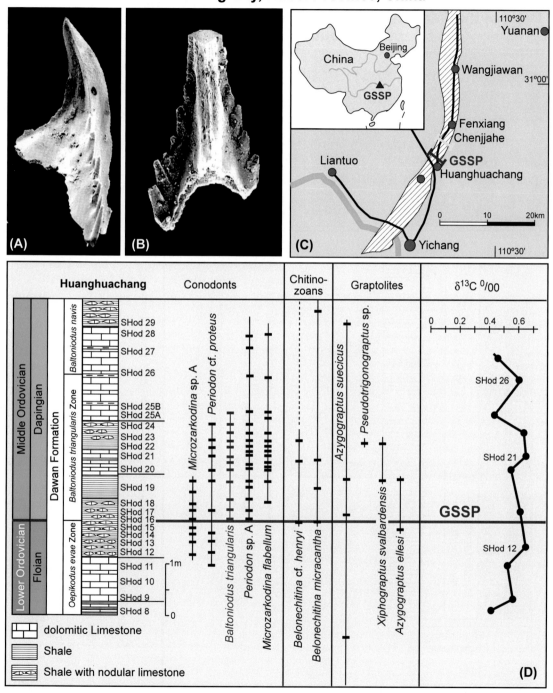

Figure 6.3 GSSP for base of the Middle Ordovician (base of Dapingian Stage) at the Huanghuachang section, Hubei, South China. The GSSP level coincides with the lowest occurrence of conodont *Baltoniodus triangularis*. Images of *Balt. triangularis*: (A) is Pa element, lateral view, 0.23 mm in height; (B) is Sa element, posterior view, 0.16 mm in height. (Set of images and stratigraphic column from Cooper et al. (2012) is partly from Wang et al. (2005; Fig. 4).)

Base of the Sandbian Stage of the Ordovician System at Fågelsång, Southern Sweden

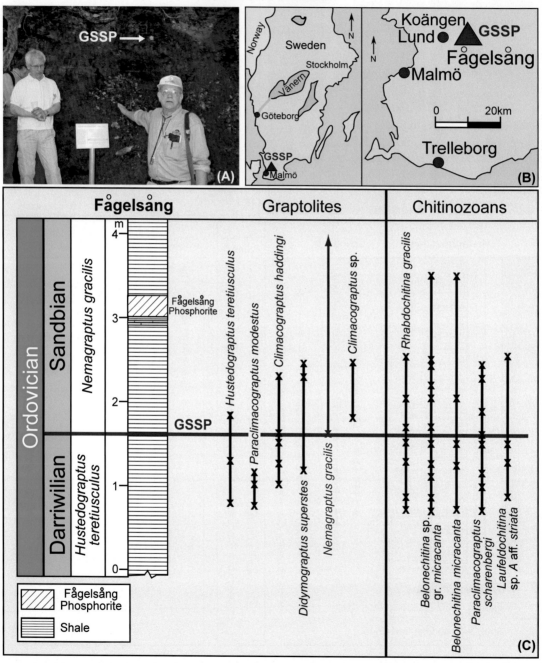

Figure 6.4 GSSP for base of the Upper Ordovician (base of Sandbian Stage) at the E14b outcrop, Fågelsång, southwestern Sweden. The GSSP level at the lowest occurrence of graptolite *Nemagraptus gracilis*. (Photograph and stratigraphic column from Cooper et al. (2012) showing graptolite and chitinozoan ranges from Bergström et al. (2000); Fig. 5.)

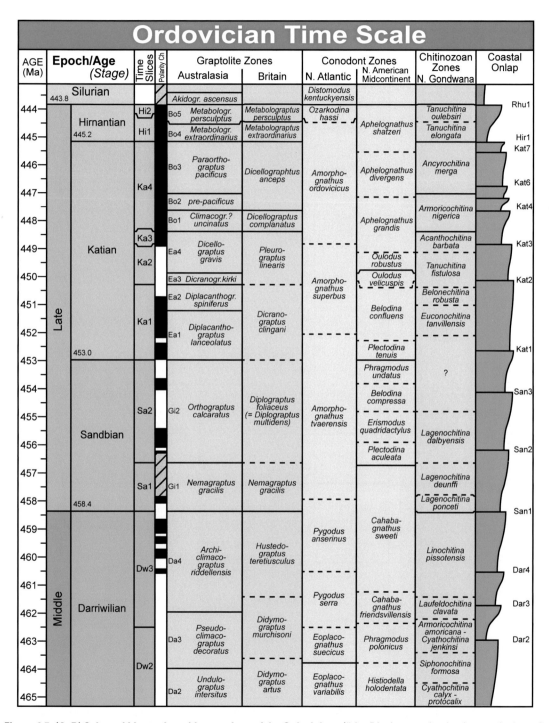

Figure 6.5 (A, B) Selected biostratigraphic zonations of the Ordovician. ("*Age*" is the term for the time equivalent of the rock-record "*stage*".) Compilation modified from Cooper et al. (2012). The Australian graptolite scale is modified from Cooper and Sadler (2004). British graptolite zones, North Atlantic and North America Midcontinent conodont zones, and North Gondwana chitinozoan zones are from Webby et al. (2004).

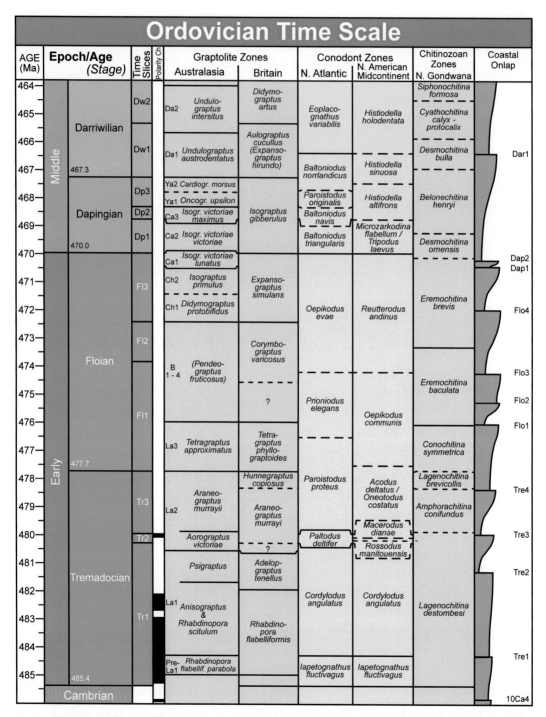

Figure 6.5 (Continued) Coastal onlap with labels for selected major sequence boundaries are modified from Haq and Schutter (2008) following advice of Bilal Haq (pers. comm., 2008). Additional zonations, biostratigraphic markers, regional stages, geochemical trends, sea-level curves, and details on calibrations are compiled in Cooper et al. (2012) and in the internal data sets within the *TimeScale Creator* visualization system (free at *www.tscreator.org*).

of widespread black-shale deposition—an important regional hydrocarbon source rock—that continued through the earliest Silurian (Melchin et al., 2013).

(2) Stable-isotope stratigraphy and selected events

A distinct break occurs in the characteristics of geochemical trends between the relatively quiet early Ordovician and the excursions and rapid rates of change in the late Ordovician (Fig. 6.1). The graptolite turnover rates also indicate a relatively steady marine environment until the Katian onset of a series of several sharp extinction episodes during the latest Ordovician through Silurian triggered by environmental crises (Cooper et al., 2012).

The carbon–isotope curve has fine-scale variations during the Early and Middle Ordovician that, when combined with biostratigraphy, are useful for high-resolution correlation (e.g., Edwards and Saltzman, 2014). Beginning with the Guttenberg positive Isotope Carbon Excursion (GICE) at the base of the Katian, a series of high-amplitude excursions affected the global carbon cycle and culminated in the major Hirnantian excursion (HICE) that accompanied the onset of the two glacial episodes (e.g., compilations by Bergström et al., 2008, 2014; Melchin et al., 2013).

Oxygen-isotope ratios in the phosphate of conodonts indicate a general long-term cooling through the Ordovician that may have played a role in increasing oceanic overturning and the Middle Ordovician surge in marine biodiversity (Trotter et al., 2008). An independent estimate of average subtropical temperatures from $\delta^{18}O$ values of carbonates that is adjusted for potential long-term trends in Phanerozoic seawater $\delta^{18}O$ (Veizer and Prokoph, 2015) yields similar trends but with significantly cooler average values (Fig. 6.1). The cooling trend culminates in

very cold oceanic temperatures during the Hirnantian glacial episodes.

Strontium $^{87}Sr/^{86}Sr$ ratios slowly decreased during the Tremadocian to mid-Darriwilian, rapidly decreased through the Sandbian, and then leveled off at a minimum during the Katian–Hirnantian before reversing to a rising trend through the entire Silurian (e.g., McArthur et al., 2012; Edwards et al., 2015). A compilation of the few magnetostratigraphic studies of Ordovician strata suggests dominance by a superchron of a reversed-polarity magnetic field during the middle Early through Middle Ordovician (Pavlov and Gallet, 2005; Pavlov et al., 2008; Fig. 6.1).

Numerical age model

GTS2012 age model and potential future enhancements

A composite sequencing of the stratigraphic ranges of over 2000 graptolite species using a database from over 500 stratigraphic sections was compiled using **CON**strained **OP**timization methods (CONOP) by Cooper and Sadler (in Cooper et al., 2012, 2014). Age models for the FADs and last-appearance datums (LADs) of the major taxa and for the Ordovician and Silurian stage boundaries calibrated to those events (placed at bases of graptolite zones) were interpolated by both a cubic spline fit and a polynomial fit to a set of 22 radioisotopic dates. The two interpolation methods generally yielded very similar estimates for the ages of Ordovician stage boundaries, except for ca. 1.3-myr younger ages from spline-fit method for the bases of the Floian and the Hirnantian (see comparison table and graphical-display comparison in Cooper et al., 2012). The spline-fit interpolations were used in GTS2012, whereas Cooper et al. (2014) used the polynomial-fit interpolation for their statistics on rates of graptolite-diversity trends.

After the publication of GTS2012, the only significant new radioisotopic dates have been mainly from the Late Ordovician. During the Sandbian and Katian stages of the Late Ordovician, there were some major explosive volcanic episodes that deposited thick ash beds over present-day eastern North America and northern Europe. One episode, the Kinnekulle bentonite of northwestern Europe, has been U-Pb redated as 454.5±0.5 Ma (Svenson et al., 2015), and extrapolation of the regional placement of the Sandbian/Katian boundary relative to this and other ash beds yielded an estimate of 451.9±0.4 Ma, which is 1-myr younger than the spline-fit interpolation of 453.0±0.7 Ma computed in GTS2012. A critique of the radioisotopic dates spanning the late Cambrian and earliest Ordovician (Landing et al., 2015) preferred using a single $^{207}Pb/^{206}Pb$ date of 486.8±2.6 Ma from detrital zircons near the Cambrian/Ordovician boundary in Wales as its best age, rather than the 485.4±1.9 Ma derived by the spline fit through this date plus adjacent dates; which highlights the need to obtain much higher precision uranium–lead isotope dilution thermal ionization mass spectrometry (U-Pb TIMS) dates from this

GSSPs of the Ordovician Stages, with location and primary correlation criteria

Stage	GSSP Location	Latitude, Longitude	Boundary Level	Correlation Events	Reference
Hirnantian	Wangjiawan North section, N of Yichang city, Western Hubei Province, China	30°59'2.68"N 111°25'10.76"E	0.39m below the base of the Kuanyinchiao Bed	Graptolite, FAD of *Metabolograptus extraordinarius*	Episodes **29/3**, 2006
Katian	Black Knob Ridge Section, Atoka, Oklahoma (USA)	34°25.829'N 96°4.473'W	4.0m above the base of the Bigfork Chert	Graptolite, FAD of *Diplacanthograptus caudatus*	Episodes **30/4**, 2007
Sandbian	Sularp Brook, Fågelsång, Sweden	55°42'49.3"N 13°19'31.8"E*	1.4m below a phosphorite marker bed in the E14b outcrop	Graptolite, FAD of *Nemagraptus gracilis*	Episodes **23/2**, 2000
Darriwilian	Huangnitang section, Changshan, Zhejiang Province, SE China	28°51'14"N 118°29'23"E*	base of Bed AEP 184	Graptolite, FAD of *Undulograptus austrodentatus*	Episodes **20/3**, 1997
Dapingian	Huanghuachang section, NE of Yichang city, Hubei Province, S China	30°51' 37.8"N 110°22' 26.5"E	10.57 m above base of the Dawan Formation	Conodont, FAD of *Baltoniodus triangularis*	Episodes **32/2**, 2009
Floian	*Diabasbrottet, Hunneberg, Sweden*	58°21'32.2"N 12°30'08.6"E	in the lower Tøyen Shale, 2.1m above the top of the Cambrian	Graptolite, FAD of *Tetragraptus approximatus*	Episodes **27/4**, 2004
Tremadocian	Green Point Section, western Newfoundland	49°40'58.5"N 57°57'55.09"W*	at the 101.8m level, within Bed 23, in the measured section	Conodont, FAD of *Iapetognathus fluctivagus*	Episodes **24/1**, 2001

* according to Google Earth

Figure 6.6 Ratified GSSPs and potential primary markers defining the Ordovician stages (*status as of early 2016*). (Details of each GSSP are available at http://www.stratigraphy.org, https://engineering.purdue.edu/Stratigraphy/gssp/, and in the *Episodes* publications.)

interval. We have retained the GTS2012 age model in this chapter.

The Ordovician time scale and scaling of biozones within stages will eventually be improved by applying cyclostratigraphy. For example, the glacial episodes during the Hirnantian appear to have been either modulated by 405-kyr eccentricity cycles superimposed on lower frequencies (eg, Melchin et al., 2013) or perhaps with a 1.2-myr modulation of obliquity (Ghienne et al., 2014). Detailed logging of Sandbian–Katian sections suggest 400-kyr modulation of 100-kyr eccentricity cycles plus 30-kyr obliquity (e.g., Ellwood et al., 2013; Elrich et al., 2013; Svensen et al., 2015; Hinnov and Diecchio, 2015).

Estimated uncertainties on assigned ages on stage boundaries

The uncertainties on stage boundaries computed by the spline-fit method in GTS2012 were largely governed by the uncertainties and spacing of the radioisotopic tie points. The uncertainties from the spline-fit statistics ranged from 1.9 myr at the base of the Ordovician, decreasing to ca. 0.7 myr at the base of the Katian, and rising to 1.5 myr at base-Hirnantian and the base of the Silurian. Reduction of these uncertainties will require reanalysis of the previously published radioisotopic ages using the EARTHTIME standards (e.g., Condon et al., 2015), adding dates within the main gaps of Tremadocian and Katian stages, and applying and verifying cyclostratigraphic scaling techniques.

Acknowledgments

This brief Ordovician summary of selected highlights and current stratigraphic issues relied heavily on the detailed overview and synthesis by Cooper et al. (2012) and updates in the *Ordovician News* newsletters of the Subcommission on Ordovician Stratigraphy (*http://ordovician.stratigraphy.org*. This review depended on the expertise of many colleagues, especially Roger Cooper, Peter Sadler, and Michael Melchin. Roger Cooper and Shanchi Peng reviewed an early version of the graphics, database, and text.

Selected publications and websites

Cited publications

Only select publications were cited in this review with an emphasis on aspects of post-2011 updates. Pre-2011 literature is well summarized in the synthesis by Cooper et al. (2012) and in some of the publications cited below.

Armstrong, H.A., Harper, D.A.T., 2014. An Earth system approach to understanding the end-Ordovician (Hirnantian) mass extinction. In: Keller, G., Kerr, A. (Eds.), *Volcanism, Impacts, and Mass Extinctions: Causes and Effects*, Geological Society of America Special Paper, **vol. 505**, pp. 287–300. http://dx.doi.org/10.1130/2014.2505(14).

Azmy, K., Stouge, S., Brand, U., Bagnoli, G., Ripperdan, R., 2014. High-resolution chemostratigraphy of the Cambrian–Ordovician GSSP: enhanced global correlation tool. *Palaeogeography, Palaeoclimatology, Palaeoecology* **409**: 135–144. http://dx.doi.org/10.1016/j.palaeo.2014.05.010.

Bergström, S.M., Finney, S.C., Xu, C., Palsson, C., Zhi-Gao, W., Grahn, Y., 2000. A proposed global boundary stratotype for the base of the Upper Series of the Ordovician System: The Fågelsång section, Scania, southern Sweden. *Episodes* **23**(2): 102–109.

Bergström, S.M., Chen, X., Gutiérrez-Marco, J.-C., Dronov, A., 2008. The new chronostratigraphic classification of the Ordovician System and its relations to major regional series and stages and to δ13C chemostratigraphy. *Lethaia* **42**: 97–107. Correlation chart available at: http://ordovician.stratigraphy.org/uploads/OrdChartHigh.jpg.

Bergström, S.M., Eriksson, M.E., Young, S.A., Ahlberg, P., Schmitz, B., 2014. Hirnantian (latest Ordovician) δ13C chemostratigraphy in southern Sweden and globally: a refined integration with the graptolite and conodont zone successions. *GFF* **136**(2): 355–386. http://dx.doi.org/10.1080/11035897.2013.851734.

Condon, D.J., Schoene, B., McLean, N.M., Bowring, S.A., Parrish, R.R., 2015. Metrology and traceability of U-Pb isotope dilution geochronology (EARTHTIME Tracer Calibration Part 1). *Geochemica et Cosmochimica Acta.* **164**: 464–480. http://dx.doi.org/10.1016/j.gca.2015.05.026.

Cooper, R.A., Nowlan, G.S., Williams, H.S., 2001. Global stratotype section and point for base of the Ordovician System. *Episodes* **24**: 19–28.

Cooper, R.A., Sadler, P.M., 2004. Ordovician. In: Gradstein, F., Ogg, J.G., Smith, A.G.(Coordinators) , *A Geologic Time Scale*. Cambridge University Press, Cambridge, pp. 165–187.

Cooper, R.A., Sadler, P.M., Gradstein, F.M., Hammer, O., 2012. The Ordovician Period. In: Gradstein, F.M., Ogg, J.G., Schmitz, M., Ogg, G. (Coordinators), *The Geologic Time Scale 2012*. Elsevier Publ, pp. 489–523 (An overview on all aspects, including graphics on the ratified GSSPs of the stages, diagrams and tables for the biostratigraphic scales, and discussion on the age models.).

Cooper, R.A., Sadler, P.M., Munnecke, A., Crampton, J.S., 2014. Graptoloid evolutionary rates track Ordovician-Silurian global climate change. *Geological Magazine* **151**: 349–364.

Edwards, C.T., Saltzman, M.R., 2014. Carbon isotope ($\delta^{13}C_{carb}$) stratigraphy of the Lower–Middle Ordovician (Tremadocian–Darriwilian) in the Great Basin, western United States: implications for global correlation. *Palaeogeography, Palaeoclimatology, Palaeoecology* **399**: 1–20.

Edwards, C.T., Saltzman, M.R., Leslie, S.A., Bergström, S.M., Sedlacek, A.R.C., Howard, A., Bauer, J.A., Sweet, W.C., Young, S.A., 2015. Strontium isotope ($^{87}Sr/^{86}Sr$) stratigraphy of Ordovician bulk carbonate: implications for preservation of primary seawater values. *Geological Society of America Bulletin* **127**: 1275–1289.

Ellwood, B.B., Brett, C.E., Tomkin, J.H., Macdonald, W.D., 2013. Visual identification and quantification of Milankovitch climate cycles in outcrop: an example from the Upper Ordovician Kope Formation, Northern Kentucky. In: Jovane, L., Herreo-Bervara, E., Hinnov, L.A., Housen, B.A. (Eds.), *Magnetic Methods and the Timing of Geological Process*. Geological Society, **vol. 373**. Special Publications, London, pp. 341–353. http://dx.doi.org/10.1144/SP373.2.

Elrich, M., Reardon, D., Labor, W., Martin, J., Descrochers, A., Pope, M., 2013. Orbital-scale climate change and glacioeustasy during the early Late Ordovician (pre-Hirnantian) determined from $\delta^{18}O$ values in marine apatite. *Geology* **41**: 775–788. http://dx.doi.org/10.1130/G34363.1.

Ghienne, J.-G., Desrochers, A., Vandenbroucke, T.R.A., Achab, A., Asselin, E., Dabard, M.-P., Farley, C., Loi, A., Paris, F., Wickson, S., Veizer, J., 2014. A Cenozoic-style scenario for the end-Ordovician glaciation. *Nature Communications* **5**: 4485. http://dx.doi.org/10.1038/ncomms5485 (9 pp.).

Harper, D.A.T., Hammarlund, E.U., Rasmussen, C.M.O., 2014. End Ordovician extinctions: a coincidence of causes. *Gondwana Research* **25**: 1294–1307. http://dx.doi.org/10.1016/j.gr.2012.12.021.

Harper, D.A.T., Zhan, R.-B., Jin, J., 2015. The Great Ordovician Biodiversification Event: Reviewing two decades of research on diversity's big bang illustrated by mainly brachiopod data. *Palaeoworld* **24**: 75–85. http://dx.doi.org/10.1016/j.palwor.2015.03.003.

Haq, B.U., Schutter, S.R., 2008. A chronology of Paleozoic sea-level changes. *Science* **322**: 64–68. http://dx.doi.org/10.1126/science.116164.

Hinnov, L.A., Diecchio, R.J., 2015. Milankovitch Cycles in the Juniata Formation, Late Ordovician, Central Appalachian Basin, USA. *Stratigraphy* **12**: 287–296.

Landing, E., Rushton, A.W.A., Fortey, R.A., Bowring, S.A., 2015. Improved geochronologic accuracy and precision for the ICS Chronostratigraphic charts: examples from the late Cambrian-Early Ordovician. *Episodes* **38**(3): 154–161.

McArthur, J.M., Howarth, R.J., Shields, G.A., 2012. Strontium isotope stratigraphy. In: Gradstein, F.M., Ogg, J.G., Schmitz, M., Ogg, G. (Coordinators), *The Geologic Time Scale 2012*. Elsevier Publ., pp. 127–144.

Melchin, M.J., Mitchell, C.E., Holmden, C., Štorch, P., 2013. Environmental changes in the Late Ordovician–early Silurian: review and new insights from black shales and nitrogen isotopes. *Geological Society of America Bulletin* **125**: 165–1670. http://dx.doi.org/10.1130/B30812.1.

Miller, J.F., Repetski, J.E., Nicoll, R.S., Nowland, G., Ethington, R., 2014. The conodont *Iapetognathus* and its value for defining the base of the Ordovician System. *GFF* **136**: 185–188. http://dx.doi.org/10.1080/11035897.2013.862851.

Ordovician News, 2015. In: Percival, I.G. (Ed.), *Subcommission on Ordovician Stratigraphy Newsletter*, **vol. 32** Available as PDF at: http://ordovician.stratigraphy.org.

Pavlov, V., Gallet, Y., 2005. A third superchron during the Early Paleozoic. *Episodes* **28**: 78–84.

Pavlov, V., Bachtadse, V., Mikhailov, V., 2008. New Middle Cambrian and Middle Ordovician palaeomagnetic data from Siberia: Llandelian magnetostratigraphy and relative rotation between the Aldan and Anabar–Angara blocks. *Earth and Planetary Science Letters* **276**: 229–242. http://dx.doi.org/10.1016/j.epsl.2008.06.021.

Scotese, C.R., 2014. *Atlas of Silurian and Middle-Late Ordovician Paleogeographic Maps (Mollweide Projection), Maps 73-80, Volume 5, The Early Paleozoic, PALEOMAP PaleoAtlas for ArcGIS, PALEOMAP Project, Evanston, IL*. https://www.academia.edu/16744278/Atlas_of_Silurian_and_Middle-Late_Ordovician_Paleogeographic_Maps.

Svenson, H.H., Hammer, Ø., Corfu, F., 2015. Astronomically forced cyclicity in the Upper Ordovician and U-Pb ages of interlayered tephra, Oslo Region, Norway. *Palaeogeography, Palaeoclimatology, Palaeoecology* **418**: 150–159.

Terfelt, F., Bagnoli, G., Stouge, S., 2011. The base of the Ordovician System – a horizon in limbo. In: Gutiérrez-Marco, J.G., Rábano, I., García-Bellido, D. (Eds.), *Ordovician of the World*, 11th International Symposium on the Ordovician System (Alcalá de Henares, Madrid, Spain, 9–13 May 2011), Guadernos del Museao Geominero (Instituto Geológico y Minero de España), **vol. 14 (682 pp.)**, p. 587 free download at: http://www.igme.es/isos11/ORDOVICIAN%20OF%20THE%20WORLD.pdf.

Terfelt, F., Bagnoli, G., Stouge, S., 2012. Re-evaluation of the conodont *Iapetognathus* and implications for the base of the Ordovician System GSSP. *Lethaia* **45**: 227–237.

Trotter, J.A., Williams, I.S., Barnes, C.R., Lécuyer, C., Nicoll, R.S., 2008. Did cooling oceans trigger Ordovician biodiversification? Evidence from conodont thermometry. *Science* **321**: 550–554.

Veizer, J., Prokoph, A., 2015. Temperatures and oxygen isotopic composition of Phanerozoic oceans. *Earth-Science Reviews* **146**: 92–104.

Wang, X., Stouge, S., Erdtmann, B.-D., Chen, X., Li, Z., Wang, C., Zeng, Q., Zhou, Z., Chen, H., 2005. A proposed GSSP for the base of the Middle Ordovician Series: the Huanghuachang section, Yichang, China. *Episodes* **28**(2): 105–117.

Wang, X., Stouge, S., Chen, X., Li, Z., Wang, C., Finney, S.C., Zeng, Q., Zhou, Z., Chen, H., Erdtmann, B.-D., 2009. The Global Stratotype Section and Point for the base of the Middle Ordovician Series and the third stage (Dapingian). *Episodes* **32**(2): 96–112.

Webby, B.D., Paris, F., Droser, M.L., Percival, I.G., 2004. *The Great Ordovician Biodiversification Event.* Columbia University Press, New York. 496 pp. (In addition to syntheses for most fossil groups, this compilation has excellent summaries of Ordovician climate, geochemistry and sea-level changes.).

Websites (selected)

Subcommission on Ordovician Stratigraphy (International Commission on Stratigraphy [ICS])—*http://ordovician.stratigraphy.org*—Ordovician News PDFs, correlation charts, details on each GSSP, and other information.

The Great Ordovician Biodiversity Event—*www.palaeocast.com/episode-19-the-great-ordovician-biodiversity-event*—Prof. David Harper (former chair of Ordovician Subcommission; and current ICS chair) leads an examination of the "causes and consequences of this complex and fascinating period" (2013).

Geobiodiversity database—*www.geobiodiversity.com*—Extensive array of stratigraphic sections (ca. 14,000 localities) with a current emphasis on Ordovician–Silurian and on China. Also includes SinoCor program for graphical correlation and other tools. Developed by Fan Junxuan (Nanjing Inst. Geology and Paleontology).

Commission Internationale Microflore Paléozoïque—*http://cimp.weebly.com*—International Commission of the Paleozoic Microflora has separate subcommissions on acritarchs, chitinozoans, and miospores; and site has archive of their newsletters and other information; although number of active newsletters diminished since 2012.

Palaeos: The Ordovician—*http://palaeos.com/paleozoic/silurian/silurian.htm*—A well-presented suite of diverse topics for a general science audience that was originally compiled by M. Alan Kazlev in 1998–2002.

Guide to the Orders of Trilobites—*http://www.trilobites.info*—award-winning site (eg, Scientific American, Geological Society of America (GSA) Geoscience Information Society, etc.) paleobiology, images, and evolutionary trees for all trilobites (plus agnostoid arthropods) and more. Developed and maintained by Sam Gon III.

7

SILURIAN

425.6 Ma Silurian

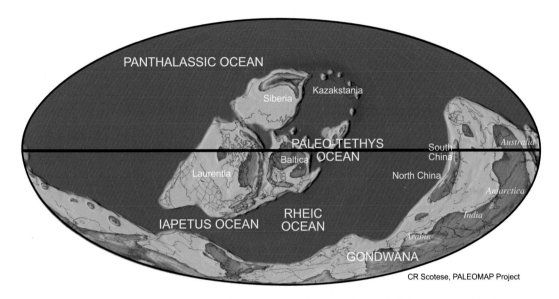

CR Scotese, PALEOMAP Project

Wenlock (middle Silurian) paleogeographic reconstruction (Sea level + 80 m). From Scotese (2014).

Basal definition and international subdivisions

The Silurian, named after the *Silures* tribe of Wales, begins during the recovery from the two-phased mass extinction that accompanied the pair of major glacial and interglacial events of the early and middle part of the Hirnantian Stage of latest Ordovician. The Global Boundary Stratotype Sections and Points (GSSPs) for the four Silurian series—Llandovery, Wenlock, Ludlow, and Pridoli (*Přídolí*)—and their stages were all placed in the United Kingdom (e.g., field guide by Ray, 2011), except for the base of the Pridoli in the

Czech Republic (Figs. 7.1 and 7.7). However, many of these GSSPs were later found deficient for precise global correlation, therefore the International Subcommission on Silurian Stratigraphy (ISSS) has working groups for the priority needs of revising the early Silurian set of the bases for Aeronian and Telychian stages in the Llandovery Series and the base for the Wenlock Series (Sheinwoodian Stage).

The GSSP for the Ordovician/Silurian boundary in Scotland coincides with the local occurrence of graptolite *Akidograptus ascensus* (Fig. 7.2). Even though it was originally intended to coincide with the base of the next higher zone of *Parakidograptus acuminatus* which was later found to begin 1.6-m higher,

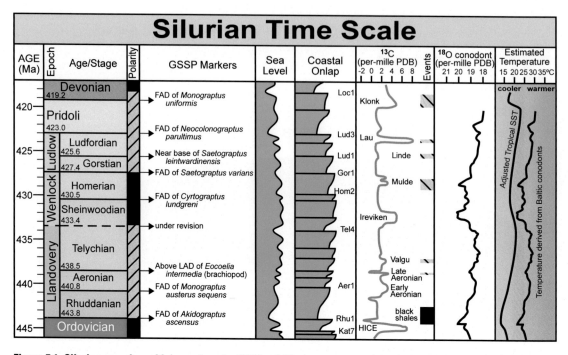

Figure 7.1 Silurian overview. Main markers for GSSPs of Silurian stages are intended to be first-appearance datums (FADs) of graptolite taxa, but some of the current GSSPs do not allow a precise calibration (e.g., Ray, 2011; Melchin et al., 2012). ("Age" is the term for the time equivalent of the rock-record "stage.") Magnetostratigraphy is essentially unstudied within the Silurian. Coastal onlap and schematic sea-level curve with labels for selected major sequence boundaries are modified from Haq and Schutter (2008) following advice of Bilal Haq (pers. comm., 2008), but there are several other published curves with different details (reviewed in Munnecke et al., 2010). The δ¹³C curve and the placement of major widespread biotic events (extinction episodes, etc.) are from Cramer et al. (2011a) (Melchin et al., 2012, 2013). The generalized δ¹⁸O curve from conodont apatite is averaged from the Baltica–Anticosti subset of Trotter et al. (2016), with schematic trends of tropical sea-surface temperatures (without adjustment for possible glacial ice). For comparison, Veizer and Prokoph (2015) derived an adjusted tropical sea-surface temperature curve from a synthesis of oxygen-18 values from carbonate fossils. The vertical scale of this diagram is standardized to match the vertical scales of the first stratigraphic summary figure in all other Phanerozoic chapters.

the current GSSP level and the implied redefinition of the basal Silurian graptolite zone was retained (Rong et al., 2008). The GSSP level is within a widespread oceanic anoxic event that continued through much of the Rhuddanian Stage and had produced extensive organic-rich shale deposits that are an important source rock for hydrocarbons (e.g., review in Melchin et al., 2013).

The GSSP for the base of the **Aeronian** Stage near the Cwm-coed-Aeron Farm in Wales that gave the name to this stage is near

the FAD of graptolite *Monograptus austerus sequens*, which was interpreted to indicate the graptolite *Demirastrites triangulatus* Zone. However, the *Mono. austerus sequens* graptolite is only found at a single other location (also in Wales), and the sparse graptolite record at the GSSP only indicates that it is at an unknown level within the *Demi. triangulates* zone (Melchin et al., 2012). The working group is focusing on alternate candidates in England, China (Shennongjia section in western Hubei Province), and

Base of the Rhuddinian Stage of the Silurian System in Dob's Linn, near Moffat in the Southern Uplands of Scotland, U.K.

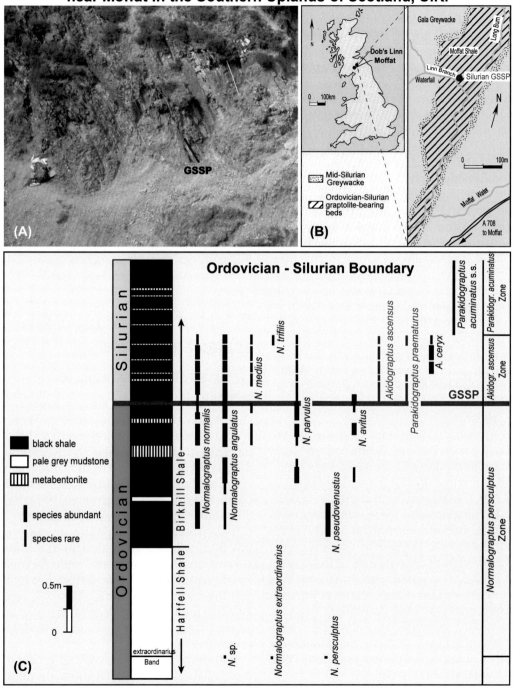

Figure 7.2 GSSP for base of the Silurian (base of Llandovery Series and base of Rhuddanian Stage) at the Dob's Linn section, southern Uplands, Scotland, United Kingdom. The GSSP level coincides with the local lowest occurrence of graptolite *Akidograptus ascensus*. Strata in outcrop are overturned. Stratigraphic columns are from Melchin et al. (2012); and photograph is by Michael Melchin.

the Prague Basin (ISSS Silurian Times, 2014, p. 19). The GSSP for the following **Telychian** Stage, named after the Pen-lan-Telych Farm in Wales, has abundant brachiopods which indicate that the GSSP level is near the last occurrence (LAD) of *Eocelia intermedia*, which elsewhere suggests a level within the upper part of the graptolite *Stimulograptus sedgwickii* Zone (Melchin et al., 2012). If the current brief *Stimul. sedgwickii* zone is further subdivided into a lower *Stimul. sedgwickii* zone and an upper *Stimul. halli* zone (corresponding to the graptolite *Lituigraptus rostrum* Zone in the Prague region), then carbon-isotope stratigraphy suggests that the GSSP level might be close to that new internal zonal boundary (Melchin et al., 2015). Potential boundary sections are in Shennongjia (Hubei, China) and Spain. A reason to have a more precise placement is that the Aeronian/Telychian boundary interval is within a time of glaciation in the South American part of Gondwana, and oscillations of sea level and carbon isotopes of ca. 100-kyr might enable precise global correlation (Melchin et al., 2015).

The current GSSP for the base of the Wenlock Series (base of Sheinwoodian Stage) in Wales at the base of the Buildwas Formation has a vague placement within biozones (Fig. 7.3). It had been intended to be near the base of the graptolite *Cyrtograptus centrifugus* Zone, but the current interpretation is that it is closer to the base of the next higher *Cyrto. murchisoni* Zone (reviewed in Melchin et al., 2012). The working group is searching for a replacement section that has correlation potential with both graptolites and conodonts, and potential sites include Gaspé (Québec, Canada), Arctic Canada, Ziyang (South China), and Gotland (Sweden) (ISSS *Silurian Times*, 2014, p. 19).

The Ludlow Series GSSP (base of Gorstian Stage) near the type area in Wales (Fig. 7.4) seems within the graptolite *Neodiversograptus*

nilssoni Zone and above the Mulde positive excursion in carbon isotopes (Fig. 7.1) (reviewed in Melchin et al., 2012).

The Pridoli (Přídolí) Series is unique in being the only chronostratigraphic series, other than Holocene, that has no stage divisions. This is partly because it was to fill a "gap" between the top of the traditional British Silurian at a Ludlow Bone Bed (top of Ludlow Series) below nonmarine strata and the Silurian/Devonian boundary as defined in 1972 with the GSSP near Prague. The GSSP in graptolite-bearing platy limestone near that Silurian/Devonian GSSP near Prague is within the graptolite *Neocolonograptus parultimus* (formerly in the *Monograptus* genus when GSSP was ratified) Zone (Fig. 7.5).

Selected main stratigraphic scales and events

(1) Biostratigraphy (marine; terrestrial)

Graptolites, the floating colonies of microscopic animals that are preserved in shales as flattened traces, recovered in diversity after their collapse at the end of the Ordovician to reach a Silurian peak during the early Telychian Stage (e.g., Cooper et al., 2014). After a secondary peak in diversity during the late Gorstian Stage, the graptolite diversity rapidly declined into the Pridoli, and only a few species continued into the early Devonian. Most graptolite taxa were widely distributed, and their relatively short stratigraphic ranges are ideal for a standardized Silurian graptolite zonation (e.g., Cramer et al., 2011a,b; Melchin et al., 2012) (Fig. 7.6).

Although graptolites are most useful in biostratigraphy of clay-rich strata, the Silurian carbonate-rich formations are commonly dated using conodonts, the phosphatic teeth and jaw elements of eel-like vertebrates.

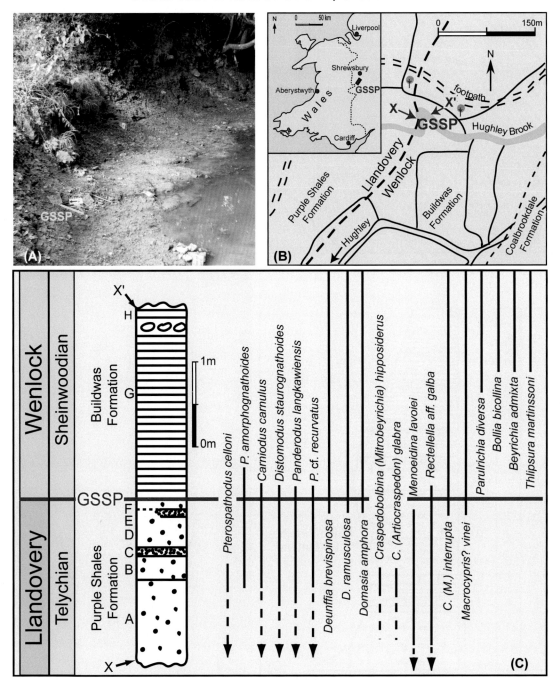

Base of the Sheinwoodian Stage (and Wenlock Series) in Hughley Brook, south-east of Leasows Farm, Great Britain

Figure 7.3 GSSP for base of the Wenlock Series (base of Sheinwoodian Stage) at the Hughley Brook section, Wales, United Kingdom. The GSSP level does not seem to coincide with any biostratigraphic or other datum that is useful for widespread correlation, other than perhaps the highest occurrence of conodont *Pterospathodus amorphognathoides*; therefore, the GSSP is under revision (see text and Melchin et al., 2012). Photograph (with 15-cm long white card for scale) is by Michael Melchin, and the stratigraphic diagram is from Melchin et al. (2012).

Base of the Gorstian Stage of the Silurian System in Pitch Coppice Quarry, United Kingdom

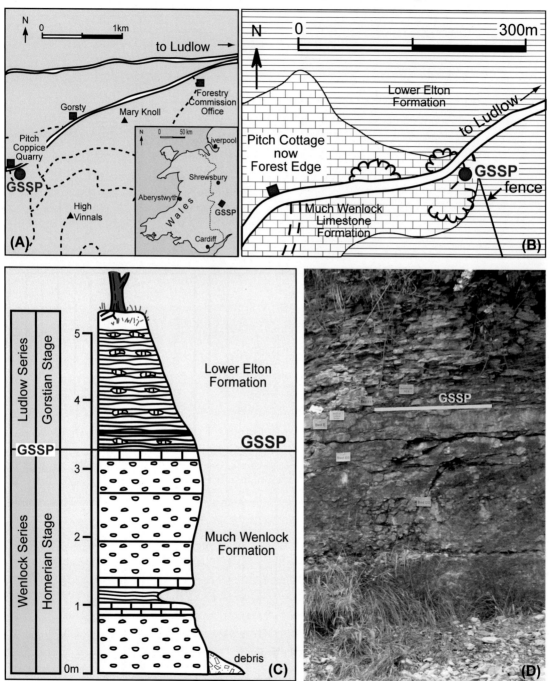

Figure 7.4 GSSP for base of the Ludlow Series (base of Gorstian Stage) at the Pitch Coppice Quarry near Ludlow, Wales, United Kingdom. The GSSP level at the base of the Lower Elton Formation seems above the local lowest occurrence of graptolite *Neodiversograptus nilssoni* and below the lowest *Saetograptus (Colonograptus) varians*, but it is uncertain how to place this GSSP within the *Neo. nilssoni* Zone (Melchin et al., 2012). Photograph (with 15-cm long white card for scale) is by Michael Melchin, and the stratigraphic diagram is from Melchin et al. (2012).

Base of the Pridolí Series of the Silurian System in the Daleje Valley, Prague, Czech Republic

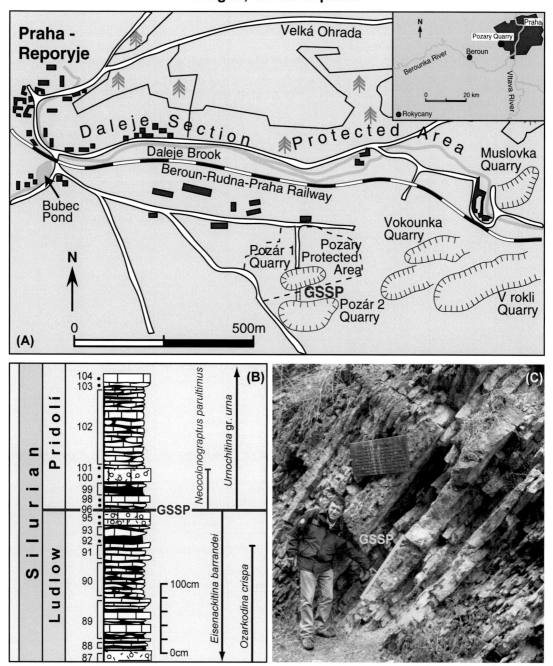

Figure 7.5 GSSP for base of the Přídolí Series at the Požáry Section, Řeporyjie, near Prague, Czech Republic. The GSSP level is above the lowest occurrence of graptolite *Neocolonograptus parultimus*, but because the immediately underlying beds are barren of graptolites, then the relative placement within that *N. parultimus* Zone is uncertain (see Melchin et al., 2012). The GSSP is just below the base of the chitinozoan *Fungochitina kosovensis* Zone. Stratigraphic column is from Melchin et al. (2012); and photograph is by Michael Melchin.

Silurian Time Scale

AGE (Ma)	Epoch/Age (Stage)	Time Slices	Polarity Ch	Graptolite Zones	Conodont Zones	Chitinozoan Zones	Spores	Vertebrates
419 — / 419.2	Devonian			*Monograptus uniformis*	*Latericriodus woschmidti - postwoschmidti*			*Katoporus timanicus - K. lithuanicus*
420 —	Pridoli	Pr2		*Monograptus transgrediens - perneri*	*Oulodus elegans detortus*	*Angochitina superba*		
421 —					*Ozarkodina eosteinhornensis* s.l. Interval Zone			*Poracanthodes punctatus*
422 —				*Monograptus bouceki*				
423.0 / 423 —		Pr1		*Monogr. lochkovensis - M. lochkov. branikensis* / *Neocolonogr. ultimus* / *Neocolonogr. parultimus*		*Margachitina elegans* / *Fungoch. kosovensis*	*Synorisporites tripapillatus - Apiculiretusispora spicula*	*Nostolepis gracilis*
	Ludfordian (Ludlow)	Lu3		*Monograptus formosus*	*Ozarkodina crispa*	*Eisenackitina barrandei*	*Lophozono-triletes? poecilomorphus - Synorisorites libycus*	*Thelodus sculptilis*
424 —		Lu2		*Neocucullogr. kozlowskii - Polonogr. podoliensis* / *Bohemograptus*	*Ozarkodina snajdri* Interval Zone			
		Lu1		*Saetograptus leintwardinensis - linearis*	*Polygnathoides siluricus*	*Eisenackitina phillipi*		*Andreolepis hedei*
425 — / 425.6					*Ancoradella ploeckensis*			*Phlebolepis elegans*
426 —	Gorstian (Ludlow)	Go2		*Lobograptus scanicus*	*Kockelella variabilis variabilis* Interval Zone	*Angochitina elongata*	*Sclya. downiei - Concen. sagittarius*	*Phlebolepis ornata*
427 — / 427.4		Go1		*Neodiversograptus nilssoni*	*Kockelella crassa*			
	Homerian (Wenlock)	Ho3		*Colonograptus ludensis* / *Colonograptus? deubeli* / *Colon? praedeubeli*	*Kockelella ortus absidata*	*Sphaerochitina lycoperdoides*		*Paralogania martinssoni*
428 —		Ho2		*Gothograptus nassa - Pristiog. dubius parvus*			*Artemopyra brevicostata - Hispanaediscus verrucatus*	
429 —		Ho1		*Cyrtograptus lundgreni*	*Ozarkodina bohemica longa*	*Conochitina pachycephala*		
430 — / 430.5					*Ozarkodina sagitta sagitta*			*Loganellia grossi*
431 —	Sheinwoodian (Wenlock)	Sh3		*Cyrtograptus rigidus - Monograptus antennularius - M. belophorus*	*Kockelella ortus ortus* / *Kockelella walliseri*	*Cingulochitina cingulata*	*Archaeozono-triletes chulus nanus - Archaeozono-triletes chulus chulus*	
432 —		Sh2			*Ozarkodina sagitta rhenana*			

Figure 7.6 (A,B) Selected marine and terrestrial biostratigraphic zonations of the Silurian. ("*Age*" is the term for the time equivalent of the rock-record "*stage*".) Compilation modified from Melchin et al. (2012), which used graptolite and conodont zonations based on Cramer et al. (2011a,b, respectively); chitinozoan zonation slightly modified from Verniers et al. (1995); spore zonation from Subcommission on Silurian Stratigraphy (1995) modified after Burgess and Richardson (1995) (Melchin et al., 2012).

Silurian Time Scale

AGE (Ma)	Epoch/Age (Stage)	Time Slices	Polarity Ch	Graptolite Zones	Conodont Zones	Chitinozoan Zones	Spores	Vertebrates
430.5	Wenlock — Homerian	Ho1		Cyrtograptus lundgreni	O. sagitta sagitta	Conochitina pachycephala		Loganellia grossi
431	Wenlock — Sheinwoodian	Sh3		Cyrtograptus rigidus - Monograptus antennularius - M. belophorus	Kockelella ortus ortus			
					Kockelella walliseri	Cingulochitina cingulata	Archaeozono-triletes chulus nanus - Archaeozono-triletes chulus chulus	
432		Sh2			Ozarkodina sagitta rhenana			
				M. riccartonensis - firmus	K. ranuliformis SZ	Margachitina margaritana		Loganellia avonia
433		Sh1		Cyrtograptus murchisoni	Pterospathodus pennatus procerus Super Zone			
433.4	Llandovery — Telychian	Te5		Cyrtograptus centrifugus / Cyrtograptus insectus				
434		Te4		Cyrtograptus lapworthi	Pterospathodus amorphognathoides amorphognathoides	Angochitina longicollis		
435								
					P. am. lithuanicus			
436		Te3		Oktavites spiralis	P. amorph. lennarti			
					Pterospathodus amorphognathoides angulatus			
				Monoclimacis crenulata				
				Monocl. griestoniensis	Pterospathodus eopennatus SuperZone		Ambitisporites avitus - Ambitisporites dilatus	Loganellia sibirica/ Loganellia scotica
437		Te2		Streptograptus crispus		Eisenackitina dolioliformis		
				Spirograptus turriculatus	Distomodus stauro-gnathoides			
438		Te1						
438.5				Spirograptus guerichi				
	Llandovery — Aeronian	Ae3		Stimulogr. sedgwickii	Pterospathodus tenuis			
439		Ae2		Lituigraptus convolutus		Conochitina alargada		
				P. leptotheca-M. argenteus				
440		Ae1		Demirastrites pectinatus - triangulatus	Aspelunda expansa	Spinachitina maennili		
440.8	Llandovery — Rhuddanian	Rh3		Coronograptus cyphus			Pseudodyado-spora sp. B - Segestrespora membranifera	
441						Conochitina electa		
442		Rh2		Cystograptus vesiculosis		Beloechitina postrobusta		Valyalepis crista
					Distomodus kentuckyensis			
443		Rh1		Parakidograptus acuminatus		Spinachitina fragilis		
443.8				Akidograptus ascensus				
444	Ordovician	Hi2		Metabolograptus persculptus	Ozarkodina hassi	Tanuchitina oulebsiri		

Figure 7.6 (Continued) And vertebrate zonation (mainly Thelodonti jawless fish with some Acanthodian jawed-fish zones in the Prídolí) is from Märss et al. (1995). Additional zonations, biostratigraphic markers, geochemical trends, sea-level curves, and details on calibrations are compiled in Melchin et al. (2012) and in the internal data sets within the *TimeScale Creator* visualization system. Free at www.tscreator.org.

The zonation in Fig. 7.6 is the semiglobal composite compiled by Cramer et al. (2011a,b), and adding a slight modification by Trotter et al. (2016, suppl.) of a narrowed span for the *Pterospathodus tenuis* Zone within the Aeronian. Organic-walled chitinozoans, which may have been microscopic floating egg capsules of a marine animal, are also widespread in marine facies. The Silurian seas hosted a variety of early forms of jawless Thelodonti fish, with jawed Acanthodian fish appearing in the Pridoli. The earliest land plants appeared during the Silurian, and their spores enable a broad correlation of terrestrial and nearshore settings.

(2) Stable-isotope stratigraphy, magnetostratigraphy, and selected events

The Silurian is characterized by a series of major oscillations in sea level, carbon isotopes, and sea-surface temperatures. Although, there is not yet agreement on the precise intercalibration and magnitude of some of these events, especially in sea-level curves (see discussions of different sea-level and sequence interpretations in Munnecke et al., 2010; Johnson et al., 2010; Melchin et al., 2012), it has become apparent that most of these events are interconnected. In general, but not always, a major wave of extinctions of graptolites and conodonts occurs during the beginning of a climate-cooling episode, a sea-level lowstand, a possible expansion of continental ice sheets, and a positive excursion in carbon isotopes. In contrast to the common association of major large igneous province eruptions with such excursions during the Late Paleozoic through Cenozoic, the triggers for the major episodes, which are spaced at approximately 5-myr intervals, remain speculative. There

are no major volcanic eruptive provinces yet identified within the Silurian (e.g., compilation by Rampino and Self, 2015). The Lau Event that begins in the uppermost part of the graptolite *Saetograptus leintwardinensis* zone of the Ludfordian Stage at 424.1 ± 0.2 Ma (Cramer et al., 2015) is the largest positive excursion of the global carbon cycle in the entire Phanerozoic.

Strontium $^{87}Sr/^{86}Sr$ ratios steadily climbed through the Silurian, with the most rapid rate of rise during the Ludlow Epoch until the Lau Event (Melchin et al., 2012; Cramer et al., 2015).

Temperature trends from $\delta^{18}O$ measurements using conodont apatite indicate that, after the rapid warming following the cold mid-Hirnantian glacial, average temperatures remained relatively warm during the Llandovery, underwent a set of cold episodes during the Wenlock, and were generally warm during the Ludlow and Pridoli (Trotter et al., 2016). The schematic temperature curve from these conodont $\delta^{18}O$ values in Fig. 7.1 illustrates the trends and relative oscillations; and actual temperatures would require adjustment for uncertain extent of continental ice sheets. An independent estimate of average subtropical temperatures from $\delta^{18}O$ values of carbonates that is adjusted for potential long-term trends in Phanerozoic seawater $\delta^{18}O$ (Veizer and Prokoph, 2015) yields significantly cooler average values with similar trends (Fig. 7.1). The major episode of widespread black-shale deposition—an important hydrocarbon source rock—at the beginning of the Silurian (late Hirnantian through early Rhuddanian) was triggered by the melting of the large Hirnantian glacial sheets combined with an influx of nutrients, and possible development of strong oceanic stratification, and then partly sustained by enhanced phosphate recycling under anoxic bottom water conditions (Melchin et al., 2013). The Silurian ends with the Klonk Event of a positive $\delta^{13}C$ excursion and rapid cooling into the basal Devonian.

Numerical age model

GTS2012 age model and potential future enhancements

A composite sequencing of the stratigraphic ranges of over 2000 graptolite species using a database from over 500 stratigraphic sections was compiled using **CON**strained **OP**timization methods (CONOP) by Cooper and Sadler (in Cooper et al., 2012, 2014). Age models for the FADs and LADs of the major taxa and for the Ordovician and Silurian stage boundaries calibrated to those events (placed at bases of graptolite zones) were interpolated by both a cubic spline fit and a polynomial fit to a set of 22 radioisotopic dates. The two interpolation methods generally yielded very similar estimates for the ages of Silurian stage boundaries, except for the bases of the Aeronian and Telychian stages, which were ca. 1.6- and 1.1-myr younger, respectively, using the spline-fit method (see comparison table in Melchin et al., 2012; and graphical-display comparison in Cooper et al., 2012). The spline-fit interpolations were used in Geologic Time Scale 2012 (GTS2012), whereas the polynomial-fit interpolation was used by Cooper et al. (2014) for their statistics on graptolite diversity trends.

There are several radioisotopic dates for the late Silurian published after GTS2012, and these are generally very close to the GTS2012 age model (e.g., Cramer et al., 2012, 2015). For example, a 422.91±0.07 Ma date from immediately below the basal graptolite zone of the Pridoli Stage (Cramer et al., 2015) is nearly identical to the interpolated

422.96 Ma age on that stage boundary. The only new date that may suggest a revised age for a stage boundary is a 431.83±0.23 Ma date from the middle of the conodont *Pseudooneotodus bicornis* subzone near the Ireviken Event in basal Sheinwoodian (base of Wenlock Series) (Cramer et al., 2012), which suggests that this yet-to-be-revised Wenlock Series boundary might be ca. 1.5-myr younger than the interpolation in GTS2012. In the Silurian scale used in Figs. 7.1 and 7.6, the GTS2012 age model was retained.

In Fig. 7.6, a suite of radioisotopic dates within the Homerian Stage including bracketing the Mulde Excursion (Cramer et al., 2012) and within the Ludfordian stage at the base of the Lau Excursion (Cramer et al., 2015) have been incorporated in adjustment of the associated graptolite- and conodont-zone boundaries within those stages.

Estimated uncertainties on assigned ages on stage boundaries

The uncertainties on stage boundaries computed by the spline-fit method in GTS2012 were largely governed by the uncertainties and spacing of the radioisotopic tie points. The uncertainties from the spline-fit statistics ranged from 1.5 myr at the base of the Silurian, decreasing to ca. 0.5 myr at the base of the Gorstian (427.4±0.5 Ma in GTS2012, which was supported by a later-published date of 427.86±0.32 Ma near that stage base by Cramer et al., 2012), and rising to 2.3 myr at base-Pridoli and 3.2 myr at the base of the Devonian. Reduction of these relatively high uncertainties will require reanalysis of the previously published radioisotopic ages using the EARTHTIME standards (e.g., Condon et al., 2015) and adding dates within the main gaps of the late Llandovery and early Ludlow epochs.

GSSPs of the Silurian Stages, with location and primary correlation criteria

Stage	GSSP Location	Latitude, Longitude	Boundary Level	Correlation Events	Reference
Přídolí (Series)	Požáry Section, Řeporyjie, Prague, Czech Republic	50°01'39.82"N 14°19'29.56"E*	within Bed 96	Graptolite, FAD of *Neocolonograptus parultimus*	Episodes 8/2, 1985; Geol. Ser., Nat. Mus. Wales **9**, 1989
Ludfordian	Sunnyhill Quarry, near Ludlow, UK	52°21'33"N 2°46'38"W*	coincident with the base of the Leintwardine Formation	Near base of Graptolite, *Saetograptus leintwardinensis*	Lethaia **14**; Episodes 5/3, 1982; Geol. Ser.,Nat. Mus. Wales **9**, 1989
Gorstian	Pitch Coppice Quarry near Ludlow, UK	52°21'33"N 2°46'38"W*	coincident with the base of the Lower Elton Formation	Graptolite, FAD of *Saetograptus (Colonograptus) varians*	Lethaia **14**; Episodes 5/3, 1982; Geol. Ser.- Nat. Mus. Wales **9**, 1989
Homerian	Sheinton Brook, Homer, UK	52°36'56"N 2°33'53"W*	within upper part of the Apedale Member of the Coalbrookdale Formation	Graptolite, FAD of *Cyrtograptus lundgreni*	Lethaia **14**; Episodes 5/3, 1982; Geol. Ser., Nat. Mus. Wales **9**, 1989
Sheinwoodian (under revision)	Hughley Brook, UK	52°34'52"N 2°38"20"W*	base of the Buildwas Formation	Imprecise, between the base of acritarch biozone 5 and LAD of conodont *Pterospathodus amorphognathoides*	Lethaia **14**; Episodes 5/3, 1982; Geol. Series, Nat. Mus. Wales **9**, 1989
Telychian	Cefn-cerig Road Section, Wales, UK	51.97°N 3.79°W**	approximately 31 m below the top of the Wormwood Formation	Just above LAD of Brachiopod *Eocoelia intermedia* and below FAD of *Eocoelia curtisi*	Episodes 8/2, 1985; Geol. Series, Nat. Mus. Wales **9**, 1989
Aeronian	Trefawr Track Section, Wales, UK	52.03°N 3.70°W**	within Trefawar Formation	Graptolite, FAD of *Monograptus austerus sequens*	Geol. Series, Nat. Mus. Wales **9**, 1989
Rhuddanian	Dob's Linn, Scotland	55.44°N 3.27°W**	1.6m above the base of the Birkhill Shale Formation	Graptolite, FAD of *Akidograptus ascensus*	Episodes 8/2, 1985; Episodes 31/3, 2008

* according to Google Earth, ** derived from map

Figure 7.7 Ratified GSSPs and potential primary markers under consideration for defining the Silurian stages (*status as of early 2016*). Details of each GSSP are available at http://www.stratigraphy.org, https://engineering.purdue.edu/Stratigraphy/gssp/, and in the *Episodes* publications.

Acknowledgments

This brief Silurian summary of selected highlights and current stratigraphic issues relied heavily on the detailed overview and synthesis by Melchin et al. (2012) and Cramer et al. (2011a,b); and updates in the *Silurian Times* newsletters of the International Subcommission on Silurian Stratigraphy (*http://unica2.unica.it/sds/*). This review depended on the expertise of many colleagues, especially Michael Melchin, Roger Cooper, and Peter Sadler. Mike Melchin reviewed an early version of the graphics and text.

Selected publications and websites

Cited publications

Only select publications were cited in this review with an emphasis on aspects of post-2011 updates. Pre-2011 literature is well summarized in the synthesis by Melchin et al. (2012) and in some of the publications cited in the following.

Burgess, N.D., Richardson, J.B., 1995. Late Wenlock to early Přídolí cryptospores and miospores from south and southwest Wales, Great Britain. *Palaeontographica Abteilung B: Palaeophytologie* **236**: 1–44.

Condon, D.J., Schoene, B., McLean, N.M., Bowring, S.A., Parrish, R.R., 2015. Metrology and traceability of U-Pb isotope dilution geochronology (EARTHTIME Tracer Calibration Part 1). *Geochemica et Cosmochimica Acta* **164**: 464–480. http://dx.doi.org/10.1016/j.gca.2015.05.026.

Cooper, R.A., Sadler, P.M., Gradstein, F.M., Hammer, O., 2012. The Ordovician Period. In: Gradstein, F.M., Ogg, J.G., Schmitz, M., Ogg, G., (Coordinators), *The Geologic Time Scale 2012.* Elsevier Publisher, pp. 489–523 (An overview on all aspects, including graphics on the ratified GSSPs of the stages, diagrams and tables for the biostratigraphic scales, and discussion on the age models.).

Cooper, R.A., Sadler, P.M., Munnecke, A., Crampton, J.S., 2014. Graptoloid evolutionary rates track Ordovician-Silurian global climate change. *Geological Magazine* **151**: 349–364.

Cramer, B.D., Brett, C.E., Melchin, M.A., Männik, P., Kleffner, M.A., McLaughlin, P.I., Loydell, D.K., Munnecke, A., Jeppsson, L., Corradini, C., Brunton, F.R., Saltzman, M.R., 2011a. Revised chronostratigraphic correlation of the Silurian System of North America with global and regional chronostratigraphic units and d^{13}C$_{carb}$ chemostratigraphy. *Lethaia* **44**: 185–202.

Cramer, B.D., Davies, J.R., Ray, D.C., Thomas, A.T., Cherns, L., 2011b. Siluria Revisited: an introduction. In: Ray, D.C. (Ed.), *Siluria Revisited, a Field Guide*, International Subcommission on Silurian Stratigraphy Field Meeting 2011, pp. 6–27.

Cramer, B.D., Condon, D.J., Söderlund, U., Marshall, C., Worton, G.J., Thomas, A.T., Calner, M., Ray, D.C., Perrier, V., Boomer, I., Patchett, P.J., Jeppsson, L., 2012. U-Pb (zircon) age constraints on the timing and duration of Wenlock (Silurian) paleocommunity collapse and recovery during the 'Big Crisis'. *Geological Society of America Bulletin* **124**: 1841–1857. http://dx.doi.org/10.1130/B30642.1.

Cramer, B.D., Schmitz, M.D., Huff, W.D., Bergström, S.M., 2015. High-precision U–Pb zircon age constraints on the duration of rapid biogeochemical events during the Ludlow Epoch (Silurian Period). *Journal of the Geological Society, London* **172**: 157–160. http://dx.doi.org/10.1144/jgs2014-094.

Haq, B.U., Schutter, S.R., 2008. A chronology of Paleozoic sea-level changes. *Science* **322**: 64–68. http://dx.doi.org/10.1126/science.116164.

International Subcommission on Silurian Stratigraphy (ISSS), 2014. (for 2014) *Silurian Times* **22**. PDFs at http://silurian.stratigraphy.org.

Johnson, M.E., 2010. Tracking Silurian eustasy: alignment of empirical evidence or pursuit of deductive reasoning. *Paleogeography, Palaeoclimatology, Palaeoecology* **296**: 276–284.

Märss, T., Fredholm, D., Talimaa, V., Turner, S., Jeppsson, L., Nowlan, G.S., 1995. Silurian vertebrate biozonal scheme. In: Lelievre, H., Wenz, S., Blieck, A., Cloutier, R. (Eds.), *Premiers vertebrés et vertebrés inférieurs* Géobios, **vol. 19**, pp. 369–372.

Melchin, M.J., Sadler, P.M., Cramer, B.D., Cooper, R.A., Gradstein, F.M., Hammer, O., 2012. The Silurian Period. In: Gradstein, F.M., Ogg, J.G., Schmitz, M., Ogg, G., (Coordinators), *The Geologic Time Scale 2012.* Elsevier Publisher, pp. 525–558 (An overview on all aspects, including graphics on the ratified GSSPs of the stages, diagrams and tables for the biostratigraphic scales, and discussion on the age models.).

Melchin, M.J., Mitchell, C.E., Holmden, C., Štorch, P., 2013. Environmental changes in the Late Ordovician–early Silurian: review and new insights from black shales and nitrogen isotopes. *Geological Society of America Bulletin* **125**: 1635–1670. http://dx.doi.org/10.1130/B30812.1.

Melchin, M.J., MacRae, K.-D., Bullock, P., 2015. A multi-peak organic carbon isotope excursion in the late Aeronian (Llandovery, Silurian): evidence from Arisaig, Nova Scotia, Canada. *Palaeoworld* **24**: 191–197.

Munnecke, A., Calner, M., Harper, D.A.T., Servais, T., 2010. Ordovician and Silurian sea-water chemistry, sea level, and climate: a synopsis. *Paleogeography, Palaeoclimatology, Palaeoecology* **296**: 389–413.

Rampino, M.R., Self, S., 2015. Large igneous provinces and biotic extinctions. In: Sigurdsson, H., Houghton, B., McNutt, S., Rymer, H., Stix, J. (Eds.), The Encyclopedia of Volcanoes, second ed., Elsevier Publications, pp. 1049–1058. http://dx.doi.org/10.1016/B978-0-12-385938-9.00061-4. Chapter 61.

Ray, D.C.(editor), 2011. *Siluria revisted: a field guide.* International Subcommission on Silurian Stratigraphy: Field Meeting 2011(ISBN: 978-0-9569190-0-7): 166p. free download at: www.igcp591.org/downloads/siluria_revisited_excursion_2011.pdf.

Rong, J., Melchin, M.J., Williams, S.H., Koren, T.N., Verniers, J., 2008. Report of the restudy of the defined global stratotype of the base of the Silurian System. *Episodes* **31**: 315–318.

Scotese, C.R., 2014. *Atlas of Silurian and Middle-Late Ordovician Paleogeographic Maps (Mollweide Projection), Maps 73–80, Volume 5, The Early Paleozoic, PALEOMAP PaleoAtlas for ArcGIS, PALEOMAP Project, Evanston, IL.* https://www.academia.edu/16744278/Atlas_of_Silurian_and_Middle-Late_Ordovician_Paleogeographic_Maps.

Trotter, J.A., Williams, I.S., Barnes, C.R., Männik, P., Simpson, A., 2016. New conodont δ^{18}O records of Silurian climate change: Implications for environmental and biological events. *Palaeogeography, Palaeoclimatology, Palaeoecology* **443**: 34–48.

Veizer, J., Prokoph, A., 2015. Temperatures and oxygen isotopic composition of Phanerozoic oceans. *Earth-Science Reviews* **146**: 92–104.

Verniers, J., Nestor, V., Paris, F., Dufka, P., Sutherland, S., van Grootel, G., 1995. A global Chininozoa biozonation for the Silurian. *Geological Magazine* **132**: 651–666.

Websites (selected)

International Subcommission on Silurian Stratigraphy—*http://silurian.stratigraphy.org*—News items, PDFs of *Silurian Times* Newsletters (vol. 14 of 2006, through present), and details (in future) on GSSPs and working groups.

Palaeos: The Silurian—*http://palaeos.com/paleozoic/silurian/silurian.htm*—A well-presented suite of diverse topics for a general science audience that was originally compiled by M. Alan Kazlev in 1998–2002.

Geobiodiversity Database—*www.geobiodiversity.com*—Extensive array of stratigraphic sections (ca. 14,000 localities) with a current emphasis on Ordovician–Silurian and on China. Also includes SinoCor program for graphical correlation and other tools. Developed by Fan Junxuan (Nanjing Inst. Geology and Paleontology).

Commission Internationale Microflore Paléozoïque—*http://cimp.weebly.com*—International Commission of the Paleozoic Microflora has separate subcommissions on acritarchs, chitinozoans, and miospores; and site has archive of their newsletters and other information; although number of active newsletters diminished since 2012.

The Silurian Reef—*http://www.mpm.edu/content/collections/learn/reef/*—An educational site by the Milwaukee Public Museum that focuses on the Silurian reef ecosystems.

8

DEVONIAN

388.2 Ma Devonian

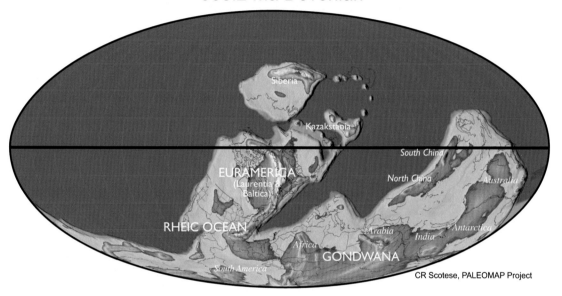

CR Scotese, PALEOMAP Project

Givetian (Middle Devonian) paleogeographic reconstruction (Sea level + 40 m, Frasnian supersequence boundary) from Scotese (2014).

Basal definition and international subdivisions

The Devonian was named after rock exposures in Devon County of southwest England. In Britain, the type Silurian is truncated by the nonmarine Old Red Sandstone that was shed from the Caledonian orogeny when the Baltica plate collided with Laurentia. The extinction of graptolites had been used as a guide to the Silurian/Devonian boundary, until it was recognized that graptolites continued much higher than the traditional lithologic placement of the boundary. Indeed, their last occurrence is now being considered as a marker for the base of the Emsian Stage at ca. 15 myr after the current base of the Devonian. The process to standardize usage of the Silurian/Devonian boundary culminated in the international ratified decision of 1972 to define a global boundary stratotype section and point (GSSP) within a graptolite-bearing succession at Klonk, near Prague in the Czech Republic. This decision was the impetus to define all geologic stages with precise GSSPs. The Subcommission on Devonian Stratigraphy completed a set of seven GSSPs in 1996 (Fig. 8.1 and Fig. 8.6), although some of these are now being reexamined.

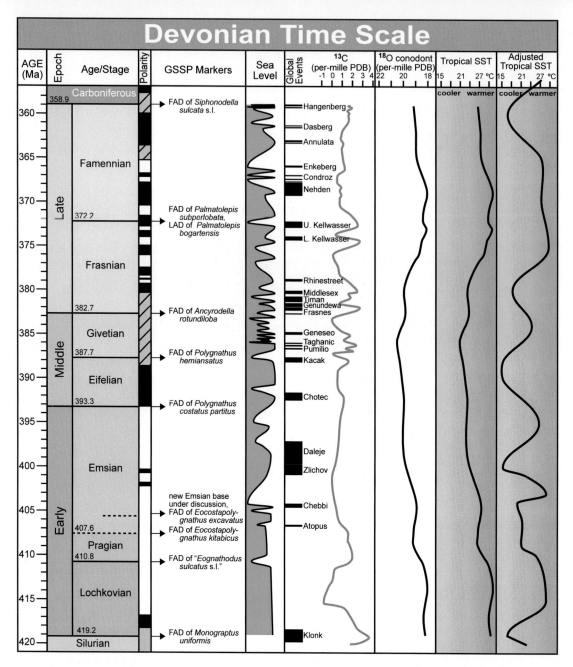

Figure 8.1 Devonian overview. Most main markers for GSSPs of Devonian stages are first-appearance datums (FADs) of conodont taxa as detailed in the text and in Fig. 8.6. ("Age" is the term for the time equivalent of the rock-record "stage".) See Carboniferous chapter for discussion on potential revised definition of the Devonian/Carboniferous boundary. Magnetic polarity scale is from Becker et al. (2012) with revised Frasnian–Famennian from Hansma et al. (2015). Schematic sea-level curve was compiled by Becker et al. (2012) from various sources. An independent interpretation of Devonian sequences by Haq and Schutter (2008) is shown in Fig. 8.5. The carbon isotope (δ^{13}C) curve is from Buggisch and Joachimski (2006) with global events (often associated with widespread anoxic events, as indicated in black) modified from Becker et al. (2012). Generalized oxygen isotope (δ^{18}O) curve and estimates of tropical sea-surface temperatures from conodont apatite are averaged from Joachimski et al. (2009). For comparison, the adjusted tropical sea-surface temperature curve that Veizer and Prokoph (2015) derived from a synthesis of oxygen-18 values from carbonate fossils is also shown. The vertical scale of this diagram is standardized to match the vertical scales of the first stratigraphic summary figure in all other Phanerozoic chapters. *PDB*, PeeDee Belemnite ^{13}C standard; *SST*, sea-surface temperature.

The GSSP for the base of the Devonian and of the Lochkovian Stage is a point in the rock record that was selected to also coincide with the FAD of the graptolite *Monograptus uniformis uniformis* (Fig. 8.2). Recognition of the boundary in carbonate strata or in sections that lack graptolites is enabled by a major turnover in conodont species and a major positive carbon-isotope excursion at the anoxic Klonk Event that straddles the boundary interval.

All other GSSPs for Devonian stages were placed to coincide with FADs of conodont taxa (reviewed by Becker et al., 2012) (Figs. 8.1 and 8.6). For example, the Lower/Middle Devonian boundary (base of Eifelian Stage) GSSP in the Eifel Mountains of Germany, near the border with Belgium, coincides with the FAD of the conodont *Polygnathus costatus partitus* (Fig. 8.3). The Middle/Upper Devonian GSSP (base of Frasnian Stage) in the Montagne Noire of southern France coincides with the FAD of the conodont *Ancyrodella rotundiloba pristina* (Fig. 8.4).

Many of the Devonian stages have commonly used substage divisions. To avoid confusion in usage, the Subcommission on Devonian Stratigraphy (SDS) is now working to formalize these substages with both agreements on stratigraphic datums (e.g., FADs of conodonts) and formal GSSP placements (e.g., Subcommission on Devonian Stratigraphy, 2014). For example, a proposed base for the Upper Famennian substage would coincide with the global *Annulata* Event of extensive anoxic conditions and faunal turnover (Fig. 8.1). Potential substage levels are shown in Fig. 8.5 and discussed by Becker et al. (2012; see his Fig. 22.1).

Continued research has indicated that some of the Devonian GSSPs were not placed at levels that were intended (e.g., see discussion on the Devonian/Carboniferous boundary in the Carboniferous chapter) or do not enable useful correlations. For example, the base of the Emsian Stage of Lower Devonian has a GSSP at Zinzil'ban Gorge in Uzbekistan that coincides with the FAD of conodont *Eocostapolygnatus* (formerly *Polygnathus*) *kitabicus*, but this level was later found to be much lower than the base of the classical Emsian in Germany and essentially transferred the upper half of the traditional Pragian into the Emsian. Therefore, in 2008, the SDS decided to shift the base of the Emsian to a higher level; and current discussions center on the FAD of the conodont *Eolinguipolygnathus excavatus* M114 (Fig. 8.5C) and on which section in eastern Uzbekistan is most suitable.

Selected main stratigraphic scales and events

Biostratigraphy (marine; terrestrial)

Conodonts, the phosphatic tooth and jaw elements of eel-like chordates, provide both a standardized global biostratigraphic framework for the Devonian and a record of oceanic oxygen-18 values for sea-surface temperatures. Graptolites, the enigmatic floating colonial clusters that are commonly preserved as elongate organic-rich films in Ordovician–Silurian shales, continued to the middle of the Early Devonian. The extinction of graptolites just predates the appearance and increased diversity of ammonoids. Ammonoid genus zones (e.g., Fig. 8.5) are used for interregional correlation (Becker et al., 2012). Images of typical Devonian ammonoid markers and selected regional species-taxa zones are compiled by Korn and Klug (2015). Dacryoconarids (small cone-shaped shells of pelagic tentaculitid animals, which are known from the Ordovician through the Devonian), pelagic ostracods, siliceous radiolarians, and organic-walled microfossils of acritarchs, and flask-shaped chitinozoans (another group of uncertain affinity) provide additional regional to global zonations (reviewed with examples of zonations in Becker et al., 2012). At the

Base of the Lochkovian Stage of the Devonian System in Klonk, Czech Republic

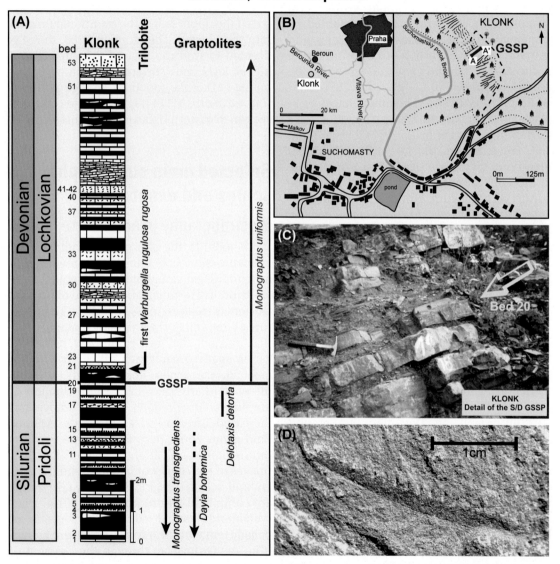

Figure 8.2 GSSP for base of the Devonian (base of Lochkovian Stage) at the Klonk section, Czech Republic. The GSSP level was to coincide with the lowest occurrence of graptolite *Monograptus uniformis uniformis*. Outcrop photograph and stratigraphic column are from Becker et al. (2012).

Base of the Eifelian Stage (Base of the Middle Devonian Series) of the Devonian System near Schönecken-Wetteldorf, Germany

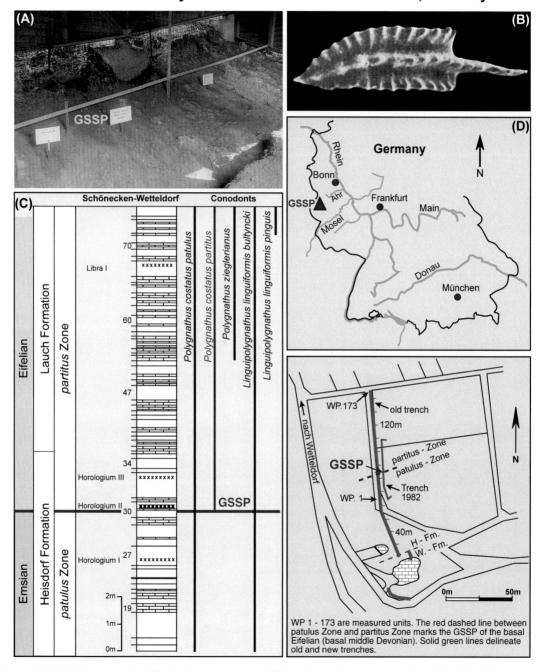

Figure 8.3 GSSP for base of the Middle Devonian (base of Eifelian Stage) in the trench at Wetteldorf, Eifel Mountains, northwest Germany. The GSSP level coincides with the lowest occurrence (FAD) of the conodont *Polygnathus costatus partitus*. Photographs (A) of the section details within the Happel Hut with marked GSSP position and (B) of the index conodont *Po. costatus partitus* by K. Weddige. Section stratigraphy is from Becker et al. (2012).

Base of the Frasnian Stage of the Devonian
System at Col du Puech de la Suque, Montagne Noire, France.

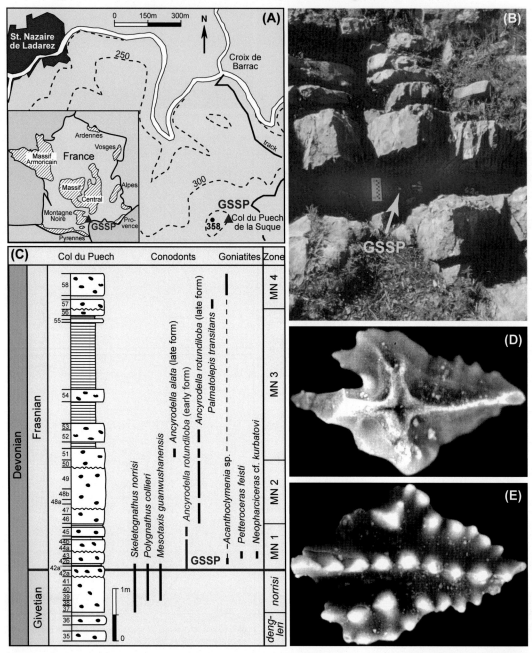

Figure 8.4 GSSP for base of the Upper Devonian (base of Frasnian Stage) Col du Puech de la Suque, Montagne Noire, southern France. The GSSP level coincides with the lowest occurrence (FAD) of conodont *Ancyrodella rotundiloba pristina* (= early morphotype). Photographs (D and E) of representative adult specimens of *An. rotundiloba* by G. Klapper. Section stratigraphy and photograph is from Becker et al. (2012).

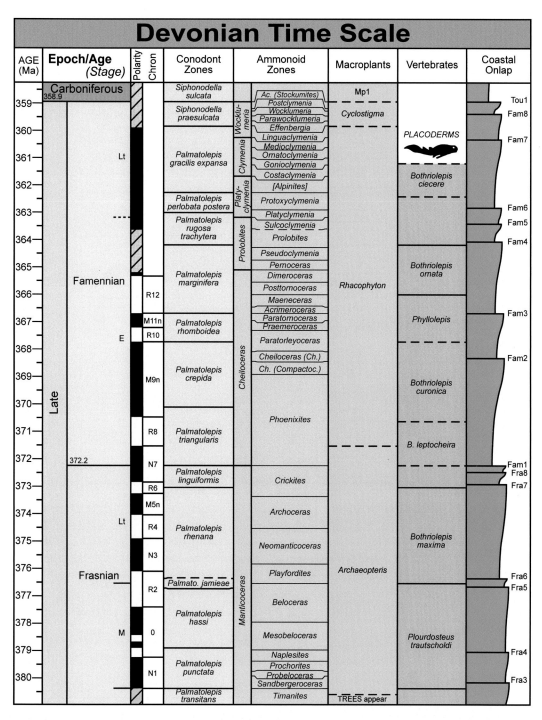

Devonian Time Scale

AGE (Ma)	Epoch/Age (Stage)	Polarity	Chron	Conodont Zones	Ammonoid Zones		Macroplants	Vertebrates	Coastal Onlap
359	Carboniferous 358.9			Siphonodella sulcata	Ac. (Stockumites)	Wocklumeria	Mp1		Tou1
					Postclymenia				Fam8
				Siphonodella praesulcata	Wocklumeria		Cyclostigma		
					Parawocklumeria			PLACODERMS	
360					Effenbergia	Clymenia			Fam7
					Linguaclymenia				
361	Lt			Palmatolepis gracilis expansa	Medioclymenia			Bothriolepis ciecere	
					Ornatoclymenia				
					Gonioclymenia				
362					Costaclymenia	Platy-clymenia			
				Palmatolepis perlobata postera	[Alpinites]				
363					Protoxyclymenia				Fam6
				Palmatolepis rugosa trachytera	Platyclymenia	Prolobites			Fam5
364					Sulcoclymenia				Fam4
					Prolobites				
					Pseudoclymenia				
365				Palmatolepis marginifera	Pernoceras		Rhacophyton	Bothriolepis ornata	
	Famennian				Dimeroceras				
366			R12		Posttornoceras				
					Maeneceras				
367			M11n	Palmatolepis rhomboidea	Acrimeroceras			Phyllolepis	Fam3
		E	R10		Paratornoceras				
					Praemeroceras				
368					Paratorleyoceras				Fam2
			M9n	Palmatolepis crepida	Cheiloceras (Ch.)	Cheiloceras			
369					Ch. (Compactoc.)			Bothriolepis curonica	
370	Late								
					Phoenixites				
371			R8	Palmatolepis triangularis				B. leptocheira	
372	372.2		N7						Fam1
				Palmatolepis linguiformis	Crickites				Fra8
373			R6						Fra7
			M5n		Archoceras				
374		Lt	R4	Palmatolepis rhenana				Bothriolepis maxima	
375			N3		Neomanticoceras				
376					Playfordites	Manticoceras	Archaeopteris		Fra6
	Frasnian		R2	Palmato. jamieae					Fra5
377					Beloceras				
378		M	0	Palmatolepis hassi	Mesobeloceras			Plourdosteus trautscholdi	
379					Naplesites				Fra4
			N1	Palmatolepis punctata	Prochorites				
380					Probeloceras				Fra3
					Sandbergeroceras				
				Palmatolepis transitans	Timanites		TREES appear		

Figure 8.5a,b,c Selected marine and terrestrial biostratigraphic zonations of the Devonian. ("*Age*" is the term for the time equivalent of the rock-record "*stage*".) See Carboniferous chapter for discussion on potential revised definition of the Devonian/Carboniferous boundary. Potential revised placement for base of Emsian Stage of Early Devonian is dashed above the current GSSP assignment.

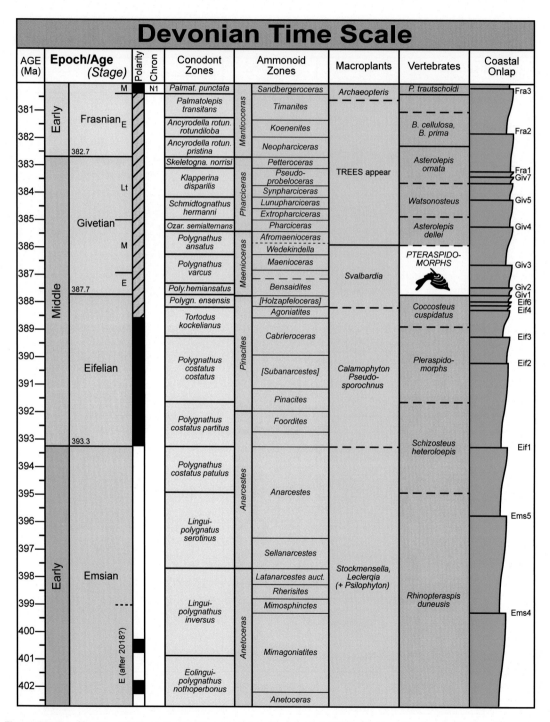

Devonian Time Scale

AGE (Ma)	Epoch/Age (Stage)			Polarity	Chron	Conodont Zones	Ammonoid Zones		Macroplants	Vertebrates	Coastal Onlap	
381	Early	Frasnian	M / E		N1	Palmat. punctata	Sandbergeroceras	Manticoceras	Archaeopteris	P. trautscholdi		Fra3
382						Palmatolepis transitans	Timanites			B. cellulosa, B. prima		Fra2
			382.7			Ancyrodella rotun. rotundiloba	Koenenites					
383						Ancyrodella rotun. pristina	Neopharciceras			Asterolepis ornata		Fra1 / Giv7
384	Middle	Givetian	Lt			Skeletogna. norrisi	Petteroceras	Pharciceras	TREES appear			
						Klapperina disparilis	Pseudo-probeloceras					
							Synpharciceras			Watsonosteus		Giv5
385						Schmidtognathus hermanni	Lunupharciceras					
							Extropharciceras			Asterolepis dellei		Giv4
386				M		Ozar. semialternans	Pharciceras		Svalbardia			
						Polygnathus ansatus	Afromaenioceras	Maenioceras		PTERASPIDO-MORPHS		Giv3
387							Wedekindella					
						Polygnathus varcus	Maenioceras					
			387.7	E		Poly. hemiansatus	Bensaidites					Giv2 / Giv1
388						Polygn. ensensis	[Holzapfeloceras]	Pinacites		Coccosteus cuspidatus		Eif6 / Eif4
						Tortodus kockelianus	Agoniatites					
389		Eifelian					Cabrieroceras		Calamophyton Pseudo-sporochnus	Pleraspido-morphs		Eif3
390						Polygnathus costatus costatus						Eif2
391							[Subanarcestes]					
392							Pinacites					
						Polygnathus costatus partitus	Foordites			Schizosteus heteroloepis		
393			393.3									Eif1
394	Early	Emsian				Polygnathus costatus patulus	Anarcestes	Anarcestes				
395												Ems5
396						Lingui-polygnatus serotinus						
397							Sellanarcestes		Stockmensella, Leclerqia (+ Psilophyton)			
398						Lingui-polygnathus inversus	Latanarcestes auct.	Anetoceras		Rhinopteraspis duneusis		
399				E (after 2018?)			Rherisites					Ems4
							Mimosphinctes					
400												
401							Mimagoniatites					
402						Eolingui-polygnathus nothoperbonus	Anetoceras					

Figure 8.5a,b,c (Continued) Magnetic polarity scale is from Becker et al. (2012) with Frasnian–Famennian from Hansma et al. (2015). Conodont, ammonoid, graptolite, macroplants, and vertebrate zones are modified from Becker et al. (2012). Coastal onlap sea-level curve with sequence boundary nomenclature is modified from Haq and Schutter (2008) following advice of Bilal Haq (pers. comm., 2008); but see Fig. 8.1 for an independent compilation by Becker et al. (2012).

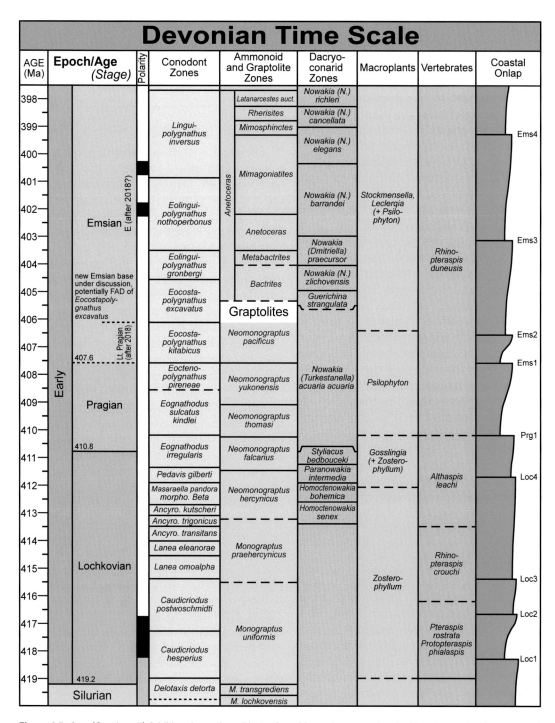

Figure 8.5a,b,c (Continued) Additional zonations, biostratigraphic markers, geochemical trends, sea-level curves, and details on calibrations are compiled in Becker et al. (2012) and in the internal data sets within the *TimeScale Creator* visualization system (free at *www.tscreator.org*).

top of the marine food chain were large sea scorpions, giant orthoconic nautiloids, and diverse fish, and their relatives. Armored fish (pteraspidomorphs in the early half of the Devonian, placoderms in the later half) were accompanied by sharks in the Middle and Late Devonian (zonations in Becker et al., 2012). There was a progressive increase in the average size of these marine vertebrates through the Devonian, until a sudden collapse in their sizes during the end-Devonian Hangenberg Event (Salim and Galimberti, 2015).

The barren land was progressively populated by vegetation through the Devonian, especially following the evolution and rapid spread of tree-sized *Archaeopteris* beginning at the end of the Middle Devonian (Gradstein and Kerp, 2012). In addition to providing macroplant and miospore zonations for regional and global correlation (example columns in Becker et al., 2012), the forestation of the landscape resulted in more mature soils with enhanced release of phosphate–nitrate nutrients which may have contributed to the organic-rich events and general cooling in the Middle and Late Devonian (e.g., Algeo et al., 2001). The earliest semiaquatic tetrapods are known from the Eifelian (Niedźwiedzki et al., 2010), but they became more widespread in the forested world of the Famennian (e.g., Clack, 2012).

Magnetostratigraphy

The generalized magnetostratigraphy synthesis in Becker et al. (2012) indicated that the Early Devonian had a dominance of reversed polarity, the Eifelian was suggested a normal polarity, and the latter Devonian was essentially uncertain; but this scale mainly reflected a dearth of research and verification in multiple sections. The Frasnian–Famennian polarity pattern has now been partly resolved from several sections in the Canning Basin of Western Australia as a series of frequent reversals (Hansma et al., 2015) (Fig. 8.5) that may enable future marine to terrestrial correlations.

Stable-isotope stratigraphy and selected events

The shallow-marine Devonian strata have a series of widespread anoxic events that often coincide with turnovers of conodonts and other marine groups (Fig. 8.1). Many of these events coincide with positive excursions in the $\delta^{13}C$ curve (e.g., Buggisch and Joachimski, 2006). Within the Mesozoic–Cenozoic, some of the widespread anoxic events and carbon-isotope excursions have been linked to eruption of large igneous provinces. It has been proposed that some of the Devonian excursions have a similar initial causation, for example the Viluy and Yakutsk traps in eastern Siberia have $^{40}Ar/^{39}Ar$ ages that cluster at 376.7 ± 1.7 Ma and 364.4 ± 1.7 Ma, which, when combined with the ±1.5-myr uncertainties of the Late Devonian age model, suggest that the two main eruptive phases might overlap the Lower Kellwasser event of latest Frasnian and the *Annulata* event of mid-Famennian (Ricci et al., 2013).

The phosphatic conodonts preserve shallow-marine $\delta^{18}O$ trends, which can be interpreted as tropical temperatures (e.g., Joachimski et al., 2009) (Fig. 8.1). A different estimate of marine $\delta^{18}O$ trends and adjusted tropical sea-surface temperature has been derived from carbonate fossils (e.g., composite by Veizer and Prokoph, 2015; and merger of selected studies in Becker et al., 2012). Although Becker et al. (2012) caution that the available reliable data is inadequate to compile a detailed temperature curve, both methods show a general trend of warmer conditions in the Early and the Late Devonian, with relatively cooler conditions during the Middle Devonian (Fig. 8.1).

In contrast to the suggested warmer conditions from these $\delta^{18}O$ trends, there is evidence from basins in the northern parts of South America, Africa, and the United States that Gondwana experienced episodic glacial conditions in latest Famennian and in the earliest Carboniferous (ca. middle of the Tournaisian Stage) (e.g., Isaacson et al., 2008).

Numerical age model

GTS2012 age model and potential future enhancements

There are very few direct radioisotopic dates on Devonian strata that are well constrained by biostratigraphy. Therefore, the scaling for the Devonian conodont standard zonation in GTS2012 used a spline fit to a schematic diagram of relative thicknesses of those zones in the rock record with tie-point constraints from the few radioisotopic dates on those zones (Becker et al., 2012). The ages assigned to stage boundaries (and all other stratigraphic events) came from their placement relative to that conodont scale. Uncertainties for those stage boundary age assignments came from the spline-fit statistics that incorporated uncertainty on the radioisotopic dates.

After the publication of GTS2012, progress has been made in obtaining cyclostratigraphic estimates for the actual durations of stages and placement of events within those stages—the Givetian (De Vleeschouwer et al., 2014), the Frasnian (De Vleeschouwer et al., 2012), and the upper part of the Famennian (De Vleeschouwer et al., 2013)—using identification of 405-kyr long–eccentricity cycles. These initial cycle-derived estimates for the Givetian (4.35 ± 0.45 myr) and the Frasnian (6.7 ± 0.6 myr) are significantly shorter than those from the spline-fit method of GTS2012 (5.0 and 10.5 myr, respectively); therefore, De Vleeschouwer and Parnell (2014) suggested compensating by adding to the yet-to-be-constrained durations of the preceding

Eifelian Stage and of the succeeding Famennian Stage. However, there is the possibility of "missed beats" in cyclostratigraphic analysis; therefore, a cycle analysis is required to justify lengthening those stages and to identify which zones span considerably more time.

The timing of anoxic events and the durations of conodont zones in both the late Frasnian (L. and U. Kellwasser) and the latest Famennian (*Annulata*–Dasberg–Hangenberg) were partly controlled by high-amplitude 405-kyr sea-level transgressions driven by postulated regional glaciations in northern South America (De Vleeschouwer et al., 2013). This aspect of cyclostratigraphy has been incorporated in Figs. 8.1 and 8.5; but all other ages and internal scaling are from GTS2012.

To improve the Devonian time scale, it is essential that more radioisotopic dates and cyclostratigraphic analyses are obtained that are precisely tied to conodont zone or magnetic zone boundaries.

Estimated uncertainties on assigned ages on stage boundaries

The uncertainties on stage boundaries computed by the spline-fit method in GTS2012 were largely governed by the uncertainties on the radioisotopic tie points, and ranged from ca. ±2.9 myr for Early Devonian stages, ca. ±1.0 myr for Middle and Late Devonian stages, and a tightly constrained ±0.4 myr for base-Carboniferous. De Vleeschouwer and Parnell (2014) applied a Bayesian statistical method to the same conodont data set with incorporation of "floating astronomical durations" for Givetian and Frasnian; and their method yielded similar uncertainties for the Early Devonian and base-Eifelian (ca. ±3 myr), and ca. ±1.5 myr on the ages for the bases of the Givetian through Famennian stages. These estimates of uncertainty by De Vleeschouwer and Parnell (2014) are preferred.

GSSPs of the Devonian Stages, with location and primary correlation criteria

Stage	GSSP Location	Latitude, Longitude	Boundary Level	Correlation Events	Reference
Famennian	Coumiac Quarry, near Cessenon, Montagne Noire, France	43°27'40.6"N 3°02'25"E*	base of Bed 32a	Conodont, FAD of *Palmatolepis subperlobata*, Conodont, LAD of *Palmatolepis bogartensis*	Episodes **16**/4, 1993
Frasnian	Col du Puech de la Suque, Montage Noire, France	43°30'11.4"N 3°05'12.6"E*	base of Bed 42a at Col du Puech de la Suque section E	Conodont, FAD of *Ancyrodella rotundiloba pristina*	Episodes **10**/2, 1987
Givetian	Jebel Mech Irdane, Morocco	31°14'14.7"N 4°21'14.8"W*	base of Bed 123	Conodont, FAD of *Polygnathus hemiansatus*	Episodes **18**/3, 1995
Eifelian	Wetteldorf, Eifel Hills, Germany	50°08'58.6"N 6°28'17.6"E*	21.25m above the base of the exposed section, base of unit WP30	Conodont, FAD of *Polygnathus costatus partitus*	Episodes **8**/2, 1985
Emsian	Zinzil'ban Gorge, Uzbekistan	39°12'N 67°18'20"E	base of Bed 9/5 in the Zinzil'ban Gorge in the Kitab State Geological Reserve	Conodont, FAD of *Eocostapoly-gnathus kitabicus.* New Emsian base under discussion, potentially FAD of *Eocostapolygnathus excavatus*	Episodes **20**/4, 1997
Pragian	Velká Chuchle, Prague, Czech Republic	50°00'53"N 14°22'21.5"E*	base of Bed 12 in Velká Chuchle Quarry	Conodont, FAD of "*Eognathodus sulcatus* s.l."	Episodes **12**/2, 1989
Lochkovian	Klonk, near Prague, Czech Republic	48.855°N 13.792°E**	within Bed 20	Graptolite, FAD of *Monograptus uniformis*	IUGS Series A, **5**, 1977

* according to Google Earth, ** derived from map

Figure 8.6 Ratified GSSPs and primary markers defining the Devonian stages. (*Status as of early 2016.*) Details of each GSSP are available at *http://www.stratigraphy.org, https://engineering.purdue.edu/Stratigraphy/gssp/*, and in the *Episodes* publications.

Acknowledgments

This brief overview of the Devonian was limited to only a few selected highlights and current stratigraphic issues. A detailed overview and synthesis is by Becker et al. (2012); and updates are at the website and newsletters of the Sub-commission on Devonian Stratigraphy (*http://unica2. unica.it/sds/*). This review depended on the expertise of many colleagues, especially Thomas Becker, who reviewed an early version of the graphics, database, and text.

Selected publications and websites

Cited publications

Only select publications were cited in this review with an emphasis on aspects of post-2011 updates. Pre-2011 literature is well summarized in the syntheses by Becker et al. (2012) and in some of the publications cited in the following.

Algeo, T.J., Scheckler, S.E., Maynard, J.B., 2001. Effects of the Middle to Late Devonian spread of vascular land plants on weathering regimes, marine biota, and global climate. In: Gensel, P.G., Edwards, D. (Eds.), *Plants Invade the Land: Evolutionary and Environmental Approaches*. Columbia University Press, New York, pp. 213–236.

Becker, R.T., Gradstein, F.M., Hammer, O., 2012. The Devonian Period. In: Gradstein, F.M., Ogg, J.G., Schmitz, M., Ogg, G., (Coordinators), *The Geologic Time Scale 2012*. Elsevier Publ., pp. 559–601 (An overview on all aspects, including graphics on the ratified GSSPs of the stages, diagrams and tables for the biostratigraphic scales, and discussion on the age models.).

Buggisch, W., Joachimski, M.M., 2006. Carbon isotope stratigraphy of the Devonian of Central and Southern Europe. *Palaeogeography, Palaeoclimatology, Palaeoecology* **240**: 68–88.

Clack, J.A., 2012. *Gaining Ground: The Origin and Evolution of Tetrapods*. Indiana University Press. 544 pp.

De Vleeschouwer, D., Parnell, A.C., 2014. Reducing time-scale uncertainty for the Devonian by integrating astrochronology and Bayesian statistics. *Geology* **42**: 491–494. http://dx.doi.org/10.1130/G35618.1.

De Vleeschouwer, D., Whalen, M.T., Day, J.E., Claeys, P., 2012. Cyclostratigraphic calibration of the Frasnian (Late Devonian) time scale (western Alberta, Canada). *Geological Society of America Bulletin* **124**: 928–942. http://dx.doi.org/10.1130/B30547.1.

De Vleeschouwer, D., Ralpconski, M., Racki, G., Bond, D.P.G., Sobien, K., Claeys, P., 2013. The astronomical rhythm of Late-Devonian climate change (Kowala section, Holy Cross Mountains, Poland). *Earth and Planetary Science Letters* **365**: 25–37. http://dx.doi.org/10.1016/j.epsl.w013.01.016.

De Vleeschouwer, D., Boulvain, F., Da Silva, A.C., Pas, D., Labaye, C., Claeys, P., 2014. The astronomical calibration of the Givetian (Middle Devonian) timescale (Dinant Synclinorium, Belgium). In: Da Silva, A.C., Whalen, M.T., Hladil, J., Chadimova, L., Chen, D., Spassov, S., Boulvain, F., De Vleeschouwer, X. (Eds.), *Magnetic Susceptibility Application: A Window onto Ancient Environments and Climatic Variations*. Geological Society, **414**. Special Publications, London. http://dx.doi.org/10.1144/SP414.3. 12 pp.

Gradstein, S.R., Kerp, H., 2012. A brief history of plants on Earth. In: Gradstein, F.M., Ogg, J.G., Schmitz, M., Ogg, G., (coordinators), *The Geologic Time Scale 2012*. Elsevier Publ., pp. 233–237. http://dx.doi.org/10.1016/B978-0-444-59425-9.00023-8.

Hansma, J., Tohver, E., Yan, M., Trinajstic, K., Roelofs, B., Peek, S., Slotznick, S.P., Kirschvink, J., Playton, T., Haines, P., Hocking, R., 2015. Late Devonian carbonate magnetostratigraphy from the Oscar and Horse Spring Ranges, Lennard Shelf, Canning Basin, western Australia. *Earth and Planetary Science Letters* **409**: 232–242.

Haq, B.U., Schutter, S.R., 2008. A chronology of Paleozoic sea-level changes. *Science* **322**: 64–68. http://dx.doi.org/10.1126/science.116164.

Isaacson, P.E., Díaz-Martínez, E., Grader, G.W., Kalvoda, J., Babek, O., Devuyst, F.X., 2008. Late Devonian–earliest Mississippian glaciation in Gondwanaland and its biogeographic consequences. *Earth and Planetary Science Letters* **268**: 126–142.

Joachimski, M.M., Breisig, S., Buggisch, W., Talent, J.A., Mawson, R., Gereke, M., Morrow, J.R., Day, J., Weddige, K., 2009. Devonian climate and reef evolution: insights from oxygen isotopes in apatite. *Earth and Planetary Science Letters* **284**: 599–609.

Korn, D., Klug, C., 2015. Paleozoic ammonoid biostratigraphy. In: Klug, C., Korn, D., De Baets, K., Kruta, I., Mapes, R.H. (Eds.), *Ammonoid Paleobiology: From Macroevolution to Paleogeography*. Topics in Geobiology, **44**. Springer Publ., pp. 299–328. http://dx.doi.org/10.1007/978-94-017-9633-0_13 (Chapter 12).

Niedźwiedzki, G., Szrek, P., Narkiewicz, K., Narkiewicz, M., Ahlberg, P.E., 2010. Tetrapod trackways from the early Middle Devonian Period of Poland. *Nature* **463**: 43–48.

Ricci, J., Quidelleur, X., Pavlov, V., Shatsillo, A., Courtillot, V., 2013. New ^{40}Ar/^{39}Ar and K–Ar ages of the viluy traps (Eastern Siberia): further evidence for a relationship with the Frasnian–Famennian mass extinction. *Palaeogeography, Palaeoclimatology, Palaeoecology* **386**: 531–540. http://dx.doi.org/10.1016/j.palaeo.2013.06.020.

Salim, L., Galimberti, A.K., 2015. Body-size reduction in vertebrates following the end-Devonian mass extinction. *Science* **350**: 812–815 plus supplementary materials.

Scotese, C.R., 2014. Atlas of Devonian Paleogeographic Maps. *PALEOMAP PaleoAtlas for ArcGIS*, **vol. 4**, The Late Paleozoic, Maps 65–72, Mollweide Projection, PALEOMAP Project, Evanston, IL. https://www.academia.edu/16711496/Atlas_of_Devonian_Paleogeographic_Maps.

Subcommission on Devonian Stratigraphy (ICS, International Commission on Stratigraphy), 2014. Annual Report 2014. Submitted to International Union of Geological Sciences (IUGS) at: http://iugs.org/uploads/ICS%202014.pdf and SDS Newsletters of 2013 and 2014 at: http://unica2.unica.it/sds/.

Veizer, J., Prokoph, A., 2015. Temperatures and oxygen isotopic composition of Phanerozoic oceans. *Earth-Science Reviews* **146**: 92–104.

Websites (selected)

Subcommission on Devonian Stratigraphy
(International Commission on Stratigraphy
[ICS])—*http://unica2.unica.it/sds/*—News items,
PDFs of recent Newsletters (vol. 21 of 2008, through
present), and details (in future) on GSSPs.

Commission Internationale Microflore Paléozoïque—
http://cimp.weebly.com—International Commission
of the Paleozoic Microflora has separate subcommissions on acritarchs, chitinizoans, and miospores; and
site has archive of their newsletters and other
information; although number of active newsletters
diminished since 2012.

Devonian Times—*http://www.devoniantimes.org*—
Extensive well-presented and illustrated site
compiled by Dennis Murphy that received Science &
Techonology award from *Scientific American* in 2005
and illustrates most aspects of Devonian life.
(However, 2006 was the last indicated update.)

Palaeos: Devonian—*http://palaeos.com/paleozoic/
devonian/devonian.htm*—A well-presented suite of
diverse topics for a general science audience that
was originally compiled by M. Alan Kazlev in
1998–2002.

9

CARBONIFEROUS

306Ma Carboniferous

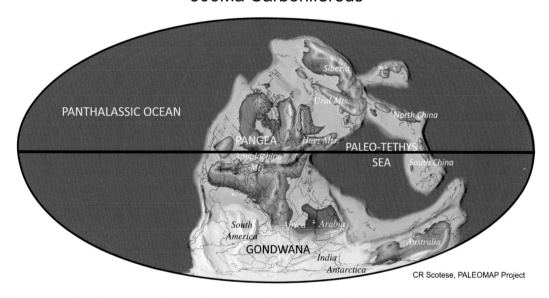

Kasimovian (late Pennsylvanian) paleogeographic reconstruction. (Sea level + 120 m maximum-flooding surface [MFS]) from Scotese (2014).

Basal definition and international subdivisions

The Carboniferous is named after its abundance of coal (carbon) deposits in Europe and North America. The Carboniferous chronostratigraphy is unique in that it has two subsystems, Mississippian and Pennsylvanian, each with a set of Lower/Middle/Upper series, but most of these series consist of a single stage (Fig. 9.1). The Global Boundary Stratotype Sections and Points (GSSPs), ratified or under discussion for these stages, are mainly placed to coincide with first-appearance datums (FADs) of conodont taxa on an evolutionary lineage, and it appears the most complete sections in deeper-water facies are in South China and Russia (Richards, 2013; Chen et al., 2014). The evolution of this international Carboniferous chronostratigraphy and its comparison to the various regional series/stage divisions are summarized by Davydov et al. (2012).

The Carboniferous began after the end-Devonian episodes of cooling, waves of extinctions, and widespread organic-rich deposits (e.g., Hangenberg Event levels in Europe; see Devonian chapter Fig. 8.1). The Devonian–Carboniferous GSSP at Bed 89 in the La Serre

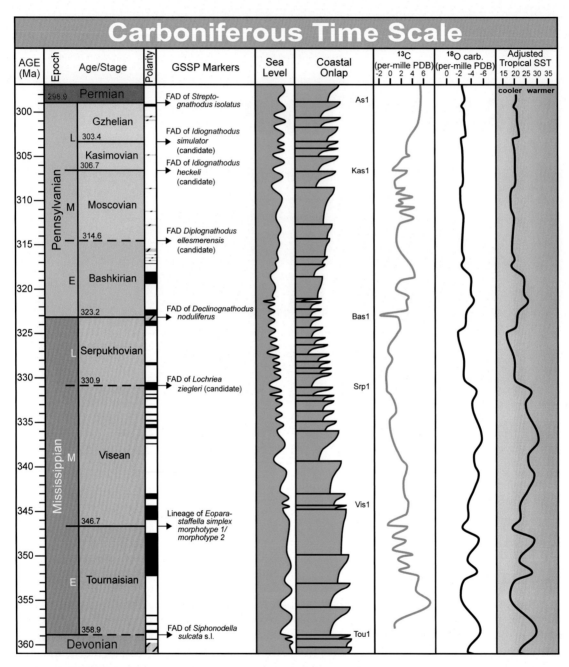

Figure 9.1 Carboniferous overview. Main markers or candidate markers for GSSPs of Carboniferous stages are first-appearance datums (FADs) of conodont taxa as detailed in the text and in Fig. 9.6. ("Age" is the term for the time equivalent of the rock-record "stage".) Magnetic polarity scale is from Davydov et al. (2012). Coastal onlap and schematic sea-level curve with labels for selected major sequence boundaries are modified from Haq and Schutter (2008) following advice of Bilal Haq (pers. comm., 2008). The δ13C curve is from Saltzman (2003). The Carboniferous δ18O curve is modified from the mean of the curve of Veizer and Prokoph (2015) with a smoothed schematic version of their adjusted estimates of tropical sea-surface temperatures derived from those oxygen-18 values. The vertical scale of this diagram is standardized to match the vertical scales of the first stratigraphic summary figure in all other Phanerozoic chapters. *PDB*, PeeDee Belemnite 13C and 18O standard; *SST*, sea-surface temperature.

section of southern France was intended to coincide with the FAD of conodont *Siphonodella sulcata* within the lineage *S. praesulcata-S. sulcata*, but this taxon was later found at a lower level that overlapped its presumed ancestor (Kaiser, 2009) (Fig. 9.2). Therefore, both the criteria for global correlation and retention of the GSSP section at La Serre are being reevaluated with a focus on a potential level within the Hangenberg Event (Aretz, 2011).

The Mississippian–Pennsylvanian boundary GSSP at Arrow Canyon, Nevada, United States, was placed to coincide with the FAD of conodont *Declinognathodus noduliferus* (sensu lato) (Fig. 9.3). This taxon is now split into subspecies; and the FAD of subspecies *D. noduliferus inaequalis* (*D. inaequalis*) is commonly used as the boundary marker (Richards, 2013).

Other than the bases of the Mississippian and the Pennsylvanian subsystems, as of February 2016 only the base of the Visean (Viséan) Stage (Middle Mississippian) has a ratified GSSP. The Tournaisian–Visean boundary interval does not have any significant events in conodont or ammonoid stratigraphy; therefore, the **Visean** GSSP at the base of Bed 83 in a stream section near the village of Pengchong (near Liuzhou city, Guangxi Province, South China) was selected to coincide with a change in morphotype within the benthic foraminifer *Eoparastaffella* genus, from *Eo. rotunda* (or the *Eo. ovalis* branch in the GSSP section*)* to *Eo. simplex typica*, and just below the FAD of regional conodont *Gnathodus homopunctatus* (Devuyst et al., 2003).

The following summary of defining the other stage GSSPs is from Richards (2013) and the Carboniferous Subcommission newsletters (www.stratigraphy.org/Carboniferous/pub/pub.asp).

Serpukhovian (Late Mississippian): The Visean/Serpukhovian boundary interval is within a major glaciation. The boundary task group is focusing on a level coincident with the FAD of conodont *Lochriea ziegleri* in the lineage *L. nodosa-L. ziegleri*. The two leading candidate sections are Naqing (Nashui) in southern Guizhou Province, South China (e.g., Qi et al., 2014), and the Verkhnyaya Kardailovka section on the Ural River in the Bashkortostan Republic of southern Russia (Sevastopulo and Barham, 2014). A volcanic ash bed in the Verkhnyaya Kardailovka section about 1.5m below the potential GSSP level gave an age of 333.87±0.08Ma (Schmitz and Davydov, 2012).

Moscovian (Middle Pennsylvanian): The FAD of conodont *Declinognathodus donetzianus* had been considered as a potential index for the base of this stage, but is not found in North America. Therefore, the FAD of the more cosmopolitan *Diplognathodus ellesmerensis*, "one of the most widely recovered conodonts in the Upper Carboniferous," in the lineage *Decl. coloradoensis-Dipl. ellesmerensis* is considered to have the greatest potential for correlation (Alekseev and Task Group, 2014). The leading candidate GSSP section is the same Naqing section in South China (e.g., Qi et al., 2016) that has been proposed for the base-Serpukhovian; and other candidates are those in the southern Urals and Donets Basin (Sungatullina, 2014) that have yielded radio-isotopic dates.

Kasimovian (Late Pennsylvanian): There is a well-dated (U-Pb) conodont-zoned cyclostratigraphy (cyclothems) across the Moscovian–Kasimovian boundary (Schmitz and Davydov, 2012), and individual cycles can be correlated between the Southern Urals–Donets Basin and North America. The boundary task group (Ueno and Task Group, 2014) is considering the FAD of conodont *Idiognathodus heckeli* in the lineage *Id. swadei—Id. heckeli—Id. turbatus* (Rosscoe and Barrick, 2013). The FAD of *Id. heckeli* in the Exline lithologic cyclothem is at the base of regional Missourian Stage of North America (Heckel, 2013b) and would be close to the base of the traditional Kasimovian Stage of Russia. This level would be slightly younger

Base of the Tournaisian Stage of the Carboniferous System in the La Serre Section, Montagne Noire, France

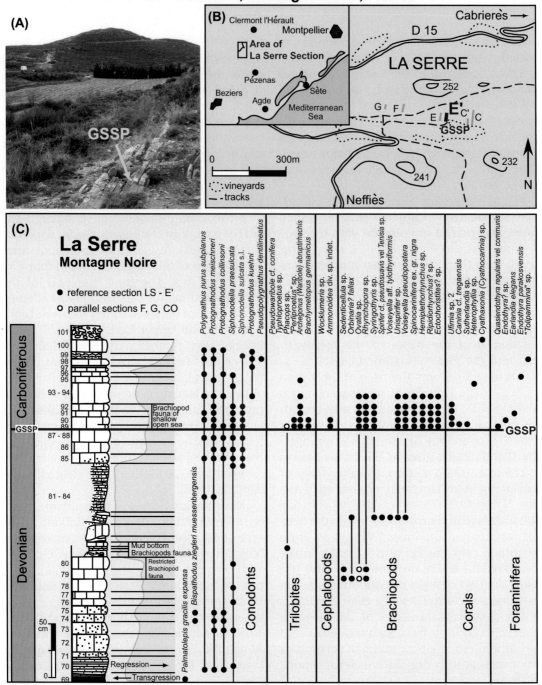

Figure 9.2 GSSP for base of the Carboniferous (base of Mississippian; base of Tournaisian Stage) at the La Serre hill section, 2.5 km southwest of the village of Cabrières, Montagne Noire, southern France. The GSSP level was intended to coincide with the lowest occurrence of conodont *Siphonodella sulcata* in the lineage *S. praesulcata-S. sulcata*, but this taxon is now found at a lower level. A revised criteria for the basal marker and potentially a different boundary section for the Devonian–Carboniferous boundary are now being sought. Outcrop photograph provided by Marus Aretz; stratigraphic column from Paproth et al. (1991).

Base of the Pennsylvanian Sub-system of the Carboniferous System at Arrow Canyon, Nevada, U.S.A.

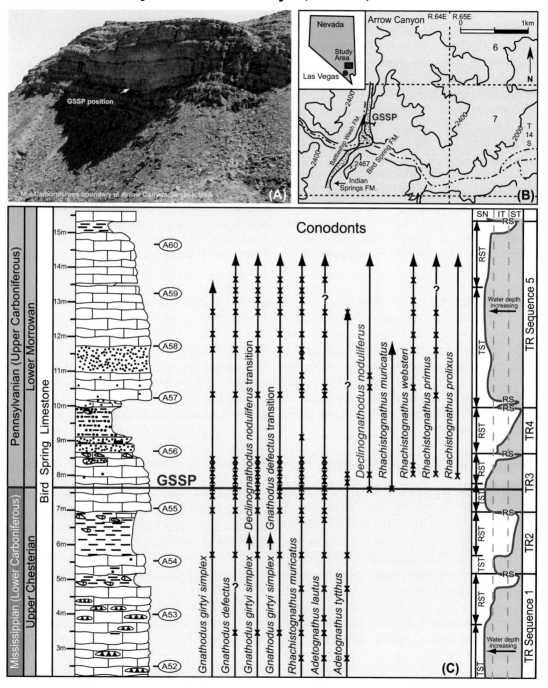

Figure 9.3 GSSP for base of the Pennsylvanian subsystem (base of Bashkirian Stage) at the Arrow Canyon section, Nevada, United States. The GSSP level coincides with the first-appearance datum (FAD) of conodont *Declinognathodus noduliferus* (sensu lato). (Photograph, stratigraphic column, and schematic sea-level trends from Davydov et al. (2012) and Lane et al. (1999).)

than the Westphalian–Stephanian regional series boundary of Western Europe. A leading potential GSSP section is the same Naqing section of South China that is the candidate for the other GSSPs.

Gzhelian (mid-Late Pennsylvanian): The FAD of conodont *Idiognathodus simulator sensu stricto* in the lineage *Id. eudoraensis - Id. simulator* was decided in 2008 as the boundary-defining event, but the decision on the GSSP section has progressed slowly even though the Usolka section in southern Urals of Russia had also been proposed in 2008 (Davydov et al., 2008). One problem is that there is a lack of a well-documented transition to the *Id. simulator*, although the high-amplitude oscillations of global sea levels during the glacial-influenced boundary interval, coupled with the local FAD of *Id. simulator*, might allow using a precise maximum-flooding level (Ueno and Task Group, 2014). Other potential GSSP sections being investigated include the Rusavkino quarry and historical stratotype of the Gzhelian Stage in the Gzhel quarry east of Moscow and the same Naqing section of South China under consideration for other Pennsylvanian stage GSSPs.

Selected main stratigraphic scales and events

Biostratigraphy (marine; terrestrial)

Marine biostratigraphy mainly relies on conodonts, benthic foraminifers (especially fusulinaceans), and ammonoids, plus brachiopods in marginal settings. The microscopic phosphatic feeding apparatuses of conodonts are more ubiquitous in different marine facies; therefore, they are especially important for interregional correlation and as primary markers for stage GSSPs. The Pennsylvanian conodont zonations of Europe and of the North American midcontinent have been precisely correlated using common high-amplitude sea-level cycles (Schmitz and Davydov, 2012;

Barrick et al., 2013; Heckel, 2013b) (Fig. 9.4). Benthic foraminifers are particularly useful in biostratigraphic correlations within each region, but there are divergent standard zonations for Eurasia and North America (e.g., Davydov et al., 2012; Wahlman, 2013). Ammonoids underwent increased provincialism during the Mississippian; therefore, they have different regional standards (e.g., Korn and Klug, 2015; Boardman and Work, 2013). At the top of the food chain in the marine realm, the average size of fish and their vertebrate cousins, which had been progressively getting larger during each succeeding Devonian stage, underwent a sharp drop across the end-Devonian Hangenberg Event. Even though a few survivors regained larger dimensions, the marine fossil record is dominated by a progressive and unexplained decrease in average size of the marine vertebrates into the Visean and continuing into the Serpukhovian (Sallan and Galimberti, 2015).

Regional palynological spore and megafloral assemblages correlate the extensive Northern Hemisphere coal deposits. The Devonian tree-sized *Archaeopteris* continued into the Mississippian, and the lycophyte *Lepidodendron* and *Sigillaria* trees dominated the Pennsylvanian coal forests until their extinction at the end of the Carboniferous (Gradstein and Kerp, 2012).

Evolution of reptiles from amphibians characterizes the top of terrestrial ecosystems (Chen et al., 2014). Archaic Devonian fish-like semiaquatic tetrapods continued into the earliest Carboniferous. However, a ca. 20-myr "Romer's Gap" without suitable preservation of terrestrial tetrapods spans the majority of the Tournaisian to mid-Visean; and, after that gap, ecosystems with reptile-like Anthracosauria amphibians flourished into the Early Permian (e.g., Clack, 2012). The sail-backed pelycosaur reptiles (e.g., *Dimetrodon* and its ancestors), the earliest known herbivorous tetrapods, developed in the late Moscovian and continued through the Early Permian.

Pennsylvanian Time Scale

AGE (Ma)	Age (Stage)	Donets Basin 400-kyr cycle	European Conodont Zones	N. American Mid-Continent Conodont Zones	N. American 400-kyr cycle	N. Am. Glacials	European Ammonoid Zones	Europen Fusulinids & Benthic Forams
298	Permian		Streptognathodus sigmoidalis - S. cristellaris	Streptognathodus isolatus		Interglacial VI ?	Svetlanoceras - Paragastrioceras	
298.9		SA1	Streptognathodus isolatus		Red Eagle 2			
299	Gzhelian	SG14-16	St. wabaunsensis - St. fissus	St. binodosus	Foraker 2	Glacial F		Daixina bosbytauensis - Globifus. robusta
		SG13	Streptognathodus simplex - St. bellus	St. farmeri	Five Point			Daixina sokensis
				St. flexuosus				
300		SG11-12		Streptognathodus bellus	Falls City			
		SG8-10			Brownville			
		SG7			Dover			
301		SG6b	Streptognathodus virgilicus		Tarkio		Vidrioceras - Shumardites	Jigulites jigulensis
		SG6a		Streptognathodus virgilicus	Howard			
302		SG5			Topeka	Interglacial V		
		SG4			Deer Creek			Rauserites rossicus - Rauserites stuckenbergi
303		SG3	Streptognathodus vitali	St. vitali	Lecompton			
303.4		SG2	Idiognathodus simulator	Idiognathodus simulator	Oread	E		
	Kasimovian	SG1		St. zethus	Stanton (& Cass)	IV		
304		SK7	Streptognathodus firmus	I. eudoraensis	Iola			Rauserites quasiarcticus
		SK6		Streptognathodus gracilis	Dewey	D		
305		SK5	I. toretzianus	I. confragus	Dennis	III	Dunbarites - Parashumardites	
		SK4	St. cancellosus	I. cancellosus	Swope	C		Montiparus paramontiparus
		SK3	Idiognathodus sagittalis	Idiognathodus turbatus	Hertha	II		
306		SK2	St. subexcelsus - Sw. makhlinae	Sw. nodocarinata	Lost Branch	B		Protriticites pseudomontiparus
306.7		SK1		Sw. neoshoensis	Altamont			
307	Moscovian	SM18			Pawnee	Interglacial I	Eoschistoceras	Fusulina cylindrica - Protriticites ovatus
		SM17		Idiognathodus delicatus	Upper Fort Scott			
		SM16	Neognathodus roundyi		Lower Fort Scott (Excello)			
308		SM15		Neognathodus roundyi	Bevier			Fusulinella bocki
		SM14			Verdigris			
309		SM13			Fleming			
		SM12			Russell Creek		Pseudopara- legoceras	Fusulinella colaniae - F. voshgalensis - Beedeina kamensis
		SM11	Idiognathodus podolskensis	Neognathodus asymmetricus	Upper Tiawah			
310		SM10			Lower Tiawah			
		SM9			Post-Wainright	Glacial A ?		
311		SM8			Inola			
		SM7	Neognathodus caudatus		Doneley		Paralegoceras / Eowellerites	Fusulinella subpulchra
		SM6	Swadelina dissectus		Spaniard			

Figure 9.4 Selected marine biostratigraphic zonations of the Pennsylvanian subperiod with 405-kyr sea-level cycles from eccentricity-forced oscillations of the Gondwana ice sheets. ("*Age*" is the term for the time equivalent of the rock-record "*stage*.") 400-kyr cycles (cyclothems) and scaling of European and North American Mid-Continent conodont zones to these cycles are a composite from Schmitz and Davydov (2012), Davydov et al. (2010, 2012), Heckel (2013b, and pers. comm., Sept. 2015) and Barrick et al. (2013).

Pennsylvanian Time Scale

AGE (Ma)	Age (Stage)	Donets Basin 400-kyr cycle	European Conodont Zones	N. American Mid-Continent Conodont Zones	N. American 400-kyr cycle	N. Am. Glacials	European Ammonoid Zones	European Fusulinids & Benthic Forams
310	Moscovian	SM11 / SM10 / SM9	Idiognathodus podolskensis	Neognathodus asymmetricus	Upper Tiawah / Lower Tiawah / Post-Wainright	Glacial A ?	Pseudopara-legoceras	Fusulinella colaniae - F. voshgalensis - Beedeina kamensis
311		SM8 / SM7			Inola / Doneley			
312		SM6 / SM5 / SM4	Swadelina dissectus	Neognathodus caudatus	Spaniard / Tamaha / McCurtain		Paralegoceras / Eowellerites	Fusulinella subpulchra / Priscoidella priscoidea
313		SM3b / SM3a	Neognathodus uralicus	Neognathodus colombiensis	unnamed		Diaboloceras - Winslowoceras	Aljutovella aljutovica
314		SM2b / SM2a	Diplognathodus ellesmerensis - Declinognathodus donetzianus					
314.6		SM1						
315	Bashkirian	upper Melekesian	Neognathodus atokaensis	Neognathodus atokaensis			Diaboloceras - Axinolobus	Verella spicata - Alj. tikhonovichi
316					Pounds			
317			Declinognathodus marginodosus				Branneroceras / Gastrioceras	Profusulinella rhombiformis
318				Neognathodus nataliae				Profusulinella primitiva - Pseudostaffella gorskyi
319			Idiognathodus sinuosus		Bostwick		Bilinguites / Cancelloceras	Staffellaeformes staffellaeformis - Pseudostaffella praegorskyi
320			Neognathodus askynensis	Neognathodus bassleri	Trace Creek			Pseudostaffella antiqua
321			Idiognathodus sinuatus	Neognathodus symmetricus	Dye - Kessler		Baschkortoceras / Reticuloceras	Semistaffella variabilis - Semistaffella minuscilaria
322				Id. sinuatus	Prairie Grove 2			
323			Declinognathodus noduliferus	D. noduliferus, Rhachisto. primus	Cane Hill		Homoceras / Hudsonoceras	Plectostaffella bogdanovkensis
323.2	Serpukhovian		G. postbilineatus	upper R. muricatus				Monotaxinoides transitorius

Figure 9.4 (Continued) The main North American cycles in the Bashkirian Stage are from Heckel (2008, 2013b, and pers. comm., Sept. 2015). Interpretation of the major glacial/interglacial intervals from the North American Mid-Continent record is from Heckel (2013b and pers. comm., Sept. 2015). European ammonoid zones and fusulinid/benthic foraminifer zones are mainly from Davydov et al. (2012) and Heckel (2008), which schematically positions these zones relative to the conodont scales. Additional zonations, biostratigraphic markers, geochemical trends, sea-level curves, and details on calibrations are compiled in Davydov et al. (2012) and in the internal data sets within the *TimeScale Creator* visualization system (free at www.tscreator.org).

Magnetostratigraphy, stable-isotope stratigraphy, and selected events

The Carboniferous magnetic polarity time scale is poorly known. The Kiaman hyperchron of late Carboniferous through middle Permian is dominated by reversed polarity. The youngest significant pre-Kiaman normal-polarity zones are upper Bashkirian. The polarity scale in Fig. 9.1 was compiled in Davydov et al. (2012).

In addition to other factors, the carbon-isotope record is influenced by the major episodes of coal burial that preferentially buried ^{12}C to enrich the mean δ^{13}C values in the ocean, but the timing and magnitude of many of the reported major fluctuations have not yet been explained (reviewed in Davydov et al., 2012). The schematic curve in Fig. 9.1 is a running average of the data from several studies of Carboniferous compiled by Saltzman and Thomas (2012), and many of the regional features require additional global verification.

The Carboniferous–Permian had the longest ice age of the Phanerozoic with a fluctuating ice sheet dominating the southern high latitudes (e.g., the chapter opening figure, "Kasimovian (late Pennsylvanian) paleogeographic reconstruction"). The contributing factors include the Gondwana paleogeography under the South Pole, carbon burial in coal deposits, atmospheric CO_2 levels, and sea-level feedbacks (e.g., synthesis by Montañez and Poulsen, 2013). The late Mississippian (Serpukhovian) through Pennsylvanian deposits on continental margins are characterized by "cyclothems" caused by major periodic sea-level changes driven by the response of the major Gondwana ice sheets to Milankovitch orbital-climate cycles. Most of these cyclothems are ca. 100-kyr short-eccentricity oscillations modulated by long-eccentricity envelopes which produce periodic (405-kyr) enhanced major lowstand and exposure surfaces flooded by major sea-level incursions. The combination of U-Pb dating with conodont datums has produced a precise time scale and correlation of these cyclothems between North America midcontinent and the southern Urals–Donets Basin (Fig. 9.4).

The relative magnitude of clusters of sea-level lowstands in the North American cyclothem record is interpreted by Heckel (2013b) as episodes of major glacial ice buildup in Gondwana (Fig. 9.4). An independent estimate of major Serpukhovian through Pennsylvanian glacial episodes is from influxes of glacial-derived sediment into the basins of eastern Australia (Fielding et al., 2008). The ages for both of these estimates of the main glacial (cooling) episodes are only partially consistent with the synthesis of a global oxygen-18 curve and the approximations of tropical sea-surface temperatures by Veizer and Prokoph (2015, shown in Fig. 9.1).

Numerical age model

GTS2012 age model

As the Gondwana ice sheets underwent regular expansion/contraction during the late Carboniferous through early Permian in response to Milankovitch orbital-climate feedbacks, a set of high-amplitude cyclic sea-level onlaps (cyclothems) were generated on continental margins. The detailed U-Pb dating of conodont-rich successions in the Donets Basin and southern Urals enabled recognition of a full set of 405-kyr long-eccentricity cycles spanning the Asselian and Sakmarian stages (Schmitz and Davydov, 2012). Those cycles also have counterparts in the North American Midcontinent conodont-zoned lithostratigraphy (Heckel, 2013b), thereby enabling a high-resolution age model for the duration of each conodont zone between the two regions (Fig. 9.4). Many of the benthic foraminifer and ammonoid zones were independently calibrated to those cycles (e.g., Heckel, 2013a) or to the conodont scale (Fig. 9.5).

Mississipian Time Scale

AGE (Ma)	Age (Stage)	European Conodont Zones	N. American Mid-Continent Conodont Zones	Ammonoid Zones	Fusulinids & Benthic Forams
323.2	Bashkirian	Declinogn. noduliferus	Decl. noduliferus, Rh. primus	Homoceras / Hudsonoceras	Plectostaffella bogdanovkensis
325	Serpukhovian	Gnathodus postbilineatus	upper Rhach. muricatus / lower Rhach. muricatus / Adetognathus unicornis		Monotaxinoides transitorius
		Gnathodus bollandensis	Cavusgnathus naviculus	Delepinoceras / Fayettevillea	Eostaffellina protvae
		Lochriea cruciformis		Cravenoceras / Uralopronorites	Neoarchaediscus postrugosus
330		Lochriea ziegleri	upper Gnathodus bilineatus		
330.9		Lochriea nodosa		Hypergoniatites / Ferganoceras	
		Lochriea mononodosa	lower Gnathodus bilineatus		Endothyranopsis crassus - Archaediscus gigas
335		Gnathodus bilineatus	Hind. scitulus, Ap. scalenus		
	Visean	Gnathodus praebilineatus		Beyrichoceras / Goniatites	Endothyranopsis compressa - Pararchaediscus kokjubensis
340			Gnathodus texanus		
345		Gnathodus texanus			Uralodiscus rotundus
346.7			Gnathodus bulbosus		Eoparastaffella simplex - Eoendothyranopsis donica
		Gnathodus pseudosemiglaber - Scaliognathus anchoralis	Eotaphrus burlingtonensis / Bactrognathus lanei / Doliognatus latus	Fascipericyclus / Ammonellipsites	Endothyra elegia - Eotextularia diversa
		Gnathodus semiglaber - Polygnathus communis	Pseudopolyg. multistriatus / Polyg. communis carinus		
350		Dollimae bouckaerti	Gnathodus punctatus	Protocanites / Pericyclus	Spinoendothyra costifera
		Gnathodus typicus - Siphonodella isosticha	Siphonodella isosticha - upper S. crenulata (upper)		Palaeospiroplectammina tchernyshinensis
355	Tournaisian	upper Siphonodella quadruplicata - Patrognathus andersoni	Siphonodella isosticha - upper S. crenulata (lower)	Protocanites / Gattendorfia	Chernyshinella disputabilis
		Siphonodella sandbergi - Siph. belkai	lower Siphonodella crenulata / Siphonodella sandbergi	Eocanites / Gattendorfia	
		Siphonodella duplicata	upper Siphonodella duplicata / lower Siphonodella duplicata		Bisphaera malevkensis - Earlandia minima
358.9		Siphonodella sulcata	Siphonodella sulcata	Acutimitoceras	T. pseudobeata - S. njumylga
	Devonian	Siphonodella praesulcata			

Figure 9.5 Selected marine biostratigraphic zonations of the Mississippian subperiod. *(Note that the age scale is more compact than in Fig. 9.4).* Conodont, ammonoid, and fusulinid/benthic foraminifer zones are from Davydov et al. (2012) with the Tournaisian–Visean boundary interval partially modified according to the Carboniferous Subcommission Correlation Chart (2014). Additional zonations, biostratigraphic markers, geochemical trends, sea-level curves, and details on calibrations are compiled in Davydov et al. (2012) and in the internal data sets within the *TimeScale Creator* visualization system (free at www.tscreator.org).

The age model for the combined Carboniferous–Permian in GTS2012 was a statistical spline fit of a collection of U-Pb ages to a composite biostratigraphic scale that had been constructed by applying CONstrained Optimization (CONOP)-9 to an extensive suite of reference sections (Davydov et al., 2012; Schmitz and Davydov, 2012). This technique yielded very similar ages for the conodont zones, but the cycle-tuned scale of Schmitz and Davydov (2012) is used here for the Pennsylvanian age model. However, the age model for the Mississippian through lower Bashkirian conodont and benthic foraminifer zones retains the GTS2012 spline fit of the CONOP scale, except for revised zonal calibrations within the Tournaisian–Visean boundary interval by the Subcommission on Carboniferous Stratigraphy (2014).

Revised ages compared to GTS2012 and potential future enhancements

The only changes in the age model for the Carboniferous are for the yet-to-be-formalized Pennsylvanian stages. The shift to slightly younger ages is from the combination of (1) enhanced high-precision U-Pb and cyclostratigraphic dating of conodont zone boundaries (Schmitz and Davydov, 2012), and (2) the conodont FADs currently preferred by the Carboniferous boundary working groups as the stage-boundary markers. However, there is the possibility that the eventual GSSPs for these stages may have different conodont FADs or other primary markers than the candidates in Figs. 9.1 and 9.6. *[Note that an external uncertainty of ca. 0.3 myr should be included if comparing these EARTHTIME-standardized dates to other dating methods, as explained in* Burgess et al. (2014).*]*

Gzhelian base (**303.4** vs 303.7 ± 0.1 Ma in GTS2012): This is the cyclothem-calibrated age for the FAD of conodont marker *Streptognathodus simulator*.

Kasimovian base (**306.7** vs 307.0 ± 0.2 Ma in GTS2012): The potential conodont marker, *Idiognathodus heckeli* in the Exline cyclothem is used in Fig. 9.4 to place the boundary.

Moscovian base (**314.6** vs 315.2 ± 0.2 Ma in GTS2012): This is the cyclothem-calibrated age for the potential conodont marker, *Diplognathodus ellesmerensis.*

Estimated uncertainties on assigned ages on stage boundaries

The high-precision radioisotopic dates with well-constrained biostratigraphic ages that constrain the Pennsylvanian portion of this Carboniferous time scale typically have a published uncertainty less than 0.2 myr (e.g., Schmitz and Davydov, 2012). However, an external uncertainty of ca. 0.3 myr should be included if comparing these Permian EARTH-TIME-standardized dates to other dating methods, as explained in Burgess et al. (2014). The duration of those stages, if computed from the dates in the same EARTHTIME data sets, have uncertainties that omit that external factor.

The Mississippian portion retained the spline fit to the CONOP scale used in GTS2012, which yielded a 0.4-myr uncertainty (Davydov et al., 2012). The Visean–Serpukhovian boundary (330.9 Ma in GTS2012 and in Fig. 9.5), has a U-Pb date of 333.87 ± 0.08 Ma just below the candidate GSSP level (Schmitz and Davydov, 2012), therefore probably has a precision similar that of the Pennsylvanian stages. The spline-fit age for the Tournaisian–Visean boundary in GTS2012 (346.7 Ma) is constrained by a 345.17 ± 0.4 Ma date in the benthic foraminifer *Uralodiscus rotundus* zone (Davydov et al., 2012; Schmitz, 2012); but because placement of that zone relative to the base-Visean is disputed (e.g., Davydov et al., 2012 chart versus the Subcommission on Carboniferous Stratigraphy website chart of 2014), then an uncertainty of ca. 1 myr for the base-Visean might

	GSSPs of the Carboniferous Stages, with location and primary correlation criteria				
Stage	**GSSP Location**	**Latitude, Longitude**	**Boundary Level**	**Correlation Events**	**Reference**
Gzhelian	*Candidates are in southern Urals or Nashui (South China)*			*Conodont, FAD of Idiognathodus simulator (candidate)*	
Kasimovian	*Candidates are in Nashui, south-central China*			*Conodont, FAD of Idiognathodus heckeli*	
Moscovian	*Candidates are in southern Urals or Nashui (South China)*			*Conodont, FAD Diplognathodus ellesmerensis (candidate)*	
Bashkirian	Arrow Canyon, Nevada	36°44'00" N, 114°46'40" W**	82.9m above the top of the Battleship Formation in the lower Bird Spring Formation	Conodont, FAD of *Declinognathodus noduliferus*	Episodes **22/4**, 1999
Serpukhovian	*Candidates are Verkhnyaya Kardailovka (Urals) or Nashui (China)*			*Conodont, FAD of Lochriea ziegleri (candidate)*	
Visean	Pengchong, South China	24°26'8.88"N, 109°27'19.49"E	base of bed 83 in the Pengchong Section	Foraminifer, FAD of *Eopara-staffella simplex*	Episodes **26/2**, 2003
Tournaisian (GSSP under reevaluation)	La Serre, France	43°33'19.9"N 3°21'26.3"E*	base of Bed 89 in Trench E' at La Serre, (but FAD now known to be at base of Bed 85)	Conodont, FAD of *Siphonodella sulcata* s.l.	Episodes **14/4**, 1991; Kölner Forum Geol. Paläont., **15**, 2006

* according to Google Earth, ** derived from map

Figure 9.6 Ratified GSSPs and potential primary markers under consideration for defining the Carboniferous stages *(status as of early 2016).* Details of each GSSP are available at http://www.stratigraphy.org, https://engineering. purdue.edu/Stratigraphy/gssp/, and in the *Episodes* publications.

be appropriate. The Devonian–Tournaisian boundary, even though the GSSP level might shift slightly to be associated with the immediately underlying Hangenberg Event, is fairly well constrained by radioisotopic dates on the transition interval as 358.9 ± 0.4 Ma (Davydov et al., 2012), which was verified by U-Pb dating that Hangenberg Event as between 358.89 ± 0.20 and 358.97 ± 0.11 Ma (Myrow et al., 2013).

Acknowledgments

This summary of the Carboniferous and its detailed cyclostratigraphy was limited to only a few aspects, selected highlights, and some current stratigraphic issues of this fascinating topic. For more information, see the detailed overviews and syntheses by Davydov et al. (2012) and Richards (2013), and on updates of the stratigraphic scales and GSSP status at the website and newsletters of the Subcommission on Carboniferous Stratigraphy (www.stratigraphy.org/Carboniferous). This review depended on the expertise of many colleagues, including (*in alphabetical order only*): Vladimir Davydov, Phil Heckel, Rich Lane, Barry Richards, and Mark Schmitz. Phil Heckel and Zhong-Qiang Chen reviewed an early version of the graphics, database, and text.

Selected publications and websites

Cited publications

Only select publications were cited in this review with an emphasis on aspects of post-2011 updates. Pre-2011 literature is well summarized in the synthesis by Davydov et al. (2012) and in some of the cited publications in the following.

Alekseev, A.S., Task Group, 2014. Report of the Task Group to establish a GSSP close to the existing Bashkirian-Moscovian boundary. *Newsletter on Carboniferous Stratigraphy* **31**: 33–36. Available at: www.stratigraphy.org/Carboniferous/pub/pub.asp.

Aretz, M., 2011. Report on the workshop of the task group for defining the Devonian-Carboniferous boundary. *Subcommission on Devonian Stratigraphy Newsletter* **26**: 18–20.

Barrick, J.E., Lambert, L.L., Heckel, P.H., Rosscoe, S.J., Boardman, D.R., 2013. Midcontinent Pennsylvanian conodont zonation. In: Heckel, P.H. (Ed.), Pennsylvanian Genetic Stratigraphy and Biostratigraphy of Midcontinent North America*Stratigraphy* **10**(1/2): 55–72.

Boardman, D.R., Work, D.M., 2013. Pennsylvanian (Desmoinesian-Virgilian) ammonoid zonation for Midcontinent North America. In: Heckel, P.H. (Ed.), Pennsylvanian Genetic Stratigraphy and Biostratigraphy of Midcontinent North America*Stratigraphy* **10**(1/2): 105–116.

Burgess, S.D., Bowring, S., Shen, S.Z., 2014. High-precision timeline for Earth's most severe extinction. *Proceedings of the National Academy of Science (PNAS)* **111**: 3316–3321.

Clack, J.A., 2012. *Gaining Ground: The Origin and Evolution of Tetrapods.* Indiana University Press, p. 544.

Chen, Z.-Q., Wang, X., Richards, B., Aretz, M., 2014. Multidisciplinary studies of global Carboniferous stage boundaries: towards a better definition and global correlations: an introduction. *Geological Magazine* **151**(2): 199–200.

Davydov, V.I., Shernykh, V.V., Chuvashov, B.I., Schmitz, M., Snyder, W.S., 2008. Faunal assemblage and correlation of Kasimovian-Gzhelian transition at Usolka section, southern Urals, Russia (a potential candidate for GSSP to define base of Gzhelian Stage). *Stratigraphy* **5**: 113–136.

Davydov, V.I., Crowley, J.L., Schmitz, M.D., Poletaev, V.I., 2010. High-precision U-Pb zircon age calibration of the global Carboniferous time scale and Milankovitch-band cyclicity in the Donets Basin, eastern Ukraine. *Geochemistry, Geophysics, Geosystems* **11**: Q0AA04. http://dx.doi.org/10.1029/2009GC002736.

Davydov, V.I., Korn, D., Schmitz, M.D., with Gradstein, F.M., Hammer, O., 2012. The Carboniferous Period. In: Gradstein, F.M., Ogg, J.G., Schmitz, M., Ogg, G., (Coordinators), *The Geologic Time Scale 2012.* Elsevier Publication, pp. 603–651. http://dx.doi.org/10.1016/B978-0-444-59425-9.00023-8 (An overview on all aspects, including graphics on the ratified GSSPs of the stages, diagrams and tables for the biostratigraphic scales, and discussion on the age models.).

Devuyst, F.-X., Hance, L., Hou, H., Wu, X., Tian, S., Coen, M., Sevastopulo, G., 2003. A proposed Global Stratotype Section and Point for the base of the Viséan Stage (Carboniferous): the Pengchong section, Guangzi, south China. *Episodes* **26**(2): 105–115.

Fielding, C.R., Frank, T.D., Birgenheier, L.P., Rygel, M.C., Jones, A.T., Roberts, J., 2008. Stratigraphic imprint of the late Paleozoic ice age in eastern Australia: a record of alternating glacial and non-glacial climate regime. *Journal of the Geological Society of London* **165**: 129–140.

Gradstein, S.R., Kerp, H., 2012. A brief history of plants on Earth. In: Gradstein, F.M., Ogg, J.G., Schmitz, M., Ogg, G., (Coordinators), *The Geologic Time Scale 2012.* Elsevier Publication, pp. 233–237. http://dx.doi.org/10.1016/B978-0-444-59425-9.00023-8.

Haq, B.U., Schutter, S.R., 2008. A chronology of Paleozoic sea-level changes. *Science* **322**: 64–68. http://dx.doi.org/10.1126/science.116164.

Heckel, P.H., 2008. Carboniferous Period. In: Ogg, J.G., Ogg, G., Gradstein, F.M., (Eds.), *The Concise Geologic Time Scale.* Cambridge University Press, pp. 73–83.

Heckel, P.H., (Ed.), 2013a. Pennsylvanian Genetic Stratigraphy and Biostratigraphy of Midcontinent North America. *Stratigraphy* **10**(1/2): 1–126.

Heckel, P.H., 2013b. Pennsylvanian stratigraphy of the northern Midcontinent shelf and biostratigraphic correlation of cyclothems. In: Heckel, P.H. (Ed.), Pennsylvanian Genetic Stratigraphy and Biostratigraphy of Midcontinent North America*Stratigraphy* **10**(1/2): 3–40.

Kaiser, S.I., 2009. The Devonian/Carboniferous boundary stratotype section (La Serre, France) revisited. *Newsletters on Stratigraphy* **43**: 195–205.

Korn, D., Klug, C., 2015. Chapter 12. Paleozoic ammonoid biostratigraphy. In: Klug, C., Korn, D., De Baets, K., Kruta, I., Mapes, R.H. (Eds.), Ammonoid Paleobiology: From Macroevolution to Paleogeography *Topics in Geobiology* 44: 299–328. http://dx.doi.org/10.1007/978-94-017-9633-0_13 Springer Publication.

Lane, H.R., Brenckle, P.L., Baesemann, J.F., Richards, B., 1999. The IUGS boundary in the middle of the Carboniferous; Arrow Canyon, Nevada, USA. *Episodes* 22: 272–283.

Montañez, I.P., Poulsen, C.J., 2013. The late Paleozoic ice age: an evolving paradigm. *Annual Review of Earth and Planetary Sciences* 41: 24.1–24.28. http://dx.doi.org/10.1146/annurev.earth.031208.1001118.

Myrow, P.M., Ramezani, J., Hanson, A.E., Bowring, S.A., Racki, G., Rakocinski, M., 2013. High-precision U-Pb age and duration of the latest Devonian (Famennian) Hangenberg Event, and its implications. *Terra Nova* 26: 222–229.

Paproth, E., Feist, R., Flajs, G., 1991. Decision on the Devonian-Carboniferous boundary stratotype. *Episodes* 14: 331–336.

Qi, Y.-P., Nemyrovska, T.I., Wang, X.-D., Chen, J., Wang, Z., Lane, H.R., Richards, B.C., Hu, K., Wang, Q.-L., 2014. Late Visean – Early Serpukhovian conodont succession at the Naqing (Nashui) section in Guizhou, South China. *Geological Magazine* 151(2): 254–268. http://dx.doi.org/10.1017/S001675681300071X.

Qi, Y.-P., Lambert, L.L., Nemyrovska, T.I., Wang, X.-D., Hu, K.-Y., Wang, Q.-L., 2016. Late Bashkirian and Early Moscovian conodonts from the Naqing section, Luodian, Guizhou, South China. *Palaeoworld*. http://dx.doi.org/10.1016/palwor.2015.02.005. http://www.sciencedirect.com/science/article/pii/s187174x15000207.

Richards, B.C., 2013. Current status of the international Carboniferous time scale. In: Lucas, S.G., et al. (Ed.), The Carboniferous-Permian Transition *New Mexico Museum of Natural History and Science Bulletin* 60: 348–353.

Rosscoe, S.J., Barrick, J.E., 2013. North American species of the conodont genus *Idiognathodus* from the Moscovian-Kasimovian boundary composite sequence and correlation of the Moscovian-Kasimovian stage boundary. In: Lucas, S.G., DiMichele, W., Barrick, J.E., Schneider, J.W., Spielmann, J.A. (Eds.), The Carboniferous-Permian Transition *New Mexico Museum of Natural History and Science, Bulletin* 60: 354–371.

Sallan, L., Galimberti, A.K., 2015. Body-size reduction in vertebrates following the end-Devonian mass extinction. *Science* 350: 812–815 (plus supplementary materials).

Saltzman, M.R., 2003. Late Paleozoic ice age: oceanic gateway or pCO$_2$? *Geology* 31: 151–154 (and on-line supplement for dataset).

Saltzman, M.R., Thomas, E., 2012. Carbon isotope stratigraphy. In: Gradstein, F.M., Ogg, J.G., Schmitz, M., Ogg, G., (Coordinators), *The Geologic Time Scale 2012*. Elsevier Publication, pp. 207–232. http://dx.doi.org/10.1016/B978-0-444-59425-9.00023-8.

Scotese, C.R., 2014. Atlas of Permo-Carboniferous Paleogeographic Maps (Mollweide Projection), Maps 53-64. PALEOMAP PaleoAtlas for ArcGIS *The Late Paleozoic.*, **vol. 4**. PALEOMAP Project, Evanston, IL. https://www.academia.edu/16664729/Atlas_of_Permo-Carboniferous_Paleogeographic_Maps.

Schmitz, M.D., 2012. Appendix 2. Radiometric ages used in GTS2012. In: Gradstein, F.M., Ogg, J.G., Schmitz, M., Ogg, G., (Coordinators), *The Geologic Time Scale 2012*. Elsevier Publication, pp. 1045–1082. http://dx.doi.org/10.1016/B978-0-444-59425-9.00023-8.

Schmitz, M.D., Davydov, V.I., 2012. Quantitative radiometric and biostratigraphic calibration of the Pennsylvanian–Early Permian (Cisuralian) time scale and pan-Euramerican chronostratigraphic correlation. *Geological Society of America Bulletin* 124: 549–577. http://dx.doi.org/10.1130/B30385.1 and data-repository item 2012020.

Sevastopulo, G.D., Barham, M., 2014. Correlation of the base of the Serpukhovian Stage (Mississippian) in NW Europe. *Geological Magazine* 151(2): 244–253.

Subcommission on Carboniferous Stratigraphy, 2014. *Stratigraphic Chart of Carboniferous Conodont, Fusulinid & Benthic Foraminifer, and Ammonoid Zones and T-r Cycles*. Available at: www.stratigraphy.org/Carboniferous (viewed Nov' 15).

Sungatullina, G., 2014. Determination of the Bashkirian–Moscovian boundary in the Volga region via conodont species *Declinognathodus donetzianus* Nemirovskaya. *Geological Magazine* 151(2): 299–310.

Ueno, K., Task Group, 2014. Report of the task group to establish the Moscovian-Kasimovian and Kasimovian-Gzhelian boundaries. *Newsletter on Carboniferous Stratigraphy* 31: 36–40. Available at: www.stratigraphy.org/Carboniferous/pub/pub.asp.

Veizer, J., Prokoph, A., 2015. Temperatures and oxygen isotopic composition of Phanerozoic oceans. *Earth-Science Reviews* 146: 92–104.

Wahlman, G.P., 2013. Pennsylvanian to Lower Permian (Desmoinesian-Wolfcampian) fusulinid biostratigraphy of Midcontinent North America. In: Heckel, P.H. (Ed.), Pennsylvanian Genetic Stratigraphy and Biostratigraphy of Midcontinent North America. *Stratigraphy* 10(1/2): 73–104.

Websites (selected)

Subcommission on Carboniferous Stratigraphy
(International Commission on Stratigraphy [ICS])—
http://www.stratigraphy.org/Carboniferous/—Details
on GSSPs, Portable Document Formats (PDFs) of the
full set of Newsletters (1992 to present).

Paleozoic Forests—*http://www.uni-muenster.de/
GeoPalaeontologie/Palaeo/Palbot/ewald.html*—
a well-presented exploration that focuses on the
Carboniferous coal forests.

Palaeos: Carboniferous—*http://palaeos.com/paleozoic/
carboniferous/carboniferous.htm*—A well-presented
suite of diverse topics for a general science audience
that was originally compiled by M. Alan Kazlev in
1998–2002.

Climate and the Carboniferous Period (by Monte Hieb,
under Plant Fossils of West Virginia)—*http://
geocraft.com/WVFossils/Carboniferous_climate.
html*—continental drift, ice ages, and coal deposit
overview.

10

PERMIAN

255.7 Ma Permian

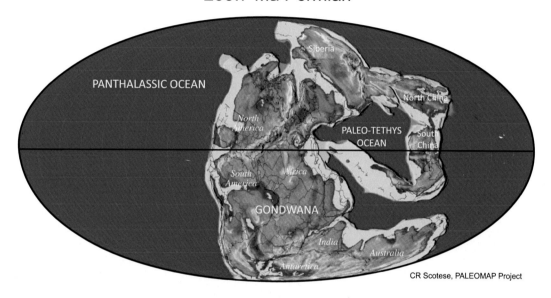

Lopingian (late Permian) paleogeographic reconstruction (Sea level–40 m) from Scotese (2014).

Basal definition and international subdivisions

The Permian, named after exposures in the southern Urals of Russia, which was within the ancient kingdom of Permia, has three named series/epochs—Cisuralian (type area is the southern Urals), Guadalupian (type area is the Guadalupe Mountain region of West Texas), and Lopingian (type area is South China) (Fig. 10.1). The ratified or candidate Global Boundary Stratotype Sections and Points (GSSPs) for the nine Permian stages (Figs. 10.1 and 10.7), all of which coincide with first-appearance datums (FADs) of conodont taxa, are mainly in these type regions (summarized in Henderson et al. (2012a) and in Shen et al. (2013b)).

The base of the Permian was originally placed at the onset of evaporites in the type area of the Urals—a level now within the Kungurian Stage—and was progressively shifted to older biostratigraphic levels with the creation of additional early Permian stages upon the distinguishing of Carboniferous versus Permian faunas. The base of the Russian Permian stabilized in 1992 at the appearance of three ammonoid families and the first inflated "*Schwagerina*" group of fusulinids (benthic foraminifers). The base of the Permian (base of Asselian Stage)

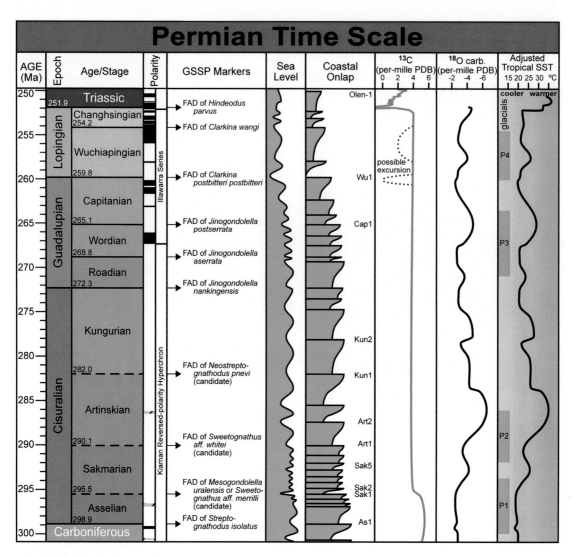

Permian Time Scale

Figure 10.1 Permian overview. Main markers or candidate markers for GSSPs of Permian stages are first-appearance datums (FADs) of conodont taxa as detailed in the text and in Fig. 10.7. ("Age" is the term for the time equivalent of the rock-record "stage.") Magnetic polarity scale is modified from Steiner (2006). Coastal onlap and schematic sea-level curve with labels for selected major sequence boundaries are modified from Haq and Schutter (2008) following advice of Bilal Haq (personal communication, 2008). The Late Permian sea-level sequences are mainly derived from South China (Chen et al., 1998). The $\delta^{13}C$ curve is from the composite by Buggisch et al. (2015), who concluded that none of the reported excursions in regional studies can yet be reliably verified on a global scale (except for the latest Permian) and that a stable mean (+4 per mil on their diagram) would currently fit those inconsistent trends. Superimposed are schematic placements of possible interregional negative excursions in latest Capitanian through Wuchiapingian (e.g., Shen et al., 2013a), and there may be another negative excursion within the early part of late Capitanian (Zhong-Qiang Chen, written communication, January 2016). The Permian $\delta^{18}O$ curve is modified from the mean of the schematic curve of Veizer and Prokoph (2015) with their adjusted estimates of tropical sea-surface temperatures derived from those oxygen-18 values. P1–P4 glacial episodes are according to the dated glacial deposits in eastern Australian basins (Frank et al., 2015; Metcalfe et al., 2015). The vertical scale of this diagram is standardized to match the vertical scales of the first stratigraphic summary figure in all other Phanerozoic chapters. *PDB*, PeeDee Belemnite ^{13}C and ^{18}O standard; *SST*, sea-surface temperature.

at the Aidaralash stream section in northwest Kazakhstan coincides with the FAD of conodont *Streptognathodus isolatus* and is 6.3 m below the appearance of the traditional "*Schwagerina*" group (Fig. 10.2). However, this main conodont marker has only rare well-documented occurrences outside of the Urals, may have taxonomic identification problems, and there is a possibility that its FAD might be diachronous relative to the *Schwagerina* secondary markers (e.g., Lucas, 2013b; with reply by Davydov, 2013).

GSSPs to define the other three stages of the Cisuralian based on evolutionary lineages of conodonts were nearing completion in 2014, but then complications arose from "disagreements on the taxonomy of conodonts and the index species for definition" of the Sakmarian and Artinskian stages (*Permophiles*, v.59, 2014) and failure to get a two-thirds majority vote for placing the Kungurian GSSP in an expanded section in Nevada. The current statuses are found at http://permian.stratigraphy.org and its post-2015 issues of *Permophiles*.

Sakmarian: GSSPs for the base and the top of the Sakmarian Stage were proposed near the health resort of Krasnousol'sky village (Bashkortostan Republic, southeast Volga district, southwestern Russia) on the western slope of the southern Urals, near the historical type-section (Chernykh et al., 2013). The candidate for the Sakmarian GSSP on the right bank of the Usolka river in Bed 25 (51.6 m above base of section) would coincide with the FAD of conodont *Mesogondolella uralensis* which evolved from *M. arcuata*. This datum is just below the FAD of the widespread conodont *Sweetognathus* aff. *merrilli* in the lineage that later evolved into *Sweetognathus* aff. *whitei*, which is proposed to coincide with the GSSP for the following Artinskian Stage [Note that the usage of "aff" (affinity) for both taxa is necessary because the lineage of the original named *Sw. merrilli* and *Sw. whitei* taxa (holotypes) were found to be restricted to the underlying Asselian Stage.]. The base

of the Sakmarian coincides with a major sea-level lowstand.

Artinskian: The candidate GSSP near the proposed Sakmarian GSSP is at 1.8 m above the base of Bed 4 as exposed in trenches along the Dal'ny Tulkus stream section on the southern end of the Usolka anticline (Chuvashov et al., 2013). This level would coincide with the evolutionary appearance of *Sweetognathus* aff. *whitei* in the lineage of *Sw.* aff. *merrilli—Sw. binodosus—Sw. anceps—Sw.* aff. *whitei*.

Kungurian: The accepted base of the Kungurian Stage is the evolution of conodont *Neostreptognathodus pnevi* from *Neostr. pequopensis*, but there are at least two competing candidates for that level. The candidate level at 26.5 m in the Rockland section near Wells, Nevada, United States (Henderson et al., 2012b) did not receive a supermajority vote from the working group. Therefore, work is continuing on a small quarry section near Mechetlino in the southern Pre-Urals, where the proposed GSSP level would be in a sandy Bed 9 (Chernykh et al., 2012). Even though the Mechetlino section is nearly an order-of-magnitude more compact than the Rockland section, the conodonts are more pristine (Henderson et al., 2012b).

The base of the Guadalupian (base of Roadian Stage) at Stratotype Canyon in West Texas coincides with the FAD of conodont *Jinogondolella nankingensis* (Fig. 10.3). The other GSSPs for stages in the Guadalupian in nearby sections are based on the evolution of that *Jinogondolella* genus.

The base of the Lopingian (base of Wuchiapingian Stage) at Penglaitan, Guangxi Province, South China coincides with the FAD of conodont *Clarkina postbitteri postbitteri* (Fig. 10.4), although the actual GSSP will need to be relocated in the near future when a new dam floods the stream outcrop. The GSSP for the base of the Changhsingian Stage is in the same exposure at Meishan in southern China as the base-Triassic GSSP that defines its top.

Carboniferous - Permian Boundary at Aidaralash near the town of Aktobe, Kazakhstan

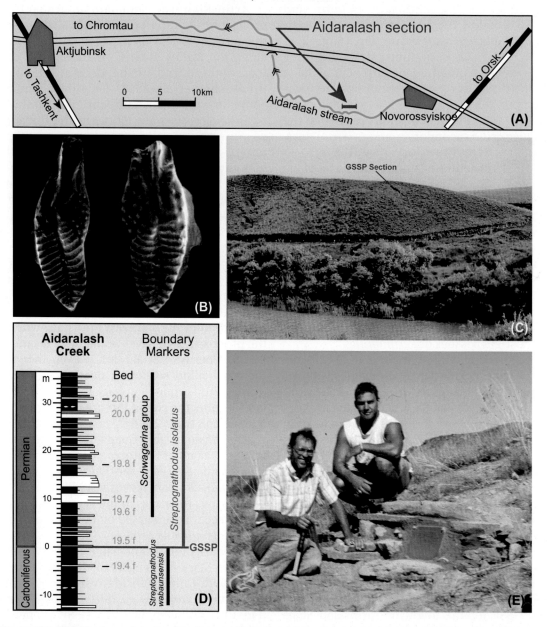

Figure 10.2 GSSP for base of the Permian (base of Cisuralian Series; base of Asselian Stage) at the Aidaralash section, southern Urals of northwest Kazakhstan. The GSSP level coincides with the lowest occurrence of conodont *Streptognathodus isolatus*. Holotype of *St. isolatus* in left image is 1.2-mm long, and in right image is 0.8-mm long. (Outcrop photograph provided by Vladimir Davydov; conodont and section photographs and stratigraphic column from Henderson et al. (2012a).)

Base of the Roadian Stage of the Guadalupian Series at Stratotype Canyon, Texas, U.S.A.

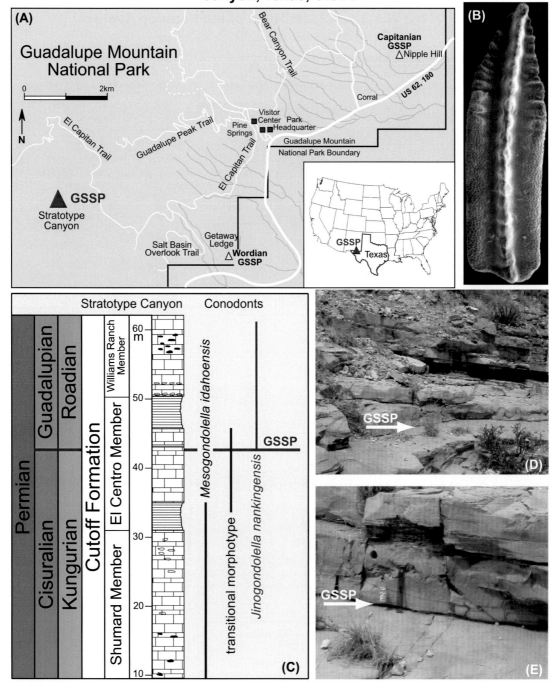

Figure 10.3 GSSP for base of the Guadalupian Series (base of middle Permian; base of Roadian Stage) at the Stratotype Canyon section, western Texas, United States. The GSSP level coincides with the lowest evolutionary occurrence of conodont *Jinogondolella nankingensis* from its ancestors *Mesogondolella idahoensis idahoensis* and *M. idahoensis lamberti*. Holotype of *J. nankingensis* in image is 1.0 mm long. (Photographs and stratigraphic column are from Henderson et al. (2012a).)

Base of the Wuchiapingian Stage of the Permian System at Penglaitan Section, South China

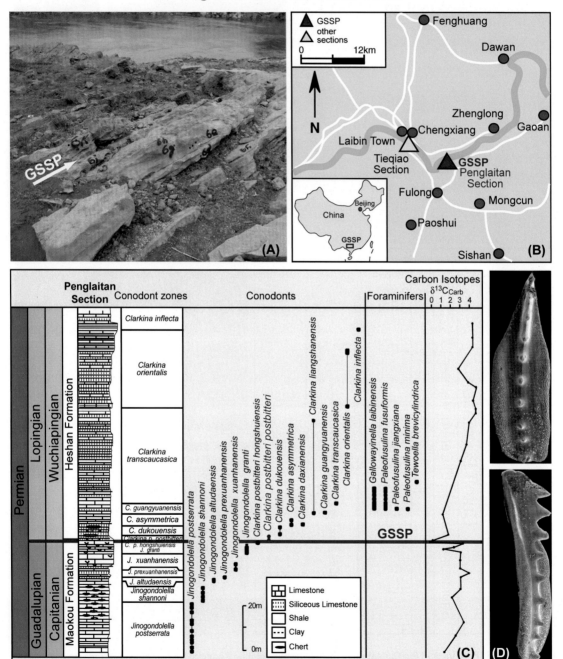

Figure 10.4 GSSP for base of the Lopingian Series (base of late Permian; base of Wuchiapingian Stage) at the Penglaitan section, Guangxi province, South China. The GSSP level coincides with the lowest occurrence of *Clarkina postbitteri postbitteri* within an evolutionary lineage from *C. postbitteri hongshuiensis* to *C. dukouensis*. Holotype of *C. postbitteri postbitteri* in image is 1.0-mm long. (Photographs and stratigraphic column are from Henderson et al. (2012a).)

Selected main stratigraphic scales and events

(1) Biostratigraphy (marine; terrestrial)

A trio of conodonts, benthic foraminifers (especially fusulinaceans) and ammonoids are the main biostratigraphic tools for correlation of marine successions. A selected subset of the regional zonations for these groups and approximate correlations in Fig. 10.5 is based on Henderson et al. (2012a) and Shen et al. (2015), who emphasized that there is not yet a precise correlation between the zonal boundaries among the different groups for most time intervals.

The phosphatic feeding apparatus of conodonts, a small, probably eel-like vertebrate, provides the main global subdivisions for middle Devonian through Triassic. All Permian conodont zones begin with the first-appearance datum (FAD) of their index taxa. The early Permian (late Asselian through Sakmarian and basal Artinskian) portion of the conodont zonation in Fig. 10.5 is for the *Streptognathodus–Sweetognathus* lineage, and there is a parallel one for *Mesogondolella* for this ca. 6-myr interval. Concentrated research on the late-Permian environmental changes and mass extinction led to a very detailed conodont zonation for the upper Changhsingian Stage through basal Triassic, especially in the important South China reference sections (e.g., Fig. 10.6). In that upper Changhsingian interval, the progressive recognition of more conodont taxa and a practice of giving a new zonal name to every consecutive FAD, while useful for detailed discussions, has resulted in some incredibly brief zones (e.g., *Clarkina meishanensis* Zone spans only 0.02 myr (Yuan et al., 2014) or 0.008 myr (Chen et al., 2015)).

Permian ammonoid zonations are not as detailed or as useful as in the Devonian or Carboniferous. A comparison of zonations with images of the main Permian index taxa is in Korn and Klug (2015). Three of the four major clades of Permian ammonoids vanished during the end-Permian mass extinction, and all Mesozoic ammonoids appear descended from the single family of morphologically simple Xenodiscidae of the Ceratitida clade (Brayard and Bucher, 2015).

The amphibian and reptile records of western United States and South Africa are used to subdivide the Permian into 10 land-vertebrate fauna-chrons named after selected reference sections (Lucas, 2006). There were slow rates of evolution of the classic sail-backed pelycosaur reptiles during the early Permian. There is a ~5-myr long "Olson's Gap" in the preserved tetrapod-evolution record which obscures the transition from that pelycosaur-dominated early Permian into the rapid initial diversification and domination of mammal-like therapsid reptiles during the middle Permian (Lucas, 2006, 2013a). The therapsids are considered the distant ancestors of modern mammals.

(2) Magnetostratigraphy

The early through middle Permian is within the long reversed-polarity Kiaman hyperchron. The first verified normal-polarity chron, named the Illawarra after its initial report from magnetostratigraphy in the Illawarra coal measures of the Sydney Basin of Australia, is in the upper part of the Capitanian Stage. Reports of an earlier normal-polarity zone within the preceding Wordian Stage are consistent with magnetostratigraphy from Japan (Kirschvink et al., 2015); and, even though the Illawarra coal measures do not extend to this older level (Metcalfe et al., 2015), the generalized name "Illawarra series" is applied. The schematic polarity pattern for the late Permian in Fig. 10.1 is based on the compilation of Steiner (2006); however, some of the magnetostratigraphic studies from this interval are difficult to intercorrelate, and, in many cases, the source data were never published.

(3) Stable-isotope stratigraphy and selected events

The calibration of the stable-isotope and glacial-episode records of the Permian has undergone major revisions since the publication of GTS2012.

The schematic carbon-isotope curve in GTS2012 was a merger of two major reference sections in slope-to-basinal carbonate successions near the Yangtze Platform of South China. These had been considered potentially representative of significant global excursions superimposed on a gradual negative trend from base-Permian to mid-Wuchiapingian—a composite of upper Carboniferous to mid-Capitanian by Buggisch et al. (2011) and of upper Capitanian through basal Triassic by Shen et al. (2010). However, Shen et al. (2013a) cautioned in their comparative study of several South China and Iranian sections that "*the extent to which the end-Guadalupian and Wuchiapingian/Changhsingian boundary excursions result from local versus global controls remains unresolved.* "Indeed, after examining additional reference sections outside of South China and compiling an extensive comparison of other published studies, Buggisch et al. (2015) concluded that "*an overall negative trend … is not obvious and negative excursions related to changes in the carbon isotope composition of the global oceanic carbon pool cannot be confirmed, except for the Permian–Triassic boundary interval.*" There are probably such carbon-isotope excursions within the Permian that will be useful as global signatures, such as one reported in late Guadalupian (the "Kamura event" of Isozaki et al., 2007), but these require improved interregional correlation for verification. Therefore, only the average value of 4‰ (per mille or per thousand) for the majority of the Permian, as diagrammed in the summary of Buggisch et al. (2015), is shown in Fig. 10.1 with dotted schematic placement of possible negative excursions.

Ice sheets in Gondwana underwent several major expansions and retreats during the Carboniferous–Permian in addition to smaller oscillations in response to Milankovitch orbital–climate feedbacks, especially as modulated by 100-kyr and 400-kyr eccentricity cycles. Major Gondwana glacial advances and retreats are recorded as diamictites and other glacial facies within the basins of eastern Australia. Using the pre-2013 calibration of Australian palynology zones to Permian stages, Fielding et al. (2008) intercalibrated the records of several basins to identify four major glacial "P" episodes within the Cisuralian and Guadalupian epochs in the eastern Australian basins. A major program to radiometrically date over 100 interbedded volcanic bentonites within these basins has significantly revised the calibration of these Australian palynology zones and revealed that the final two glacial episodes are significantly younger (e.g., Metcalfe et al., 2015; Nicoll et al., 2015). The earliest and most intense glacials, P1 and P2, are biostratigraphically calibrated in the lithologic record of eastern Australian basins as spanning the end-Carboniferous to middle Artinskian with a relatively brief nonglacial in the middle Sakmarian (Fielding et al., 2008; Frank et al., 2015). Constraints from U-Pb dating indicate that the lesser glacials of P3 (271–263.5 Ma) spans the Roadian to mid-Capitanian, and P4 (260–254.5 Ma) begins with the end-Capitanian mass extinction and continues to the middle of the Wuchiapingian (Metcalfe et al., 2015). The glacials of P3 may, in part, have accounted for the global nutrient-rich oceans in that time (Large et al., 2015; Shi et al., 2016). The cooling of sea-surface temperatures would have accelerated oceanic circulation, thereby increasing the return of nutrients from the deep water to surface oceans, to stimulate a high primary productivity by the blooming of plankton (Shi et al., 2016).

However, revised ages of these main glacial episodes are only partially consistent with the

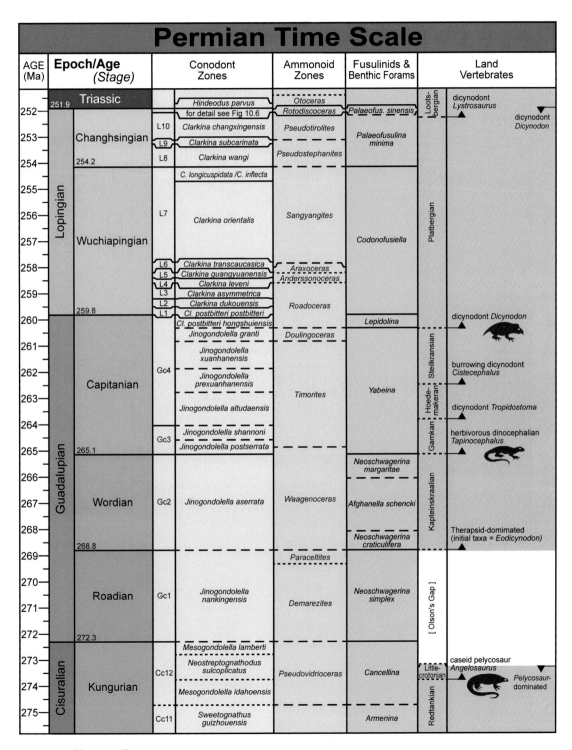

Permian Time Scale

AGE (Ma)	Epoch/Age *(Stage)*		Conodont Zones		Ammonoid Zones	Fusulinids & Benthic Forams		Land Vertebrates
251.9 252—		Triassic	Hindeodus parvus		Otoceras	Palaeofus. sinensis	Loots-bergian	dicynodont Lystrosaurus
			for detail see Fig 10.6		Rotodiscoceras			dicynodont Dicynodon
253—	Changhsingian		L10	Clarkina changxingensis	Pseudotirolites	Palaeofusulina minima		
			L9	Clarkina subcarinata				
254—	254.2		L8	Clarkina wangi	Pseudostephanites			
255—				C. longicuspidata /C. inflecta				
256—	Lopingian	Wuchiapingian	L7	Clarkina orientalis	Sangyangites	Codonofusiella	Platbergian	
257—								
258—			L6	Clarkina transcaucasica	Araxoceras			
			L5	Clarkina guangyuanensis	Anderssonoceras			
259—			L4	Clarkina leveni				
			L3	Clarkina asymmetrica	Roadoceras			
	259.8		L2	Clarkina dukouensis				
260—			L1	Cl. postbitteri postbitteri		Lepidolina		dicynodont Dicynodon
				Cl. postbitteri hongshuiensis				
				Jinogondolella granti	Doulingoceras			
261—		Capitanian		Jinogondolella xuanhanensis			Steilkransian	burrowing dicynodont Cistecephalus
262—			Gc4	Jinogondolella prexuanhanensis		Yabeina		
263—				Jinogondolella altudaensis	Timorites		Hoede-makeran	dicynodont Tropidostoma
264—				Jinogondolella shannoni			Gamkan	herbivorous dinocephalian Tapinocephalus
265—			Gc3	Jinogondolella postserrata				
	265.1					Neoschwagerina margaritae		
266—	Guadalupian	Wordian	Gc2	Jinogondolella aserrata	Waagenoceras	Afghanella schencki	Kapteinskraalian	
267—								
268—						Neoschwagerina craticulifera		Therapsid-domimated (initial taxa = Eodicynodon)
	268.8				Paraceltites			
269—								
270—		Roadian	Gc1	Jinogondolella nankingensis	Demarezites	Neoschwagerina simplex	[Olson's Gap]	
271—								
272—		272.3						
273—				Mesogondolella lamberti		Cancellina	Little-crotonian	caseid pelycosaur Angelosaurus
	Cisuralian	Kungurian	Cc12	Neostreptognathodus sulcoplicatus	Pseudovidrioceras			Pelycosaur-dominated
274—				Mesogondolella idahoensis			Redtankian	
275—			Cc11	Sweetognathus guizhouensis		Armenina		

Figure 10.5 *(Continued)*

Permian Time Scale

AGE (Ma)	Epoch/Age *(Stage)*		Conodont Zones		Ammonoid Zones	Fusulinids & Benthic Forams		Land Vertebrates
275	Cisuralian	Kungurian	Cc11	*Sweetognathus guizhouensis*		Armenina	Redtankian	
276								
277				*Neostreptognathodus prayi*	Uraloceras	Misellina		captorhinid *Labidosaurus*
278							Mitchellcreekian	
279			Cc10	*Neostreptognathodus pseudoclinei*		Brevaxina		varanopid pelycosaur *Mycterosaurus*
280								
281				*Neostreptognathodus pnevi*			Seymourian	seymouriamorph *Seymouria*
282		282.0						
283		Artinskian	Cc9	*Neostreptognathodus pequopensis*		Pamirina		
284								
285					Aktubinskia	Chalaroschwagerina		
286			Cc8	*Sweetognathus clarki*	Artinskia	Parafusulina solidissima		
287								
288					Popanoceras	*Pseudo. juresanensis*	Coyotean	
289				*Sweetognathus aff. whitei*		*Pseudofusulina pedissequa*		
290		290.1			Uraloceras			
291		Sakmarian	Cc6 - Cc7	*Sweetognathus anceps*		Leeina urdalensis		Dimetrodon
292					Metalegoceras			
293					Properrinites			
294				*Sweetognathus binodosus*	Sakmarites	Leeina vernueili		
295				*Sweetognathus aff. merrilli - Mesogondolella uralensis*		Sakmarella moelleri		
296		295.5 Asselian	Cc5	*Streptognathodus barskovi*	Juresanites	Sphaeroschwagerina sphaerica		
			Cc4	*Streptognathodus postfusus*		*Pseudoschwagerina uddeni*		
297			Cc3	*Streptognathodus fusus*		*Sphaeros. moelleri*		
			Cc2	*Streptognathodus constrictus*		*Globifusulina nux*		
298			Cc1	*Streptognath. sigmoidalis* *Streptognath. cristellaris*	Svetlanoceras	Sphaeroschwagerina fusiformis		
299		298.9 Carboniferous	Pc21	*Streptognathodus isolatus* *St. wabaunsensis - St. fissus*	*Shumardites / Vidrioceras*	*Daixina bosbytauensis - Globifusulina robusta*		

◀ **Figure 10.5 Selected marine and terrestrial biostratigraphic zonations of the Permian.** ("*Age*" is the term for the time equivalent of the rock-record "*stage.*") Scaling of Lopingian conodont zones from South China in Changhsingian is from Yuan et al. (2014; see Fig. 10.6 for details in latest Permian), and in Wuchiapingian is from Shen et al. (2013a). Scaling of Guadalupian and Cisuralian conodont zones is from Henderson et al. (2012a). Ammonoid zones and fusulinid (fusulinacean)/benthic foraminifer zones are mainly derived from the chart without explicit zonal boundaries compiled by Shen et al. (2015), who seem to schematically estimate the bases of the index taxa relative to the conodont zonal scales. Land-vertebrate zones and markers are from Lucas (2006, 2013a). Additional zonations, biostratigraphic markers, geochemical trends, sea-level curves, regional stages, and details on calibrations are compiled in Henderson et al. (2012a) and in the internal data sets within the *TimeScale Creator* visualization system (free at *www.tscreator.org*).

current syntheses of a global oxygen-18 curve and the estimates of tropical sea-surface temperatures (e.g., the schematic composites of Veizer and Prokoph, 2015; shown in Fig. 10.1). For example, in South China, the oxygen isotopic data from conodonts indicate a steady rise of surface seawater temperatures through the Roadian–Capitanian interval (Chen et al., 2013), and thus questions the influence of the Glacial P3. Similarly, sea-surface temperatures remained high through the earliest to middle Wuchiapingian (Chen et al., 2013), equivalent to the Glacial P4 period. During the late Cenozoic, a comparative oxygen-isotope record from surface and deeper water fauna is used to determine the timing and magnitudes of ice-volume changes, but this type of record has not yet been obtained for the glacial record of the Permian.

There was a continuous decline in the $^{87}Sr/^{86}Sr$ ratio of seawater through the Cisuralian and Guadalupian that reached a minimum in latest Capitanian. That minimum, the lowest $^{87}Sr/^{86}Sr$ value for the entire Phanerozoic, coincides with the greatest global regression at the end of the Guadalupian (Chen et al., 1998, 2009). An extensive set of high-precision U-Pb dates indicates that the steep decline during the Asselian–Sakmarian was essentially linear; therefore, $^{87}Sr/^{86}Sr$ from carbonates in this interval can be used to assign their ages. This linear slope may continue into the Guadalupian. During the Lopingian and into the Early Triassic, the $^{87}Sr/^{86}Sr$ values rose at a rate that was more rapid than their previous decrease.

A significant mass extinction of marine fauna occurred at the end of the Guadalupian.

This episode coincides with the initiation of the redated Glacial P4 episode, the reversal in the trend of seawater $^{87}Sr/^{86}Sr$ ratios, and the eruption of the Emeishan large igneous province in China at ca. 260 Ma (e.g., Wignall et al., 2009, 2015; Metcalfe et al., 2015; Yang et al., 2015). This end-Guadalupian episode may also be associated with a potential global major negative excursion in carbon-isotopes (e.g., Shen et al., 2013a). The Permian ends with a catastrophic mass extinction, a major negative carbon-isotope excursion, and the eruption of the Siberian Traps large igneous province (e.g., Shen et al., 2013a; see next chapter on Triassic).

Numerical age model

GTS2012 age model

The age model for the combined Carboniferous–Permian in GTS2012 was a statistical spline-fit of a collection of U-Pb ages to a composite biostratigraphic scale that had been constructed by applying CONstrained Optimization (CONOP)-9 to an extensive suite of reference sections (Davydov et al., 2012; Schmitz and Davydov, 2012). This yielded age estimates for each conodont zone boundary. All other stratigraphic and geochemical events were calibrated to that primary conodont scale. The main U-Pb constraints were several dated ash beds within the Asselian to early Artinskian (later published by Schmitz and Davydov, 2012) and within the Lopingian of South China (e.g., Shen et al., 2010). There

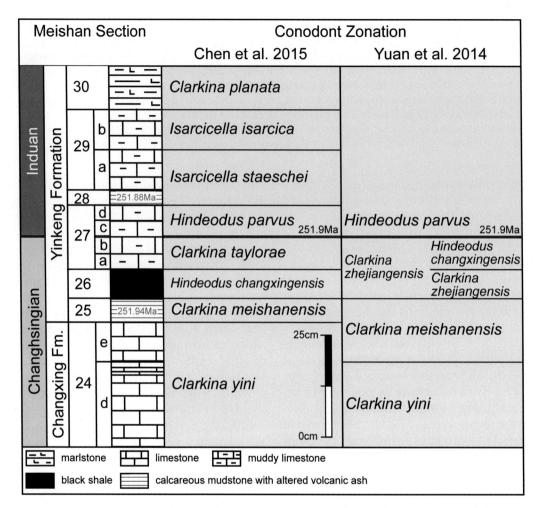

Figure 10.6 High-resolution conodont zonation and radioisotopic dating constraints across Permian–Triassic boundary GSSP at Meishan, South China. This Meishan section also has the Changhsingian GSSP; therefore, it has been extensively studied as a standard for the Changhsingian Stage. Radioisotope dates on altered volcanic ash components within Bed 25 (251.941 ± 0.037 Ma) and Bed 28 (251.880 ± 0.031 Ma) are by Burgess et al. (2014). The main end-Permian mass extinction event is at the base of Bed 25. The base of the *Clarkina yini* zone is approximately 1 m below the base of this diagram, with an extrapolated basal age of 252.06 Ma (Yuan et al., 2014). There are minor differences between recent studies, including (1) rare *Clarkina meishanensis* are reported slightly lower by Yuan et al. (2014), (2) conodonts assigned as *Clarkina taylorae* are classified as *Clarkina zhejiangensis* in this section by Yuan et al. (2014), who also shift the *Hindeodus changxingensis* zone to be topmost Permian, and (3) a brief *Clarkina planata* zone within the lower Induan at Meishan is proposed by Chen et al. (2015), which truncates the *Isarcicella isarcica* zone as used in previous publications. *Isarcicella staeschei* is now assigned to "Morphotype 2" of the broader *Isarcicella isarcica* taxon concept used in older publications (e.g., Triassic Fig. 11.2 from Yin et al., 2001; see discussion in Jiang et al., 2007). Yuan et al. (2014) did not study conodonts above the basal *Hindeodus parvus* zone. Details of conodont taxonomy, definitions of zones, comparisons to other conodont zonations, and other biostratigraphic and geochemical features of this important reference section are in Chen et al. (2015) and Yuan et al. (2014).

were no radio-isotopic dates to constrain the scaling of the composite standard in the ca. 25-myr interval between the mid-Artinskian and late Wordian; and only a single Wordian age constrained the scaling of the Guadalupian Epoch.

Revised ages compared to GTS2012 and potential future enhancements

Between GTS2012 and the preparation of this Permian chapter at the end of 2015, three major publications or revisions have appeared of radioisotopic dating constraints or within-stage scaling of biostratigraphic zones that are incorporated in the diagrams of Fig. 10.5. As the Gondwana ice sheets underwent regular expansion/contraction during the late Carboniferous through early Permian in response to Milankovitch orbital-climate feedbacks, a set of high-amplitude cyclic sea-level onlaps (cyclothems) were generated on continental margins. The detailed U-Pb dating of conodont-rich successions in the Donets Basin and southern Urals enabled recognition of a full set of 405-kyr long eccentricity cycles spanning the Asselian and Sakmarian stages (Schmitz and Davydov, 2012). Those cycles also have counterparts in the North American Midcontinent conodont-zoned lithostratigraphy, thereby enabling a high-resolution age model for the duration of each conodont zone between the two regions. This cycle scaling of conodont zones has been incorporated in Fig. 10.5.

In the Lopingian (Late Permian) of South China, the durations of conodont zones are constrained by enhanced U-Pb dating and cyclostratigraphy of reference sections. The Wuchiapingian conodont scale in Fig. 10.5 is the well-dated Shangsi section in Sichuan (Shen et al., 2013a). The Changhsingian conodont scale (enlarged in Fig. 10.6) from the Changhsing Formation between its bounding GSSPs at the Meishan section (Yuan et al., 2014) has durations similar to those

interpreted by cyclostratigraphy (Wu et al., 2013).

The main differences in the age models between GTS2012 and Fig. 10.5 for the Permian stage boundaries are:

Kungurian base (**282.0** vs. 279.3 Ma in GTS2012): GTS2012 compared two methods for the Permian scaling of the CONOP-composite. A linear-fit yielded 282.0 Ma for the base-Kungurian, but a spline fit yielded 279.3 Ma. Both were constrained by only a ca. 288.3 Ma U-Pb date in lower Artinskian and a ca. 266 Ma U-Pb date in upper Wordian. It seems that the degree of "smoothed-curve bending" in the spline-fit version, hence the curve intersection with the base-Kungurian conodont zone, may have been distorted by the uncertain placement of the 288 Ma date relative to the CONOP composite, because that dated level had come from a broad biostratigraphic zone with no upper bound. Therefore, the linear-fit version of 282.0 Ma (Schmitz and Davydov, 2012) is used here, which is closer to an independent estimate of 283.5 Ma for the base of Kungurian derived by assuming that the slope of the linear segment of the $^{87}Sr/^{86}Sr$ curve of the Asselian–Sakmarian continued with the identical slope through the ca. 7 myr of the Artinskian (Henderson et al., 2012b; which they humorously had referred to as a "smoke on the water" method).

Changhsingian base (**254.15** vs. 254.2 Ma in GTS2012): This is a very minor adjustment based on Yuan et al. (2014), and is hidden in the rounding to one decimal.

Triassic base (Base of Induan) (**251.902 ± 0.024** vs. 252.16 ± 0.2 Ma in GTS2012): The uranium–lead isotope dilution–thermal ionization mass spectrometry (U-Pb TIMS) dating of volcanic ash beds and associated interpolated base-Triassic boundary age from the Meishan GSSP section reported in Shen et al. (2010) was reanalyzed by them

(Burgess et al., 2014) using revised EARTH-TIME standards, which shifted the boundary age younger by about 0.26 myr. [*Note that an external uncertainty of ca. 0.29 myr should be included if comparing these earliest Triassic EARTHTIME-standardized dates to other dating methods, as explained in* Burgess et al. (2014).]

The main requirement for improving the Permian age model and the scaling of events within stages is the acquisition of radioisotopic dates within the ca. 30-myr span of

Stage	GSSP Location	Latitude, Longitude	Boundary Level	Correlation Events	Reference
GSSPs of the Permian Stages, with location and primary correlation criteria					
Changhsingian	Meishan, Zhejiang Province, South China	31° 4'55"N 119°42'22.9"E	base of Bed 4a-2, 88 cm above the base of Changxing Limestone at the Meishan D Section	Conodont, FAD of *Clarkina wangi*	Episodes **29**/3, 2006
Wuchiapingian	Penglaitan, Guangxi Province, South China	23°41"43"N 109°19"16"E	base of Bed 6k in the Penglaitan Section	Conodont, FAD of *Clarkina postbitteri postbitteri*	Episodes **29**/4, 2006
Capitanian	Nipple Hill, SE Guadalupe Mountains, Texas, U.S.A	31°54'32.8"N 104°47'21.1"W	4.5m above the base of the outcrop section of the Pinery Limestone Mbr of the Bell Canyon Formation	Conodont, FAD of *Jinogondolella postserrata*	
Wordian	Guadalupe Pass, Texas, U.S.A	31°51'56.9"N 104°49'58.1"W	17.6m above the base of the Getaway Ledge outcrop section of the Getaway Limestone Mbr of the Cherry Canyon Formation	Conodont, FAD of *Jinogondolella aserrata*	
Roadian	Stratotype Canyon, Texas, U.S.A	31°52'36.1"N 104°52'36.5"W	42.7m above the base of the Cutoff Formation	Conodont, FAD of *Jinogondolella nankingensis*	
Kungurian	*candidate Pequop Mtns., Nevada, U.S.A*			*Near FAD of conodont Neostreptognathodus pnevi*	
Artinskian	*candidate Usolka section, Russia*			*FAD of conodont Sweetognathus aff. whitei*	
Sakmarian	*candidate Usolka section, Russia*			*Near FAD of conodont Mesogondolella uralensis or Sweetognathus aff. merrilli*	
Asselian	Aidaralash Creek, Kazakhstan	50°14'45"N 57°53'29"E*	27m above the base of Bed 19	Conodont, FAD of *Streptognathus isolatus*	Episodes **21**/1, 1998

* according to Google Earth

Figure 10.7 Ratified GSSPs and potential primary markers under consideration for defining the Permian stages. (*status as of early 2016*). (Details of each GSSP are available at http://permian.stratigraphy.org, https://engineering.purdue.edu/Stratigraphy/gssp/, and in the *Episodes* publications.)

the mid-Artinskian through mid-Capitanian stages. If the volumes of the Permian glacial ice episodes fluctuated in response to Milankovitch cycles, then the intervals with P2 (Artinskian), P3 (Guadalupian), and P4 (lower Wuchiapingian Stage) may yield a 405-kyr and 100-kyr cyclostratigraphy to scale the majority of the within-stage events, similar to what has been accomplished during P1 (e.g., Schmitz and Davydov, 2012). The P3–P4 global climate cycles, coupled with a better-calibrated Lopingian magnetostratigraphy, will also enable a better correlation of Late Permian terrestrial deposits and the evolution of tetrapods.

Estimated uncertainties on assigned ages on stage boundaries

The high-precision radioisotopic dates with well-constrained biostratigraphic ages that constrain the Permian timescale typically have a published uncertainty less than 0.2 myr (e.g., Schmitz and Davydov, 2012). However, an external uncertainty of ca. 0.29 myr should be included if comparing these Permian EARTHTIME-standardized dates to other dating methods, as explained in Burgess et al. (2014). Those are the minimum uncertainties that apply to the bases of the Asselian, Sakmarian, and Artinskian stages of the Cisuralian and to the bases of the Wuchiapingian and Changhsingian stages of the Lopingian. The duration of those stages, if computed from the dates in the same EARTHTIME data sets, have uncertainties that omit that external factor.

In contrast, the extrapolated ages for the Kungurian, Roadian, Wordian, and Capitanian stages probably have a 2-myr uncertainty, based on the comparison given previously of the different published methods for estimating the base-Kungurian age. That empirical uncertainty of 2 myr is higher than the ca. 0.5-myr uncertainties computed by the simple statistics of the spline-fit technique in GTS2012.

Acknowledgments

This brief summary of the Permian gave only a few of the main aspects, selected highlights, and some of the current stratigraphic issues. A detailed overview and synthesis is by Henderson et al. (2012a), and updates are at the website and the *Permophiles* newsletters of the Subcommission on Permian Stratigraphy (*permian.stratigraphy.org*). This review would not have been possible without the education and advice over the past two decades from Vladimir Davydov, Charles Henderson, Lance Lambert, John Laurie, Daniel Mantle, Bob Nicoll, Mark Schmitz, Shu-zhong Shen, and Bruce Wardlaw (*in alphabetical order only*); and from field experience with Nanjing and Wuhan colleagues studying the Permian of South China. Zhong-Qiang Chen reviewed and greatly enhanced the graphics and text.

Selected publications and websites

Cited publications

Only select publications were cited in this review with an emphasis on aspects of post-2011 updates. Pre-2011 literature is well summarized in the synthesis by Henderson et al. (2012) and in some of the publications cited in the following.

Brayard, A., Bucher, H., 2015. Chapter 17. Permian-Triassic extinctions and rediversifications. In: Klug, C., Korn, D., De Baets, K., Kruta, I., Mapes, R.H. (Eds.), *Ammonoid Paleobiology: From Macroevolution to Paleogeography*. Topics in Geobiology, **44**. Springer Publications, pp. 465–473. http://dx.doi.org/10.1007/978-94-017-9633-0_13.

Buggisch, W., Wang, X.D., Alekseev, A.S., Joachimski, M.M., 2011. Carboniferous-Permian carbon isotope stratigraphy of successions from China (Yangtze platform), USA (Kansas) and Russia (Moscow Basin and Urals). *Palaeogeography, Palaeoclimatology, Palaeoecology* **301**: 18–38.

Buggisch, W., Krainer, K., Schaffhauser, M., Joachimski, M., Korte, C., 2015. Late Carboniferous to Late Permian carbon isotope stratigraphy: a new record from post-Variscan carbonates from the Southern Alps (Austria and Italy). *Palaeogeography, Palaeoclimatology and Palaeoecology* **433**: 174–190.

Burgess, S.D., Bowring, S., Shen, Z.Q., 2014. High-precision timeline for Earth's most severe extinction. *Proceedings of the National Academy of Science (PNAS)* **111**: 3316–3321.

Chen, B., Joachimski, M.M., Shen, S.Z., Lambert, L.L., Lai, X.L., Wang, X.D., Chen, J., Yuan, D.X., 2013. Permian ice volume and palaeoclimate history: oxygen isotope proxies revisited. *Gondwana Research* **24**: 77–89.

Chen, Z.Q., Jin, Y.G., Shi, G.R., 1998. Permian transgression-regression sequences and sea-level changes of South China. *Proceedings of the Royal Society of Victoria* **110**: 345–367.

Chen, Z.Q., George, A.D., Yang, W.R., 2009. Effects of Middle-Late Permian sea-level changes and mass extinction on the formation of skeletal mound from Laibin area. South China. *Australian Journal of Earth Sciences* **56**: 745–763.

Chen, Z.Q., Yang, H., Luo, M., Benton, J.J., Kaiho, K., Zhao, L., Huang, Y., Zhang, K., Fang, Y., Jiang, H., Qiu, H., Li, Y., Tu, C., Shi, L., Zhang, L., Feng, X., Chen, L., 2015. Complete biotic and sedimentary records of the Permian–Triassic transition from Meishan section, South China: Ecologically assessing mass extinction and its aftermath. *Earth-Science Reviews* **149**: 67–107. http://dx.doi.org/10.1016/j.earscirev.2014.10.005.

Chernykh, V.V., Chuvashuv, B.I., Shen, S.-Z., Henderson, C.M., 2013. Proposal for the Global Stratototype Section and Point (GSSP) for the base-Sakmarian Stage (lower Permian). *Permophiles* **58**: 16–26. Available at: http://permian.stratigraphy.org.

Chernykh, V.V., Chuvashuv, B.I., Davydov, V.I., Schmitz, M.D., 2012. Mechetlino section: a candidate for the Global Stratototype Section and Point (GSSP) of the Kungurian stage (Cisuralian, lower Permian). *Permophiles* **58**: 16–26. Available at: http://permian.stratigraphy.org.

Chuvashuv, B.I., Chernykh, V.V., Shen, S.Z., Henderson, C.M., 2013. Proposal for the Global Stratototype Section and Point (GSSP) for the base-Sakmarian Stage (lower Permian). *Permophiles* **58**: 26–34. Available at: http://permian.stratigraphy.org.

Davydov, V.I., Korn, D., Schmitz, M.D., Gradstein, F.M., Hammer, O., 2012. The Carboniferous Period. In: Gradstein, F.M., Ogg, J.G., Schmitz, M., Ogg, G.M. (Coordinators), *The Geologic Time Scale 2012*. Elsevier Publications, pp. 603–651. http://dx.doi.org/10.1016/B978-0-444-59425-9.00023-8.

Davydov, V.I., 2013. The GSSP at the Aidaralash section is solid and has no alternative. *Permophiles* **58**: 13–15. see also the immediately following article: Lucas, S.G., The Aidaralash GSSP – Reply to Davydov. *Permophiles*, **58**: 15–16. Available at: http://permian.stratigraphy.org.

Fielding, C.R., Frank, T.D., Birgenheier, L.P., Rygel, M.C., Jones, A.T., Roberts, J., 2008. Stratigraphic imprint of the Late Palaeozoic Ice Age in eastern Australia: a record of alternating glacial and non-glacial climate regime. *Journal of the Geological Society of London* **165**: 129–140.

Frank, T.D., Schultis, A.I., Fielding, C.R., 2015. Acme and demise of the late Palaeozoic ice age: a view from the southeastern margin of Gondwana. *Palaeogeography, Palaeoclimatology and Palaeoecology* **418**: 176–192.

Haq, B.U., Schutter, S.R., 2008. A chronology of Paleozoic sea-level changes. *Science* **322**: 64–68. http://dx.doi.org/10.1126/science.116164.

Henderson, C.M., Davydov, V.I., Wardlaw, B.R., 2012a. The Permian Period. In: Gradstein, F.M., Ogg, J.G., Schmitz, M., Ogg, G.M. (Coordinators), *The Geologic Time Scale 2012*. Elsevier Publication, pp. 653–679. http://dx.doi.org/10.1016/B978-0-444-59425-9.00023-8 (An overview on all aspects, including graphics on the ratified GSSPs of the stages, diagrams and tables for the biostratigraphic scales, and discussion on the age models).

Henderson, C.M., Wardlaw, B.R., Davydov, V.I., Schmitz, M.D., Schiappa, T.A., Tierney, K.E., Shen, S., 2012b. Proposal for base-Kungurian GSSP. *Permophiles* **56**: 8–21. Available at: http://permian.stratigraphy.org.

Isozaki, Y., Kawahata, H., Ota, A., 2007. A unique carbon isotope record across the Guadalupian–Lopingian (Middle–Upper Permian) boundary in mid-oceanic paleoatoll carbonates: the high-productivity "Kamura event" and its collapse in Panthalassa. *Global and Planetary Change* **55**: 21–38.

Jiang, H.S., Lai, X., Luo, G., Aldridge, R., Zhang, K., Wignall, P.B., 2007. Restudy of conodont zonation and evolution across the P/T boundary at Meishan section, Changxing, Zhejiang, China. *Global and Planetary Change* **55**: 39–55.

Kirschvink, J.L., Isozaki, Y., Shibuya, H., Otofuji, Y., Raub, T.D., Hilburn, I.A., Kasuya, T., Yokoyama, M., Bonifacie, M., 2015. Challenging the sensitivity limits of paleomagnetism: magnetostratigraphy of weakly magnetized Guadalupian–Lopingian (Permian) limestone from Kyushu, Japan. *Palaeogeography, Palaeoclimatology, Palaeoecology* **418**: 75–89.

Korn, D., Klug, C., 2015. Chapter 12. Paleozoic ammonoid biostratigraphy. In: Klug, C., Korn, D., De Baets, K., Kruta, I., Mapes, R.H. (Eds.), *Ammonoid Paleobiology: From Macroevolution to Paleogeography*. Topics in Geobiology, **44**. Springer Publications, pp. 299–328. http://dx.doi.org/10.1007/978-94-017-9633-0_13.

Large, R.R., Halpin, J.A., Lounejeva, E., Danyushevsky, L.V., Malsennikov, V.V., Gregory, D., Sack, P.J., Haines, P.W., Long, J.A., Makoundi, C., Stepamov, S., 2015. Cycles of nutrient trace elements in the Phanerozoic ocean. *Gondwana Research* **28**: 1282–1293.

Lucas, S.G., 2006. Global Permian tetrapod biostratigraphy and biochronology. In: Lucas, S.G., Cassinis, G., Schneider, J.W. (Eds.), *Non-Marine Permian Biostratigraphy and Biochronology* Geological Society of London Special Publication, **263**, pp. 65–93.

Lucas, S.G., 2013a. No gap in the Middle Permian record of terrestrial vertebrates: COMMENT. *Geology* **41**: e293. http://dx.doi.org/10.1130/G33734C.1 (online comment/reply).

Lucas, S.G., 2013b. We need a new GSSP for the base of the Permian. *Permophiles* **58**: 8–13. Available at: http://permian.stratigraphy.org.

Metcalfe, I., Crowley, J.L., Nicoll, R.S., Schmitz, M., 2015. High-precision U-Pb CA-TIMS calibration of Middle Permian to Lower Triassic sequences, mass extinction and extreme climate-change in eastern Australian Gondwana. *Gondwana Research* **28**: 61–81.

Nicoll, R.S., McKellar, J., Ayaz, S.A., Laurie, J.R., Esterlie, J., Crowley, J.L., Wood, G., Bodorkos, S., 2015. CA-IDTIMS dating of tuffs, calibration of palynostratigraphy and stratigraphy of the Bowen and Galilee basins. In: *Bowen Basin Symposium, Brisbane, 2015, Conference Volume*, p. 211.

Schmitz, M.D., Davydov, V.I., 2012. Quantitative radiometric and biostratigraphic calibration of the Pennsylvanian-Early Permian (Cisuralian) time scale, and pan-Euramerican chronostratigraphic correlation. *Geological Society of America Bulletin* **124**: 549–577. http://dx.doi.org/10.1130/B30385.1.

Scotese, C.R., 2014. Atlas of middle & late Permian and Triassic paleogeographic maps (Mollweide projection), maps 43–48, volume 3, Jurassic and Triassic, and maps 49–52 volume 4, the late Paleozoic. In: *PALEOMAP PaleoAtlas for ArcGIS (Late Paleozoic)*. PALEOMAP Project, Evanston, IL. https://www.academia.edu/11300143/Atlas_of_Middle_and_Late_Permian_and_Triassic_Paleogeographic_Maps.

Shen, S.Z., Henderson, C.M., Bowring, S.A., Cao, C.Q., Wang, Y., Wang, W., Zhang, H., Zhang, Y.C., Mu, L., 2010. High resolution Lopingian (late Permian) timescale of South China. *Geological Journal* **45**: 122–134.

Shen, S.-Z., Cao, C.-Q., Zhang, H., Bowring, S.A., Henderson, C.M., Payne, J.L., Davydov, V.I., Chen, B., Yuan, D.-X., Zhang, Y.-C., Wang, W., Zheng, Q.-F., 2013a. High-resolution δ13C$_{carb}$ chemostratigraphy from latest Guadalupian through earliest Triassic in South China and Iran. *Earth and Planetary Science Letters* **375**: 156–165.

Shen, S.-Z., Schneider, J.W., Angiolini, L., Henderson, C.M., 2013b. The international Permian timescale: March 2013 update. In: Lucas, S.G., DiMichele, W., Barrick, J.E., Schneider, J.W., Spielmann, J.A. (Eds.), *The Carboniferous-Permian Transition* New Mexico Museum of Natural History and Science Bulletin, **60**, pp. 411–416.

Shen, S.-Z., Henderson, C.M., et al., June 2015. Permian timescale. Synthesis diagram in: *Permophiles*. **375**: 40. http://permian.stratigraphy.org/files/20150630111341036.pdf, most recent version at: http://permian.stratigraphy.org/per/per.asp.

Shi, L., Feng, Q., Shen, J., Ito, T., Chen, Z.Q., 2016. Proliferation of shallow-water radiolarians coinciding with enhanced oceanic productivity in reducing conditions during the Middle Permian, South China: evidence from the Gufeng Formation of western Hubei Province. *Palaeogeography, Palaeoclimatology, Palaeoecology* **444**: 1–14.

Steiner, M.B., 2006. The magnetic polarity time scale across the Permian-Triassic boundary. In: Lucas, S.G., Cassinis, G., Schneider, J.W. (Eds.), *Non-Marine Permian Biostratigraphy and Biochronology* Geological Society of London Special Publication, **263**, pp. 15–38.

Veizer, J., Prokoph, A., 2015. Temperatures and oxygen isotopic composition of Phanerozoic oceans. *Earth-Science Reviews* **146**: 92–104.

Wignall, P.B., Sun, Y., Bond, D.P.G., Izon, G., Newton, R.J., Védrine, S., Widdowson, M., Ali, J.R., Lai, X., Jiang, H., Cope, H., Bottrell, S.H., 2009. Volcanism, mass extinction, and carbon isotope fluctuations in the Middle Permian of China. *Science* **324**: 1179–1182.

Wignall, P., 2015. *The Worst of Times: How Life on Earth Survived Eighty Million Years of Extinctions*. Princeton University Press. 224 pp.

Wu, H., Zhang, S., Hinnov, L.A., Jiang, G., Feng, Q., Li, H., Yang, T., 2013. Time-calibrated Milankovitch cycles for the late Permian. *Nature Communications* **4**: 2452. http://dx.doi.org/10.1038/ncomms3452.

Yang, J., Cawood, P.A., Due, Y., 2015. Voluminous silicic eruptions during late Permian Emeishan igneous province and link to climate cooling. *Earth and Planetary Science Letters* **432**: 166–175.

Yin, H., Zhang, K., Tong, J., Yang, Z., Wu, S., 2001. The Global Stratotype Section and Point (GSSP) of the Permian–Triassic boundary. *Episodes* **24**(2): 102–114.

Yuan, D.-X., Shen, S.-Z., Henderson, C.M., Chen, J., Zhang, H., Feng, H.-Z., 2014. Revised conodont-based integrated high-resolution timescale for the Changhsingian Stage and end-Permian extinction interval at the Meishan sections, South China. *Lithos* **204**: 220–245.

Websites (selected)

Subcommission on Permian Stratigraphy (International Commission on Stratigraphy [ICS])—*http://permian.stratigraphy.org*—Details on GSSPs and their conodont taxa used for global correlation; plus the *Permophiles* newsletter (Portable Document Formats [PDFs] of the full set, 1978 to present).

Palaeos: Permian—*http://palaeos.com/paleozoic/permian/permian.htm*—A well-presented suite of diverse topics for a general science audience that was originally compiled by M. Alan Kazlev in 1998–2002.

11

TRIASSIC

222.6 Ma Triassic

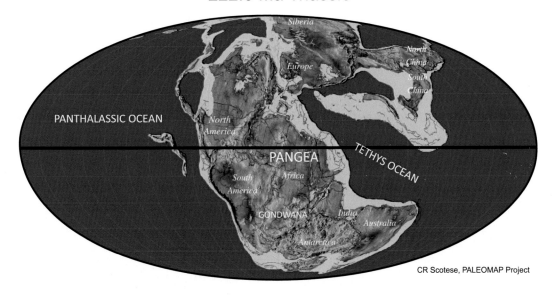

CR Scotese, PALEOMAP Project

Carnian paleogeographic reconstruction (Sea level + 40 m) from Scotese (2014).

Basal definition and international subdivisions

The Mesozoic Era is divided into the Triassic, Jurassic, and Cretaceous periods. The Triassic was named from a trio (*trias*) of widespread terrestrial to shallow-marine formations in Germany; however, the majority of its stages and commonly used substages were originally defined in the uplifted Tethyan ammonite-rich oceanic sediments exposed in the Alpine–Himalaya chain (e.g., Tozer, 1984). International research during the past decade has established a high-resolution time scale and revealed major global disruptions of the

Earth system through the Triassic. Only some select examples will be briefly cited here.

The Triassic is bounded by mass extinctions that coincide with enormous outpourings of volcanic flood basalts. The bases of the Triassic and of the Jurassic are defined by the initial recovery of marine life from each of those environmental catastrophes. The Subcommission on Triassic Stratigraphy of the International Commission on Stratigraphy (ICS) has adopted seven international Triassic stages. Except for the Rhaetian, these stages have a semistandardized nomenclature of substages. The traditional definitions for these stages (and substages) were based on lowest occurrences of ammonoid genera within exposures in the Alpine, Mediterranean, or Himalayan exposures. However, some candidates for the unassigned Global Boundary Stratotype Sections and Points (GSSPs) for the global stages utilize the phosphatic teeth of the enigmatic conodonts, which provide a more widespread method of correlation (Figs. 11.1 and 11.7). However, the possible apparent diachroneity in interregional appearances of these conodont markers, their possible offsets from the traditional lowest ammonoid zones of substages, and the taxonomic disagreements with the potential marker conodonts are among the factors that have contributed to the delays in formalizing the GSSPs for half of the Triassic stages and standardized recognition of substages. Other factors include difficulties in finding suitable sections that have multiple methods for global correlation and in achieving reliable correlations between tropical Tethyan and cooler Boreal marine realms (e.g., summaries in Konstantinov and Klets, 2009). As with most of the other stage boundaries in the geologic time scale that have not been formally defined, the controversies lie in the details on how to actually achieve reliable global correlations.

The Permian ended in a mass extinction of up to 80% of marine genera; and this catastrophe is generally considered to be caused by the initial phases of the eruption of the Siberian Trap volcanic complex (e.g., reviews by Erwin, 2006; Chen et al., 2014; Wignall, 2015). The GSSP for the base of the Triassic (base of **Induan** stage, Griesbachian substage) was selected in the Meishan section of Zhejiang Province in South China to coincide with the lowest occurrence of the conodont, *Hindeodus parvus* (e.g., Yin et al., 2001; and see historical review by Baud, 2014) (Fig. 11.2). This level followed a pronounced brief negative excursion in carbon isotopes (up to -6 $\delta^{13}C_{carb}$ relative to the late-Changhsingian *Clarkina yini* conodont zone) and is within a

rapid ~9°C increase in local sea-surface temperature (reviewed by Chen et al., 2015) (see Permian Fig. 10.1). This GSSP level is bracketed by volcanic ash beds, thereby enabling precise radioisotopic ages for both the main end-Permian mass extinction (251.950 ± 0.042 Ma for the abrupt decline in $\delta^{13}C_{carb}$) and the base of the Triassic (251.902 ± 0.024 Ma) (Burgess et al., 2014; however, an external uncertainty of ca. 0.3 myr should be included if comparing to non-EARTHTIME U-Pb dating methods). The Meishan section also hosts the GSSP for the underlying Changhsingian Stage of uppermost Permian, and is within a special GeoPark that includes a museum of Earth's history.

As of 2015, the following stages await international definition.

Olenekian (mid-Lower Triassic): The traditional Induan–Olenekian boundary was at a relatively sharp change to ammonoid assemblages with *Hedenstroemia*, *Euflemingites*, *Flemingites*, and other genera. This turnover in ammonoids is at approximately the lowest occurrence of the conodont *Novispathodus waageni* (its *Nov. eowaageni* subspecies), and seems to be at or just below an upward change from a long-duration reversed-polarity to a brief normal-polarity magnetic zone. Candidates include West Pingdingshan section near Chaohu town in Anhui Province of South China (first proposed in 2004; and used here to make the age estimate based on its cyclomagnetostratigraphy) and Muth section of Spiti region of the Himalayas in northwest India (proposed in 2005). However, a decision on the base-Olenekian definition awaits how these Tethyan-based events can be correlated to the Boreal realm (Triassic Subcommission section of ICS annual report, 2014).

Anisian (Lower-Middle Triassic boundary): The Olenekian–Anisian boundary in ammonoid zonations was placed at the appearance of *Paracrochordiceras* (closely followed by *Japonites*) and *Karangatites* genera. In the candidate GSSP at Deşli-Caira Hill in North Dobrogea of Romania, this level is slightly

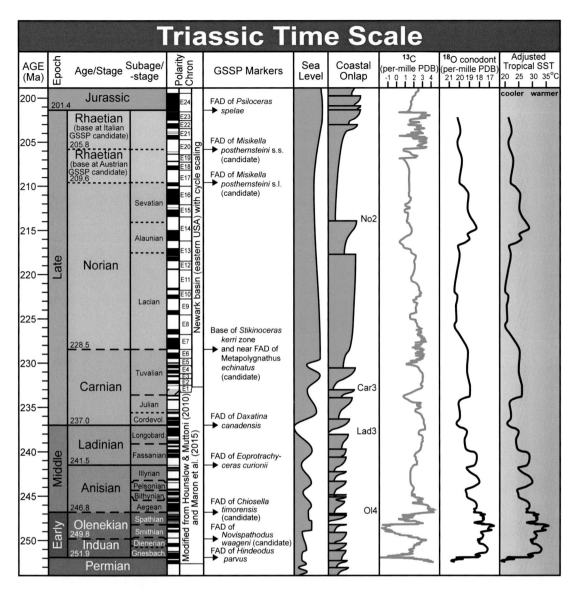

Figure 11.1 Triassic overview. Main markers or candidate markers for GSSPs of Triassic stages are a first-appearance datum (FAD) of ammonite taxa as detailed in the text and in Fig. 11.7. ("Age" is the term for the time equivalent of the rock-record "stage.") See text for Rhaetian-stage boundary options. Magnetic polarity scales are a composite of Induan through early Carnian from Hounslow and Muttoni (2010) and of the Newark Basin (Kent and Olsen, 1999) with estimated calibrations to late Carnian through Rhaetian following the "long-Rhaetian" option (Ogg, 2012; Ogg et al., 2014) as partly recalibrated by Maron et al. (2015). These magnetic polarity scales are enlarged and detailed in Figs. 11.4–11.6. Coastal onlap and schematic sea-level curve with labels for selected major sequence boundaries are modified from Haq and Al-Qahtani (2005) following advice of Bilal Haq (pers. comm., 2006); it is possible that the very long sequences of lower Norian and uppermost-Norian through early Rhaetian contain other significant sea-level excursions. The δ13C curve is a merger of generalized trends with relative magnitudes of Early Triassic to earliest Ladinian from Sun et al. (2012) and late Ladinian through Rhaetian from Muttoni et al. (2014). The Triassic δ18O curve from conodont apatite (inverted scale) has the Early Triassic from Sun et al. (2012) modified to smoothly fit the Middle and Late Triassic trend from Trotter et al. (2015); and the tropical sea-surface temperatures derived from those oxygen-18 values are adjusted following Veizer and Prokoph (2015). The vertical scale of this diagram is standardized to match the vertical scales of the first stratigraphic summary figure in all other Phanerozoic chapters. *PDB*, PeeDee Belemnite 13C and 18O standard; *SST*, sea-surface temperature.

Base of the Induan Stage of the Triassic System at Meishan, China

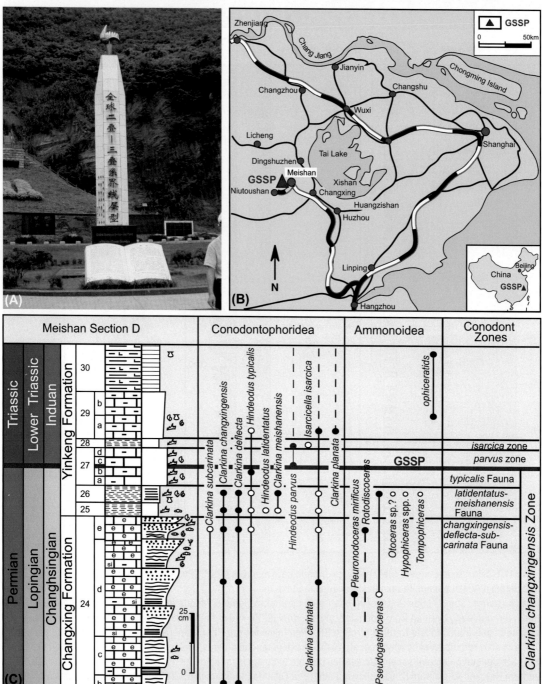

Figure 11.2 GSSP for base of the Triassic (base of Mesozoic; base of Induan Stage) at the Meishan section, South China. The GSSP level coincides with the lowest occurrence of conodont *Hindeodus parvus*, and is within a major negative excursion in carbon isotopes. Note that vertical scale is in centimeters to show details of this boundary interval. Conodont ranges are from the ratified GSSP document (Yin et al., 2001), but later studies have revised the assignments (see Chen et al., 2015; and Fig. 10.6 in the Permian chapter). Zircons from the altered volcanic ash components in Bed 25 and Bed 28 have been dated (eg, Burgess et al., 2014). The main end-Permian mass extinction event is at the base of Bed 25. The GSSP is preserved within a GeoPark, with the GSSP site located at the top of the stairs on the right side of the decorative wall.

above the lowest occurrence of conodont *Chiosella timorensis* and is close to the base of a brief normal-polarity zone ("MT1n" of Hounslow and Muttoni, 2010) (Gradinăru et al., 2007). However, the Romania section might have condensation in the boundary interval. If the lowest occurrence of *Ch. timorensis* is selected as the global marker, then the uppermost part of the ammonoid *Neopopanoceras haugi* Zone "latest Olenekian" in Tethyan realm continues into basal Anisian. In South China, the succession of *Ch. timorensis* and other conodont datums within the boundary interval are interbedded with volcanic ash beds. Although the individual zircons within each of those beds have a wide range of U-Pb radioisotopic dates, the suites enable estimates of the approximate ages for the various datums with a mean near 247 Ma (Lehrmann et al., 2015; Ovtcharova et al., 2015).

The Dolomites of northern Italy host the ratified **Ladinian** GSSP at Bagolino, where it coincides with the base of the *Eoprotrachyceras curionii* Zone (lowest occurrence of *Eoprotrachyceras* ammonoid genus, which is the onset of the Trachyceratidae ammonoid family). The Dolomites also host the ratified **Carnian** GSSP, which coincides with the lowest occurrence of *Daxatina* ammonoid genus (Fig. 11.3).

Norian (Upper Triassic): The Carnian–Norian boundary interval has no recognized major events of global significance. The boundary is traditionally placed at the base of the ammonoid *Guembelites jandianus* Zone in the Tethyan realm and at the base of the ammonoid *Stikinoceras kerri* Zone in western Canada. These levels may be approximately coincident (e.g., Jenks et al., 2015). The lowest occurrences of the conodont group of *Metapolygnathus communisti* and range of *Meta. echinatus* (reclassified as *Meta. parvus* by Orchard, 2014) are calibrated to ammonoid zones at the candidate GSSP at Black Bear Ridge in British Columbia, where the traditional base of the *S. kerri* Zone is aligned with

the top of the conodont *Meta. parvus* subzone (Orchard, 2010, 2014, pers. comm. January 2016). The other leading candidate is Pizzo Mondello in Sicily, which also has a detailed magnetostratigraphy (e.g., Balini et al., 2012; Hounslow and Muttoni, 2010). However, the conodont faunas at these two candidates appear different and difficult to correlate, so potentially the lowest occurrence of bivalve *Halobia austriaca* will be used as a primary marker, if it can be reliably correlated between the two sections (Triassic Subcommission section of ICS annual report, 2014).

Rhaetian (uppermost Triassic): The Norian–Rhaetian boundary working group selected the lowest occurrence of the conodont *Misikella posthernsteini* as the preferred marker for the Rhaetian GSSP. This conodont has a transition from *Misikella hernsteini* and continues upward to nearly the end of the Rhaetian. Even though this morphogenesis is seen worldwide, it appears that the distinctions between the component taxa are not standardized (e.g., discussions in Rigo et al., 2016; Lucas, 2016; Orchard, 2016). In the candidate GSSP section of Steinbergkogel in Austria, the interpreted base of *Mi. posthernsteini* is just above a change from a major normal-polarity magnetozone upward to a reversed-polarity-dominated magnetozone. In contrast, at the second candidate section of Pignola–Abriola in Sicily, the interpreted base of the same *Mi. posthernsteini* taxon is very high within a reversed-polarity-dominated magnetozone (Maron et al., 2015; Rigo et al., 2016). It appears that the finer features of both of these reversed-dominated magnetozones can be reliably correlated between Austria and Sicily. Therefore, the proposed correlation of these Rhaetian magnetostratigraphies to the astronomical-cycle-scaled magnetic polarity reference pattern of the Newark Basin implies nearly a 4-myr offset of the interpreted bases of *Mi. posthernsteini* between those two candidate GSSPs (Maron et al., 2015; Rigo et al., 2016). To explain this

Base of the Carnian Stage of the Triassic System in the Prati di Stuores/Stuores Wiesen Section, near San Cassiano, Italy

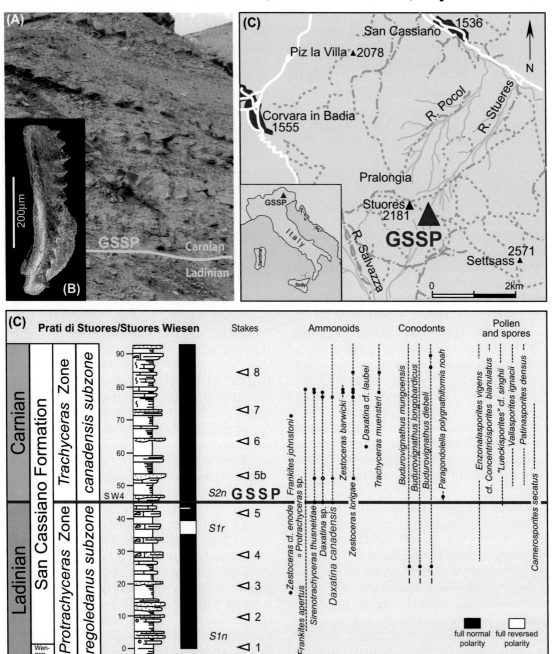

Figure 11.3 GSSP for base of the Upper Triassic (base of Carnian Stage) at the Prati di Stuores (Stuores Wiesen) section, Dolomites region of north Italy. The GSSP level coincides with the lowest occurrence of the cosmopolitan ammonoid *Daxatina* (base of *Daxatina canadensis* Subzone, lowest subzone of *Trachyceras* Zone), and is near the lowest occurrence of conodont *Paragondolella polygnathiformis* (inset "B," which is reassigned to *Quadralella* genus by Orchard [pers. comm., January 2016]). (Photographs of outcrop and conodont provided by Manuel Rigo.)

discrepancy, Maron et al. (2015) proposed that the interpreted earliest form of *Mi. posthernsteini* at the Austria GSSP candidate is an initial transitional form ("*sensu lato*") at ca. 209.5 Ma, whereas the interpreted lowest *Mi. posthernsteini* at the Sicily GSSP is the developed form ("*sensu stricto*") with an estimated age of 205.7 Ma. They consider that the higher level (and shorter Rhaetian) is more consistent with the traditional recognition of the base of the Rhaetian in other regions. Both options are shown on the figures in this chapter (Figs. 11.1 and 11.5).

Selected main stratigraphic scales and events

(1) Biostratigraphy (marine; terrestrial)

Detailed discussions of each major Triassic biostratigraphic scale are in the different chapters authored by specialists in *The Triassic Timescale* (Lucas, 2010a). Ammonoid zonations from the Alps, Canada, and Siberia have been the primary standards for subdividing the Triassic, and the correlation of these regional zonations using other stratigraphic tools has been a focus of Triassic workers (e.g., Jenks et al., 2015; Konstantinov and Klets, 2009) (Figs. 11.4–11.6).

Conodonts are generally more widespread than ammonites both in paleogeography and in different marine facies; and are approaching a well-defined taxonomy and biostratigraphy (e.g., Anisian-through-Carnian synthesis by Chen et al., 2016; discussions between Orchard, 2016; and Lucas, 2016). However, as indicated by the examples given previously for potential GSSPs, the lack of standardization of taxonomic concepts among conodont specialists for some taxa (e.g., Fig. 10.6 in Permian chapter), rarity of some primary markers, and possible apparent diachroneity among their preserved records

can be problems for high-resolution biostratigraphic correlation and for comparison to older literature.

The distinctive thin-shelled bivalves of the *Daonella*, *Halobia*, and *Monotis* genera are important for subdividing the Middle and Late Triassic (McRoberts, 2010). Radiolarians and ostracods are underutilized biostratigraphic tools; and calcareous nannofossils have great potential after their initial appearance in late Carnian.

During much of the Triassic, the sedimentary record across the Pangea supercontinent was dominated by terrestrial deposits, therefore widespread conchostracan (clam shrimp), ostracod, spore pollen, and tetrapod remains are important for global correlation, although their precise correlation to the marine-based subdivisions is a work in progress. The major dinosaur lineages became established in the Carnian; and these became the dominant land reptiles in the Norian (e.g., Lucas, 2010b; Benton et al., 2014).

(2) Magnetic stratigraphy

The Triassic magnetic polarity reference scales proposed for biostratigraphically dated marine successions (e.g., composite synthesis by Hounslow and Muttoni, 2010) and for astronomically scaled terrestrial basins (e.g., Early Triassic by Szurlies, 2007; Late Triassic by Kent and Olsen, 1999) have been verified by extensive conodont-dated magnetostratigraphy in European and Chinese sections (e.g., Muttoni et al., 2014; Maron et al., 2015; Lehrmann et al., 2015) including cycle scaling of significant intervals (e.g., Li et al., 2016; Zhang et al., 2015). As noted previously, the comparison between the magnetostratigraphies of marine sections and of cycle-scaled terrestrial sections appear to have resolved some of the uncertainties about the age models for the Late Triassic and for the Early Triassic and Anisian. For example, the "long Rhaetian-short Tuvalian" option that was preferred for

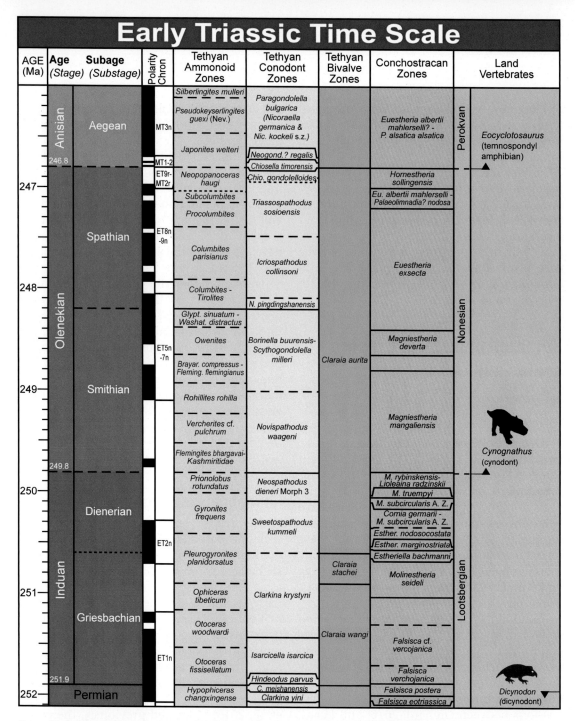

Figure 11.4 Selected marine and terrestrial biostratigraphic zonations of the Early Triassic. ("*Age*" is the term for the time equivalent of the rock-record "*stage*".) Magnetic polarity zones with nomenclature of Hounslow and Muttoni (2010) are scaled to astronomical cycles (e.g., Szurlies, 2004, 2007; Li et al., 2016) and calibrated to conchostracan zones of the Germanic Basin (Kozur and Weems, 2010, 2011) and Tethyan conodont zones as used in South China (e.g., Orchard in Lehrmann et al., 2015; with their general Tethyan zones are shown here). [*Clarkina krystyni* is classified as *Neoclarkina krystyni* by some experts (M. Orchard, pers. comm., 2016)]. Tethyan ammonoid zones are from Jenks et al. (2015), and Tethyan bivalve zones are from McRoberts (2010). Land vertebrate zones and markers are from Lucas (2010b) and Lucas and Tanner (2015). Additional zonations, biostratigraphic markers, geochemical trends, sea-level curves, and details on calibrations are compiled in the internal data sets within the *TimeScale Creator* visualization system (free at www.tscreator.org).

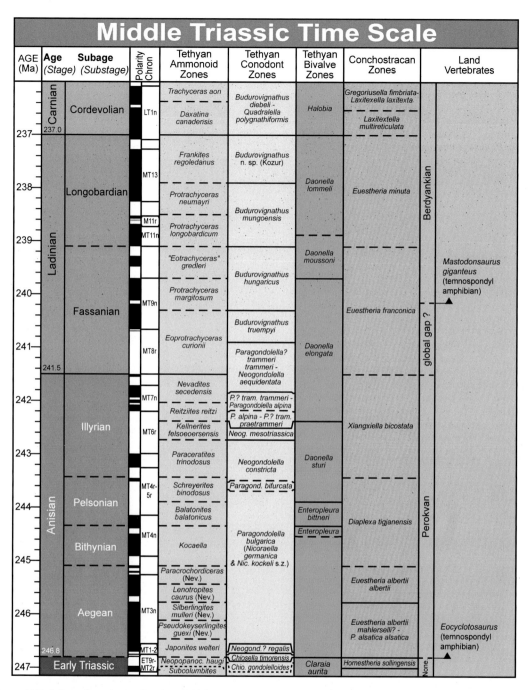

Figure 11.5 Selected marine and terrestrial biostratigraphic zonations of the Middle Triassic. *(Note that this scale is more compact than in* Fig. 11.4.*)* Magnetic polarity zones with nomenclature of Hounslow and Muttoni (2010) are scaled to astronomical cycles within Anisian (Li et al., in prep.), and are scaled relative to an arbitrary quasiequal-ammonite-subzone scaling for the Ladinian. Tethyan ammonoid zones are from Jenks et al. (2015); Tethyan conodont zones are modified from Orchard (in Lehrmann et al., 2015) and Kozur (2003). [Conodont *Budurovignathus praehungaricus* appears near base of Ladinian (M. Orchard, pers. comm., 2016); and *Bud. supralongobardica* of Kozur (2003) was reclassified by him as "n. sp. Kozur" pending full description (H. Kozur, pers. comm., 2006).] Tethyan bivalve zones are from McRoberts (2010). Conchostracan zones of the Germanic Basin are from Kozur and Weems (2010, 2011), and land vertebrate zones and markers are from Lucas (2010b) and Lucas and Tanner (2015). Additional zonations, biostratigraphic markers, geochemical trends, sea-level curves, and details on calibrations are compiled in the internal data sets within the *TimeScale Creator* visualization system (free at www.tscreator.org).

Late Triassic Time Scale

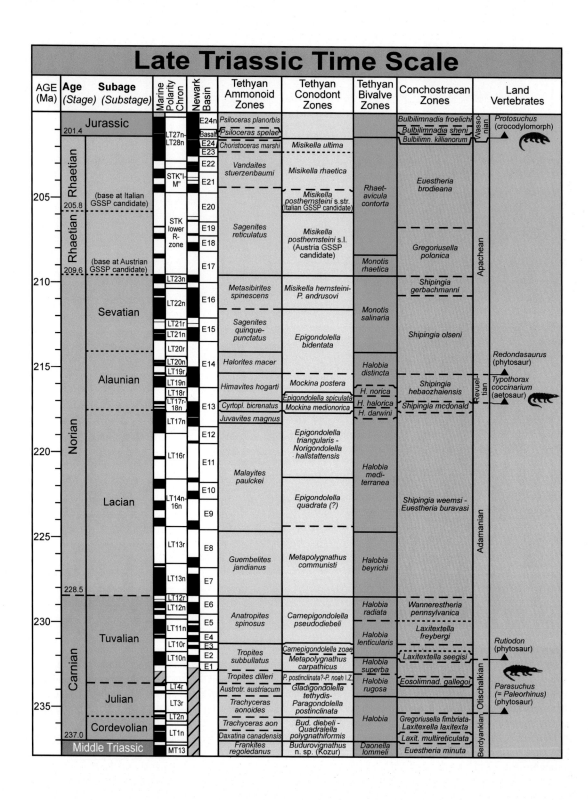

AGE (Ma)	Age (Stage)	Subage (Substage)	Marine Polarity Chron	Newark Basin	Tethyan Ammonoid Zones	Tethyan Conodont Zones	Tethyan Bivalve Zones	Conchostracan Zones		Land Vertebrates
	Jurassic		LT27n–LT28n	E24n Basalt E24 E23	Psiloceras planorbis / Psiloceras spelae / Choristoceras marshi	Misikella ultima		Bulbilimnadia froelichi / Bulbilimnadia sheni / Bulbilimn. killianorum	Vassonian	Protosuchus (crocodylomorph)
205	Rhaetian		STK"I-M" STK lower R-zone	E22 E21 E20 E19 E18 E17	Vandaites stuerzenbaumi / Sagenites reticulatus	Misikella rhaetica / Misikella posthernsteini s.str. (Italian GSSP candidate) / Misikella posthernsteini s.l. (Austria GSSP candidate)	Rhaet-avicula contorta / Monotis rhaetica	Euestheria brodieana / Gregoriusella polonica	Apachean	
205.8		(base at Italian GSSP candidate)								
209.6		(base at Austrian GSSP candidate)								
210	Norian	Sevatian	LT23n LT22n LT21r LT21n LT20r LT20n	E16 E15	Metasibirites spinescens / Sagenites quinque-punctatus	Misikella hernsteini-P. andrusovi / Epigondolella bidentata	Monotis salinaria	Shipingia gerbachmanni / Shipingia olseni		
215		Alaunian	LT19r LT19n LT18r LT17r-18n LT17n	E14 E13	Halorites macer / Himavites hogarti / Cyrtopl. bicrenatus / Juvavites magnus	Mockina postera / Epigondolella spiculata / Mockina medionorica	Halobia distincta / H. norica / H. halorica / H. darwini	Shipingia hebaozhaiensis / Shipingia mcdonald	Revueltian	Redondasaurus (phytosaur) / Typothorax coccinarium (aetosaur)
220		Lacian	LT16r LT14n-16n	E12 E11 E10 E9	Malayites paulckei	Epigondolella triangularis - Norigondolella hallstattensis / Epigondolella quadrata (?)	Halobia medi-terranea	Shipingia weemsi - Euestheria buravasi	Adamanian	
225			LT13r LT13n	E8 E7	Guembelites jandianus	Metapolygnathus communisti	Halobia beyrichi			
228.5			LT12r LT12n LT11n LT10r LT10n	E6 E5 E4 E3 E2 E1	Anatropites spinosus / Tropites subbullatus / Tropites dilleri	Carnepigondolella pseudodiebeli / Carnepigondolella zoae / Metapolygnathus carpathicus / P. postinclinata?-P. noah I.Z.	Halobia radiata / Halobia lenticularis / Halobia superba / Halobia rugosa	Wanneresheria pennsylvanica / Laxitextella freybergi / Laxitextella seegisi / Eosolimnad. gallegoi	Otischalkian	Rutiodon (phytosaur) / Parasuchus (= Paleorhinus) (phytosaur)
230	Carnian	Tuvalian								
235		Julian	LT4r LT3r LT2r		Austrotr. austriacum / Trachyceras aonoides	Gladigondolella tethydis-Paragondolella postinclinata		Gregoriusella fimbriata-Laxitexella laxitexta	Berdyankian	
237.0		Cordevolian	LT1n MT13		Trachyceras aon / Daxatina canadensis / Frankites regoledanus	Bud. diebeli - Quadralella polygnathiformis / Budurovignathus n. sp. (Kozur)	Halobia / Daonella lommeli	Laxit. multireticulata / Euestheria minuta		
	Middle Triassic									

◀ **Figure 11.6 Selected marine and terrestrial biostratigraphic zonations of the Late Triassic.** (*Note that this scale is more compact than in* Figs. 11.4 or 11.5). Magnetic polarity zones with nomenclature of Hounslow and Muttoni (2010) of late Carnian through Rhaetian are shown with their estimated calibrations to the astronomical-scaled polarity chrons of the Newark Basin (Kent and Olsen, 1999) using the "long-Rhaetian" option (Ogg, 2012; Ogg et al., 2014) as partly recalibrated by Maron et al. (2015). [Note, the durations of Newark Basin polarity zones E1 through E8, which are within noncyclic fluvial sediments, were projected by Kent and Olsen (1999) based on the average accumulation rate for the overlying polarity zones E9 through E14 preserved in lacustrine sediments.] The magnetic polarity zones of lower Carnian are as scaled by cycle stratigraphy in South China (Zhang et al., 2015). Tethyan ammonoid zones are from Jenks et al. (2015), Tethyan conodont zones are modified from Kozur (2003), and Tethyan bivalve zones are from McRoberts (2010). Conchostracan zones of the Germanic Basin are from Kozur and Weems (2010, 2011) and land vertebrate zones and markers are from Lucas (2010b) and Lucas and Tanner (2015). Additional zonations, biostratigraphic markers, geochemical trends, sea-level curves, and details on calibrations are compiled in the internal data sets within the *TimeScale Creator* visualization system (free at www.tscreator.org).

the GTS2012 scale (Ogg, 2012; Ogg et al., 2014) is consistent with those later compilations; however, Rhaetian appears to correspond to the "*sensu lato*" interpretation of the marker conodont taxon (Maron et al., 2015).

(3) Stable-isotope stratigraphy and selected events

In addition to the major disruptions of climate accompanying the end-Permian and end-Triassic mass extinctions, there are at least two major climatic events indicated by simultaneous excursions in oxygen (temperature) and carbon isotopes (e.g., compilations and reviews by Preto et al., 2010; Muttoni et al., 2014; Trotter et al., 2015). The anomalous "lethal" tropical temperatures of the Early Triassic (Sun et al., 2012) include an abrupt drop in ammonoid and conodont diversity coincident with a negative excursion in $\delta^{13}C_{carb}$ near the end of the Smithian substage (Fig. 11.1). A dramatic event in the middle of the Carnian stage that is considered "the most distinctive climate change within the Triassic" (Preto et al., 2010) was a global disruption of the Earth's land–ocean–biological system. This "Carnian Humid Episode", "Carnian Pluvial Event" or "Middle Carnian Wet Intermezzo", was marked by sudden warming and associated increased rainfall in many continental regions, the widespread termination of

tropical carbonate platforms, and a brief major negative excursion in $\delta^{13}C_{carb}$ (e.g., Ogg, 2015; Ruffell et al., 2015). Both of these environmental perturbations may have been induced by volcanic activity—a late-stage phase in the Siberian Traps and the eruption of the Wrangellia large igneous province, respectively.

Numerical age model

The merger of astronomical cycles with magnetostratigraphy is the framework for scaling the majority of the Triassic. For the Early Triassic through early Anisian, the magnetic polarity pattern that was calibrated to short-eccentricity (100-kyr) changes in monsoon intensity recorded by clastic cycles in the Germanic Basin (e.g., Szurlies, 2007) has been duplicated in the magnetostratigraphy of conodont-bearing calcareous deposits offshore of the Yangtze Platform of South China that record both long- and short-eccentricity oscillations in relative clay content (Li et al., 2016). This cycle-magnetic compilation of South China includes the Induan (base Triassic) GSSP, and normal-polarity zone of lowest Griesbachian substage at Meishan with their high-precision U-Pb dating (Burgess et al., 2014), the section with the candidate GSSP for the Olenekian at West Pingdingshan (Chaohu), and the potential GSSP candidate for the Anisian at Guandao (Guizhou province).

The magnetostratigraphy and interpreted 405-kyr eccentricity-driven cycles at the Guandao section continue through the Anisian to lowest Ladinian (Lehrmann et al., 2015; Li et al., in prep.). This enables calculation of the ages of events by their cumulative cycle offset in South China relative to the base of the Triassic at 251.9 Ma. This age model was applied to the candidate GSSPs, to the bases of substages of Induan through Anisian using their ammonoid-based correlations to magnetic polarity zones from Europe (e.g., Hounslow and Muttoni (2010), which are different in some cases from their conodont-based placements relative to polarity zones in South China, e.g., by Lehrmann et al. (2015)), and to the conodont datums and isotopic excursions as recorded in South China in these sections (Figs. 11.1, 11.4, and 11.5).

A similar cycle scaling of magnetic polarity patterns in the Newark Basin of eastern North America relative to the U-Pb-dated basalt flows that cap the succession (e.g., Kent and Olsen, 1999) has been used by Muttoni et al. (2014) and Maron et al. (2015) to calibrate their composite magnetostratigraphies of upper Carnian through Rhaetian marine strata. Their compilation includes the candidate GSSPs for Norian and Rhaetian and a detailed carbon-isotope reference. Muttoni et al. (2014) assigned the normal-polarity-dominated interval at the base of the proposed Norian (Pizzo Mondelo [PM]5n magnetozone) to the "E8n" magnetozone of the Newark reference scale, therefore an age of ca. 227 Ma. However, this lowest Norian normal-polarity zone could be a slightly condensed merger of magnetozones "E7n-E8n" of the Newark scale. This alternative of the base of the Norian as the base of magnetozone E8n is adopted here implying a Carnian–Norian boundary age of 228.45 Ma. This slightly older age assignment was partly influenced by the uranium–lead chemical abrasion–isotope dilution–thermal ionization mass

spectrometry (U-Pb CA-TIMS) radio-isotopic dates of 223.81 ± 0.78 and 224.52 ± 0.22 Ma reported from volcanic tuffs in British Columbia that bracket the lower/middle Norian substage boundary according to adjacent conodont assemblages as used in North America (Diakow et al., 2011, abstract; and Mike Orchard, pers. comm., Sept 2015).

The age model for the early Carnian is partly constrained by cycle scaling of its polarity zones in marine carbonates of South China (Zhang et al., 2015). There is no verified cyclostratigraphy calibrated to ammonoid/conodont biostratigraphy for the underlying Ladinian stage. Therefore, until additional constraints become accepted, a schematic display of Tethyan ammonoid zones within each interval was incorporated in GTS2012, in which their relative durations within the span of the Ladinian (241.5 to 237 Ma) were partly apportioned according to their relative number of ammonite subzones or allocating 1.5 "subzonal units" for undivided zones.

Revised ages compared to GTS2012 and potential future enhancements

Induan base (Base of Triassic) (**251.902 ± 0.024** vs 252.16 ± 0.2 Ma in GTS2012): The U-Pb TIMS dating of volcanic ash beds and associated interpolated base-Triassic boundary age from the Meishan GSSP section reported in Shen et al. (2010) was reanalyzed by them (Burgess et al., 2014) using revised EARTHTIME standards, which shifted the boundary age younger by about 0.26 myr. [*Note that an external uncertainty of ca. 0.29 myr should be included if comparing these earliest Triassic EARTH-TIME-standardized dates to other dating methods, as explained in* Burgess et al. (2014).]

Olenekian base (**249.8** vs 250.0 Ma in GTS2012) and **Anisian** base (**246.8** vs 247.1 Ma in

GTS2012): The age model for the Early Triassic is based on astronomical-tuning (eccentricity orbital–climate cycles) relative to the base of the Triassic. Therefore, when that base-Triassic age was shifted younger by 0.26 myr (Burgess et al., 2014), then the ages of those other stage boundaries are also shifted younger by a similar amount. In addition, there is now astronomical scaling of the proposed *Chiosella timorensis* conodont marker for base-Anisian (Li et al., 2016). U-Pb-dated volcanic ash beds bracketing this base-Anisian marker by Ovtcharova et al. (2015) are interpreted by them to imply a slightly older age (ca. 247.3 Ma), but they caution that there are inconsistencies in the progression of interpreted dates from the zircon populations in successive layers.

Carnian base (**237.0** Ma in GTS2012 is retained): The Carnian is now defined by GSSP in the Southern Alps of Italy which is one ammonite subzone above a level dated at 237.77 ± 0.14 Ma (Mietto et al., 2012). Applying the average duration (0.6 myr) of ammonite subzones in the Ladinian suggests that the rounded age in GTS2012 is appropriate.

Rhaetian base: The correlations of the magnetostratigraphy of the boundary interval to the cycle-scaled Newark Basin magnetic polarity reference pattern indicates a correlation to either 209.5 Ma (GTS2012 using proposed GSSP in Austria) or 205.7 Ma (proposed GSSP in Sicily) that appears to depend on interpreted taxonomy of the earliest forms of the proposed conodont marker *Misikella posthernsteini* (summarized by Maron et al., 2015). Therefore, both options for defining the Rhaetian are shown on the figures in this chapter.

Jurassic base: The dating of the bracketing ash beds by Schoene et al. (2010) was revised by them (in Wotzlaw et al., 2014)

using updated EARTHTIME tracers. This changed the interpolated boundary age from 201.31 ± 0.18 Ma (as used in GTS2012) to a slightly older **201.36 ± 0.17 Ma**.

The main future enhancements to the age model for the Triassic stages will be the decisions on GSSP definition for the Olenekian, Anisian, Norian and Rhaetian. Other essential developments will be the verification of the cycle scaling and interregional correlation of the magnetostratigraphy of each stage, and obtaining an astronomical tuning for the zonations within the Ladinian and Carnian stages. It is anticipated that the integration of magnetostratigraphy, astronomical cycles coupled with periodic major changes in sea level, and distinctive stable isotopic excursions will enable a more reliable interregional correlation of biostratigraphic datums for both marine (Tethyan through Boreal realms) and terrestrial settings.

Estimated uncertainties on assigned ages on stage boundaries

The few high-precision radioisotopic dates with well-constrained biostratigraphic ages that constrain the Triassic time scale typically have a published uncertainty less than 0.2 myr. However, an external uncertainty of ca. 0.29 myr should be included if comparing these Triassic EARTHTIME-standardized dates to other dating methods, as explained in Burgess et al. (2014). Such dates anchor the cyclostratigraphic scaling for much of the Early and the Late Triassic (and "current working definitions" for their stages), but there is probably an additional ca. 0.1-myr uncertainty in the assignment of any event within a 405-kyr long eccentricity cycle. Those are the minimum uncertainties, but the age estimates for the bases of Ladinian and Carnian require additional extrapolation from the nearest radioisotopic date.

GSSPs of the Triassic Stages, with location and primary correlation criteria

Stage	GSSP Location	Latitude, Longitude	Boundary Level	Correlation Events	Reference
Rhaetian	Candidates are Pizzo Mondello, Sicily, Italy, and Steinbergkogel, Austria			Near FADs of conodont Misikella posthernsteini *s.s. or,* Misikella post-hernsteini *s.l.*	
Norian	Candidates are Black Bear Ridge in British Columbia (Canada) and Pizzo Mondello, Sicily, Italy			Base of Stikinoceras kerri *ammonoid zone and near FAD of conodont* Metapoly-gnathus communisti *group, or top of* M. parvus subzone, *or FAD of bivalve* Halobia austriaca	
Carnian	Section at Prati di Stuores, Dolomites, Italy	46°31'37"N 11°55'49"E	GSSP is base of marly limestone bed SW4, 45m from base of San Cassiano Formation	FAD of *ammonoid* Daxatina canadensis, *conodont* "Paragondolella" polygnathiformis an Halobia *bivalves*	Episodes **35**/3, 2012
Ladinian	Bagolino, Province of Brescia, Northern Italy	45°49'09.5"N 10°28'15.5"E	base of a 15 – 20cm thick limestone bed overlying a distinctive groove ("Chiesense groove") of limestone nodules in a shaly matrix, located about 5m above the base of the Buchenstein Beds	Ammonoid, FAD of *Eoprotrachyceras curionii and near FAD of conodont* Budurovignathus praehungaricus	Episodes **28**/4, 2005
Anisian	Candidate section at Desli Caira (Dobrogea, Romania); significant sections in Guizhou Province (China) and South Primorye (Russia)			FAD of conodont *Chiosella timorensis or Magnetic - base of magnetic normal-polarity chronozone MT1n*	
Olenekian	Candidates are Chaohu, China and Mud (Muth) village, Spiti valley, India			FAD of conodont Novispathodus waageni, *just above base of* Rohillites rohilla *ammonite zone*	
Induan	Meishan, Zhejiang Province, China	31° 4'47.28"N 119°42'20.90"E	base of Bed 27c in the Meishan Section	Conodont, FAD of *Hindeodus parvus*	Episodes **24**/2, 2001

Figure 11.7 Ratified GSSPs and potential primary markers under consideration for defining the Triassic stages (*status as of early 2016*). Details of each GSSP are available at http://www.stratigraphy.org, https://engineering. purdue.edu/Stratigraphy/gssp/, and in the *Episodes* publications.

Acknowledgments

The Triassic compilation benefited enormously from a 3-day working visit to the home of the late Heinz Kozur in 2010, and sessions with Leopold Krystyn, Spencer Lucas, Michael Orchard, and Paul Olsen. Two graduate students from Wuhan, Mingsong Li and Yang "Wendy" Zhang, gave preprints of their important cycle scaling of Early Triassic, Anisian and early Carnian stratigraphy. My colleagues in Wuhan, especially Chunju Huang, Zhong-Qiang Chen, Hongfu Yin, and Haishui Jiang, were instrumental in my participation in their work on Triassic timescales and events. None of them entirely agree with the selected taxonomic nomenclature for zones used in the figures or with the age models, but all agree that further international efforts will soon resolve all of the disputed calibrations. Mike Orchard and Maureen Steiner reviewed an early draft, and Zhong-Qiang Chen enhanced the text and graphics.

Selected publications and websites

Cited publications

Only select publications were cited in this review with an emphasis on aspects of post-2011 updates. Pre-2010 literature is well summarized in the suite of syntheses in Lucas (editor, 2010a) and in some of the publications cited in the following.

Balini, M., Krystyn, L., Levera, M., Tripodo, A., 2012. Late Carnian-early Norian ammonoids from the GSSP candidate section Pizzo Mondello (Sicani Mountains, Sicily). *Rivista Italiana di Paleontologia e Stratigrafia* **118**: 47–84.

Baud, A., 2014. The global marine Permian-Triassic boundary: over a century of adventures and controversies (1880-2001). *Albertiana* **42**: 1–21. Available at: http://paleo.cortland.edu/albertiana/.

Benton, M.J., Forth, J., Langer, M.C., 2014. Models for the rise of the dinosaurs. *Current Biology* **24**: R87–R95. http://dx.doi.org/10.1016/j.cub.2013.11.063.

Burgess, S.D., Bowring, S., Shen, Z.Q., 2014. High-precision timeline for Earth's most severe extinction. *Proceedings of the National Academy of Science (PNAS)* **111**: 3316–3321.

Chen, Z.Q., Algeo, T.J., Bottjer, D.J., 2014. Global review of the Permian–Triassic mass extinction and subsequent recovery: part I. *Earth Science Reviews* **137**: 1–5.

Chen, Z.Q., Yang, H., Luo, M., Benton, J.J., Kaiho, K., Zhao, L., Huang, Y., Zhang, K., Fang, Y., Jiang, H., Qiu, H., Li, Y., Tu, C., Shi, L., Zhang, L., Feng, X., Chen, L., 2015. Complete biotic and sedimentary records of the Permian–Triassic transition from Meishan section, South China: ecologically assessing mass extinction and its aftermath. *Earth-Science Reviews.* **149**: 67–107. http://dx.doi.org/10.1016/j.earscirev.2014.10.005.

Chen, Y.L., Krystyn, L., Orchard, M.J., Lai, X.L., Richoz, S., 2016. A review of the evolution, biostratigraphy, provincialism and diversity of Middle and early Late Triassic conodonts. *Papers in Paleontology.* 1–29. http://dx.doi.org/10.1002/spp2.1038 [and data at Dryad Digital Repository, http://dx.doi.org/10.5061/dryad.34r55].

Diakow, L., Orchard, M.J., Friedman, R., 2011. Absolute ages for the Norian Stage: a contribution from southern British Columbia, Canada. In: *21st Canadian Paleontological Conference, University of British Columbia, 19–20 Aug 2011. [Abstracts and Additional Details Sent by M.J. Orchard to J. Ogg, July 2011 and Sept 2015].*

Erwin, D.H., 2006. *Extinction: How Life on Earth Nearly Ended 250 Million Years Ago.* Princeton University Press, Princeton. 320 pp.

Gradinăru, E., Orchard, M.J., Nicora, A., Gallet, Y., Besse, J., Krystyn, L., Sobolev, E.S., Atudorei, N.-V., Ivanova, D., 2007. The Global Boundary Stratotype Section and Point (GSSP) for the base of the Anisian Stage: Desli Caira Hill, North Dobrogea, Romania. *Albertiana* **36**: 54–71. Available at: http://paleo.cortland.edu/albertiana/.

Haq, B.U., Al-Qahtani, A.M., 2005. Phanerozoic cycles of sea-level change on the Arabian Platform. *GeoArabia* **10**(2): 127–160.

Hounslow, M.K., Muttoni, G., 2010. The geomagnetic polarity timescale for the Triassic: linkage to stage boundary definitions. In: Lucas, S.G. (Ed.), *The Triassic Timescale.* The Geological Society, London, Special Publication, **334**, pp. 61–102.

ICS (International Commission on Stratigraphy), 2014. *Annual Report 2014.* Submitted to International Union of Geological Sciences (IUGS). at http://iugs.org/uploads/ICS%202014.pdf.

Jenks, J.F., Monnet, C., Balini, M., Brayard, A., Meier, M., 2015. Chapter 13. Biostratigraphy of Triassic ammonoids. In: Klug, C., Korn, D., De Baets, K., Kruta, I., Mapes, R.H. (Eds.), *Ammonoid Paleobiology: From Macroevolution to Paleogeography.* Topics in Geobiology, **44**. Springer Publ, pp. 329–371. http://dx.doi.org/10.1007/978-94-017-9633-0_13.

Kent, D.V., Olsen, P.E., 1999. Astronomically tuned geomagnetic polarity timescale for the Late Triassic. *Journal of Geophysical Research* **104**: 12831–12841. Web page update (2002) posted at Newark Basin Coring Project website: http://www.ldeo.columbia.edu/%7Epolsen/nbcp/nbcp.timescale.html (accessed 03.07.16.).

Konstantinov, A.G., Klets, T.V., 2009. Stage boundaries of the Triassic in Northeast Asia. *Stratigraphy and Geological Correlations* **17**: 173–191. http://dx.doi.org/10.1134/S0869593809020063.

Kozur, H.W., Weems, R.E., 2010. The biostratigraphic importance of conchostracans in the continental Triassic of the northern hemisphere. In: Lucas, S.G. (Ed.), *The Triassic Timescale*. The Geological Society, **334**. Special Publication, London, pp. 315–417.

Kozur, H.W., Weems, R.E., 2011. Detailed correlation and age of continental late Changhsingian and earliest Triassic beds: implications for the role of the Siberian Trap in the Permian-Triassic biotic crisis. *Palaeogeography, Palaeoclimatology, Palaeoecology.* **308**: 22–40. http://dx.doi.org/10.1016/j.palaeo.2011.02.020.

Kozur, H.W., 2003. Integrated ammonoid, conodont and radiolarian zonation of the Triassic. *Hallesches Jahrbuch fur Geowissenschaften* **B25**: 49–79.

Lehrmann, D.J., Stepchinski, L., Altiner, D., Orchard, M.J., Montgomery, P., Enos, P., Ellwood, B.B., Bowring, S.A., Ramezani, J., Wang, H., Wei, J., Yu, M., Griffiths, J.D., Minzoni, M., Schaal, E.K., Li, X., Meyer, K.M., Payne, J.L., 2015. An integrated biostratigraphy (conodonts and foraminifers) and chronostratigraphy (paleomagnetic reversals, magnetic susceptibility, elemental chemistry, carbon isotopes and geochronology) for the Permian-Upper Triassic strata of Guandao section, Nanpanjiang Basin, South China. *Journal of Asian Earth Sciences.* **108**: 117–135. http://dx.doi.org/10.1016/j.jseaes.2015.04.030.

Li, M., Ogg, J.G., Zhang, Y., Huang, C., Hinnov, L., Chen, Z.-Q., Zou, Z., 2016. Astronomical tuning of the end-Permian extinction and the Early Triassic Epoch of South China and Germany. *Earth and Planetary Science Letters*, **441**: 10–25.

Lucas, S.G., Tanner, L.H., 2015. Triassic timescale based on tetrapod biostratigraphy and biochronology. In: Rocha, R., et al. (Eds.), *STRATI 2013*. Springer Geology, pp. 1013–1016. http://dx.doi.org/10.1007/978-3-319-04364-7_192.

Lucas, S.G. (Ed.), 2010a. *The Triassic Timescale*. The Geological Society, **334**. Special Publication, London. 500 pp [An invited suite of authoritative compilations on all aspects of Triassic stratigraphy; which provided the foundation for most of the zonal scales in GTS2012.].

Lucas, S.G., 2010b. The Triassic timescale based on nonmarine tetrapod biostratigraphy and biochronology. In: Lucas, S.G. (Ed.), *The Triassic Timescale*. The Geological Society, **334**. Special Publication, London, pp. 447–500.

Lucas, S.G., 2016. Conodonts, Triassic chronostratigraphy and the GSSP for the base of the Rhaetian Stage. *Albertiana* **43**. in press. Available at: http://paleo.

cortland.edu/albertiana/ ([and his "Conodonts, Triassic chronostratigraphy and the GSSP for the base of the Rhaetian Stage: Reply to Orchard" in the same volume]) .

Maron, M., Rigo, M., Bertinelli, A., Katz, M.E., Godfrey, L., Zaffani, M., Muttoni, G., 2015. Magnetostratigraphy, biostratigraphy, and chemostratigraphy of the Pignola-Abriola section: new constraints for the Norian-Rhaetian boundary. *Geological Society of America Bulletin* **127**: 962–974. http://dx.doi.org/10.1130/B31106.1.

McRoberts, C.A., 2010. Biochronology of Triassic pelagic bivalves. In: Lucas, S.G. (Ed.), *The Triassic Timescale*. The Geological Society, **334**. Special Publication, London, pp. 201–219.

Mietto, P., Manfrin, S., Preto, N., Rigo, M., Roghi, G., Furin, S., Gianolla, P., Posenato, R., Muttoni, G., Nicora, A., Buratti, N., Cirilli, S., Spöti, C., Ramezani, J., Bowring, S.A., 2012. The Global Boundary Stratotype Section and Point (GSSP) of the Carnian Stage (Late Triassic) at Prati di Stuores/Stuores Wiesen section (Southern Alps, NE Italy). *Episodes* **35**: 414–430.

Muttoni, G., Mazza, M., Mosher, D., Katz, M.E., Kent, D.V., Balini, M., 2014. A Middle-Late Triassic (Ladinian-Rhaetian) carbon and oxygen isotope record from the Tethyan ocean. *Palaeogeography, Palaeoclimatology, Palaeoecology* **399**: 246–259 (and on-line data repository item 2015069).

Ogg, J.G., Huang, C., Hinnov, L., 2014. Triassic timescale status: a brief overview. *Albertiana* **41**: 3–30 [A slightly revised summary of the status of the GSSPs and age models of GTS2012.].

Ogg, J.G., 2012. Triassic. In: Gradstein, F.M., Ogg, J.G., Schmitz, M., Ogg, G. (Coordinators), *The Geologic Time Scale 2012*. Elsevier Publ., pp. 681–730 [An overview, especially with extensive details and graphics on the GSSPs of the stages, the different marine microfossil and terrestrial vertebrate biostratigraphic scales, and the astronomical tuning of the magnetic polarity scale.].

Ogg, J.G., 2015. The mysterious Mid-Carnian "Wet Intermezzo" global event. *Journal of Earth Science* **26**(2): 181–191.

Orchard, M.J., 2010. Triassic conodonts and their role in stage boundary definitions. In: Lucas, S.G. (Ed.), *The Triassic Timescale*. The Geological Society, **334**. Special Publication, London, pp. 139–161.

Orchard, M.J., 2014. Conodonts from the Carnian-Norian boundary (Upper Triassic) of Black Bear Ridge, northeastern British Columbia, Canada. *New Mexico Museum of Natural History and Sciences Bulletin* **64** 139 pp.

Orchard, M.J., 2016. Base of the Rhaetian and a critique of Triassic conodont-based chronostratigraphy: comment. *Albertiana* **43**. in press. Available at: http://paleo.cortland.edu/albertiana/.

Ovtcharova, M., Goudemand, N., Hammer, Ø., Guodun, K., Cordey, F., Galfetti, T., Schaltegger, U., Bucher, H., 2015. Developing a strategy for accurate definition of a geological boundary through radio-isotopic and biochronological dating: the Early-Middle Triassic boundary (South China). *Earth-Science Reviews*. **146**: 65–76. http://dx.doi.org/10.1016/j.earscirev.2015.03.006.

Preto, N., Kustatscher, E., Wignall, P.B., 2010. Triassic climates – state of the art and perspectives. In: Kustatscher, E., Preto, N., Wignall, P. (Eds.), *Triassic Climates. Palaeogeography, Palaeoclimatology, Palaeoecology*, **290**, pp. 1–10.

Rigo, M., Bertinelli, A., Concheri, G., Gattolin, G., Godfrey, L., Katz, M.E., Maron, M., Mietto, P., Muttoni, G., Sprovieri, M., Stellin, F., Zaffani, M., 2016. The Pignola-Abriola section (southern Apennines, Italy): a new GSSP candidate for the base of the Rhaetian Stage. *Lethaia*. http://dx.doi.org/10.111/let.12145 in press .

Ruffell, A., Simms, M.J., Wignall, P.B., 2015. The Carnian Humid Episode of the late Triassic: a review. *Geological Magazine*, **153**(2): 1–14. http://onlinelibrary.wiley.com/doi/10.1111/let.12145/full in press.

Schoene, B., Guex, J., Bartolini, A., Schaltegger, U., Blackburn, T.J., 2010. Correlating the end-Triassic mass extinction and flood basalt volcanism at the 100 ka level. *Geology* **38**: 387–390.

Scotese, C.R., 2014. *Atlas of Middle & Late Permian and Triassic Paleogeographic Maps, Maps 43-48 From Volume 3 of the PALEOMAP Atlas for ArcGIS (Jurassic and Triassic) and Maps 49-52 from Volume 4 of the PALEOMAP PaleoAtlas for ArcGIS (Late Paleozoic), Mollweide Projection*. PALEOMAP Project, Evanston, IL. https://www.academia.edu/11300143/Atlas_of_Middle_and_Late_Permian_and_Triassic_Paleogeographic_Maps.

Shen, S.-Z., Henderson, C.M., Bowring, S.A., Cau, C.-Q., Wang, Y., Wang, W., Zhang, H., Zhang, Y.-C., Mu, L., 2010. High-resolution Lopingian (Late Permian) timescale of South China. *Geological Journal* **45**: 122–134.

Sun, Y., Joachimski, M.M., Wignall, P.B., Yan, C., Chen, Y., Jiang, H., Wang, L., Lai, X., 2012. Lethally hot temperatures during the early Triassic greenhouse. *Science* **338**: 366–370.

Szurlies, M., 2004. Magnetostratigraphy: the key to global correlation of the classic Germanic Trias case study Volpriehausen Formation (Middle Buntsandstein), Central Germany. *Earth and Planetary Science Letters*, **227**: 395–410.

Szurlies, M., 2007. Latest Permian to Middle Triassic cyclo-magnetostratigraphy from the Central European Basin, Germany: implications for the geomagnetic polarity timescale. *Earth and Planetary Science Letters* **261**: 602–619.

Tozer, E.T., 1984. The Trias and its ammonoids: the evolution of a time scale. *Geological Survey of Canada Miscellaneous Report* **35** 171 pp [A fascinating historical view of the personalities, debates, and gradual understanding of the Triassic system.].

Trotter, J.A., Williams, I.S., Nicora, A., Mazza, M., Rigo, M., 2015. Long-term cycles of Triassic climate change: a new $\delta^{18}O$ record from conodont apatite. *Earth and Planetary Science Letters* **415**: 165–174.

Veizer, J., Prokoph, A., 2015. Temperatures and oxygen isotopic composition of Phanerozoic oceans. *Earth-Science Reviews* **146**: 92–104.

Wignall, P., 2015. *The Worst of Times: How Life on Earth Survived Eighty Million Years of Extinctions*. Princeton University Press. 224 pp.

Wotzlaw, J.-F., Guex, J., Bartolini, A., Gallet, Y., Krystyn, L., McRoberts, C.A., Taylor, D., Schoene, B., Schaltegger, U., 2014. Towards accurate numerical calibration of the Late Triassic: high-precision U-Pb geochronology constraints on the duration of the Rhaetian. *Geology* **42**: 571–574.

Yin, H., Zhang, K., Tong, J., Yang, Z., Wu, S., 2001. The Global Stratotype Section and Point (GSSP) of the Permian–Triassic boundary. *Episodes* **24**(2): 102–114.

Zhang, Y., Li, M., Ogg, J.G., Montgomery, P., Huang, C., Chen, Z.-Q., Shi, Z., Enos, P., Lehrmann, D.J., 2015. Cycle-calibrated magnetostratigraphy of middle Carnian from South China: implications for Late Triassic time scale and termination of the Yangtze platform. *Palaeogeography, Palaeoclimatology, Palaeoecology*. **436**: 135–166. http://dx.doi.org/10.1016/j.palaeo.2015.05.033.

Websites (selected)

Subcommission on Triassic Stratigraphy (ICS)—*http://paleo.cortland.edu/sts/* with *Albertiana* (official journal of the Subcommission) at *http://paleo.cortland.edu/albertiana/*. The main subcommission site has basic information; and the *Albertiana* journal site has PDFs from 2001 to present. The *Albertiana* journal is the primary publication for GSSP discussions, biostratigraphy and inter-regional correlations.

Palaeos: Triassic—*http://palaeos.com/mesozoic/triassic/triassic.htm*—A well-presented suite of diverse topics for a general science audience that was originally compiled by M. Alan Kazlev in 1998–2002.

12

JURASSIC

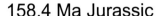

158.4 Ma Jurassic

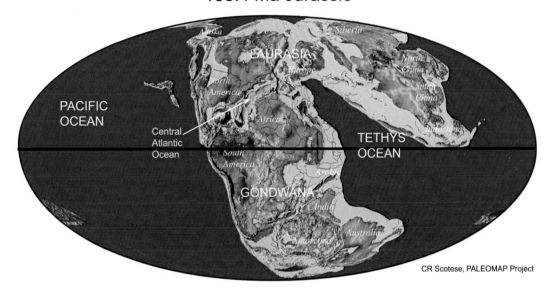

CR Scotese, PALEOMAP Project

Oxfordian paleogeographic reconstruction (Sea level + 80 m) from Scotese (2014).

Basal definition and international subdivisions

The Jurassic, named after the exposures in the Jura Mountains along the Swiss–French border, begins with the recovery from the major end-Triassic mass extinction (Fig. 12.1). This end-Triassic event was particularly catastrophic to marine fauna, including the conodonts, the distinctive phosphatic jaw elements which are used for biostratigraphy from uppermost Cambrian through Triassic, and nearly all of the ammonoid groups.

The primary marker for the base of the Jurassic is the lowest occurrence of the ammonite genus, *Psiloceras*. The earliest form, *Psiloceras spelae,* first occurs during a global sea-level fall and extended hiatus in shallow seas; therefore, this taxon is only found in rare complete sections (e.g., Peru, Alps, Nevada) (Hillebrandt and Krystyn, 2009). The GSSP for the base of the Jurassic (base of Hettangian Stage) was ratified in the Kuhjoch section within the Northern Calcareous Alps of Austria (Hillebrandt et al., 2013) (Fig. 12.2).

The lowest *Psiloceras spelae* horizon as dated by bracketing volcanic ash beds in Peru occurs at 201.36 ± 0.17 Ma, which is about 0.15 myr after the onset (201.51 ± 0.15 Ma) of a carbon-isotope excursion that coincides with

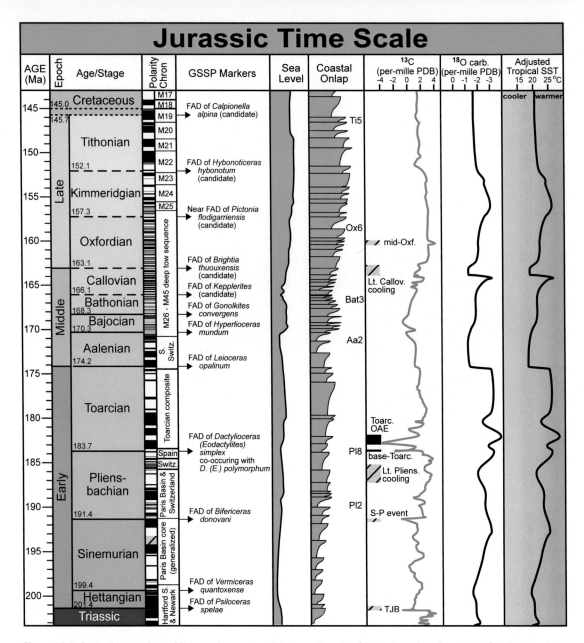

Figure 12.1 Jurassic overview; Main markers or candidate markers for Global Boundary Stratotype Sections and Points (GSSPs) of Jurassic stages are first-appearance datums (FADs) of ammonite taxa as detailed in Fig. 12.5. See text and Fig. 13.2 for Jurassic–Cretaceous boundary options. Magnetic polarity scale is a composite as compiled in Ogg (2012) and Ogg et al. (2012), and is enlarged in Fig. 12.4. Coastal onlap and schematic sea-level curve are modified from Hardenbol et al. (1998). The δ13C curve (with widespread excursion events; "OAE" = oceanic anoxic event) represents generalized trends from synthesis by Katz (in supplement to Muttoni et al., 2014) and by Jenkyns (2002), with additional schematic enhancements for the Triassic–Jurassic boundary interval (Hillebrandt et al., 2013), for the late Sinemurian through early Toarcian (Kemp et al., 2005; Bodin et al., 2010; Korte and Hesselbo, 2011), for the Callovian through early Oxfordian (Pellenard et al., 2004b), and for the middle Oxfordian through early Kimmeridgian of Central Europe (Wierzbowski, 2015). However, the δ13C values are usually systematically offset among different regions. The simplified Jurassic δ18O curve (scale is inverted) and adjusted estimates of tropical sea-surface temperatures (SST) derived from those oxygen-18 values are modified from Veizer and Prokoph (2015) with details of trends for the Sinemurian through Aalenian (Korte and Hesselbo (2011) and Korte et al. (2015)), and the Callovian through Kimmeridgian of Central Europe (e.g. Pellenard et al. (2004b) and Wierzbowski (2015)). *PDB*, PeeDee Belemnite 13C and 18O standard.

Base of the Hettangian Stage of the Jurassic System at Kuhjoch, Austria

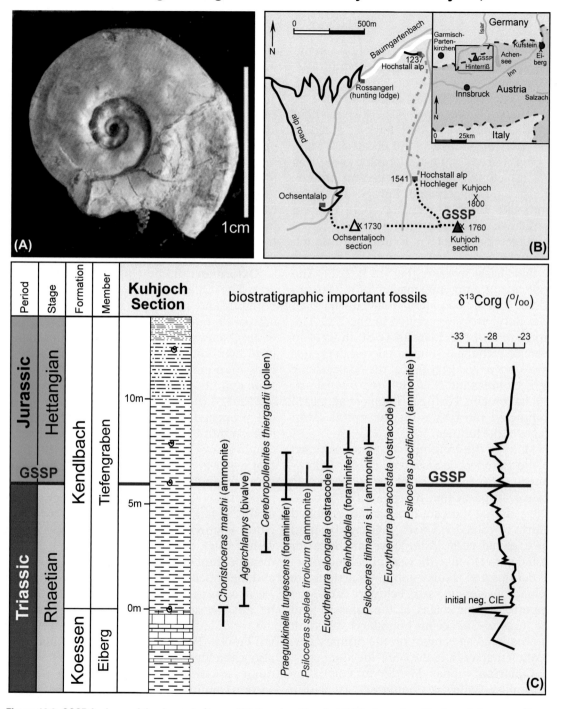

Figure 12.2 GSSP for base of the Jurassic (base of Hettangian Stage) at Kuhjoch section, Northern Calcareous Alps, Austria. The GSSP level coincides with the lowest occurrence of the ammonite *Psiloceras spelae* (subsp. *tirolicum*), marking the beginning of a new marine ecosystem after the major end-Triassic extinctions. The brief negative excursion in carbon isotopes is approximately simultaneous with those end-Triassic extinctions and main eruption phase of the Central Atlantic magmatic province. (Photograph provided by Axel von Hillebrandt.)

the main end-Triassic mass extinction and the massive volcanism of the Central Atlantic magmatic province (CAMP) at 201.48 ± 0.02 Ma. These dates are from Schoene et al. (2010, as recalibrated by Wotzlaw et al., 2014, who note that the same lower CAMP flow was dated by Blackburn et al., 2013, as 201.566 ± 0.031 Ma).

The concept of biostratigraphic zones was first developed in the succeeding ammonite-rich Jurassic strata in Europe, and these ammonite zones remain the main method of relative dating and definition of Jurassic stages. GSSPs for the Hettangian through Bathonian stages have ammonite horizons as their basal markers (Fig. 12.1; 12.5). The GSSP and its ammonite marker for the Lower–Middle Jurassic boundary (Toarcian–Aalenian boundary) are diagrammed in Fig. 12.3. The ratification of the Toarcian GSSP in 2014 to coincide with the first occurrence of the genus *Dactylioceras (Eodactylites)* at the co-occurring *D. (E.) simplex* and *D. (E.) polymorphum* horizon in Peniche, Portugal (Rocha et al., 2016) completed the GSSPs for the initial seven stages of the Jurassic.

Many of the ammonite taxa are restricted to the European exposures of Sub-Mediterranean or of Sub-Boreal paleogeographic realms. In particular, a pronounced provinciality during the Late Jurassic has inhibited efforts to establish GSSPs that can be used on a global scale. Alternative methods for global correlation with widespread microfossil datums and stable-isotope excursions are being considered. As of February 2016, the uppermost four stages of the Jurassic (and the Jurassic–Cretaceous boundary) still await international agreement on the primary correlation markers and the GSSP location:

Callovian (upper Middle Jurassic): The Bathonian–Callovian boundary is within an interval of rising sea level and is rarely preserved, although it has been traditionally placed at the lowest occurrence of the ammonite genus *Kepplerites (Kosmoceratidae)* at the base of the *K. keppleri* Subzone

of *Macrocephalites herveyi* Zone in the Sub-Boreal province (Great Britain to southwest Germany). A proposed GSSP in the Swabian Alb of southwest Germany has both an apparently complete succession at the resolution level of ammonites and a magnetostratigraphy interpreted as correlating the boundary interval to within marine magnetic anomaly M39n, but is in a very condensed iron-oolite-rich calcareous facies (e.g., Callomon and Dietl, 2000; Gipe, 2013). The Callovian stage task group is also considering potential relatively expanded sections in East Greenland (Mönnig, 2014).

Oxfordian (Middle–Upper Jurassic boundary): The Callovian–Oxfordian boundary in both England and France is at the base of the ammonite *Cardioceras scarburgense* Subzone of the *Quenstedtoceras mariae* Zone, but there have been different opinions of which specific ammonite horizon to use for that subzone definition. In Geologic Time Scale 2012 (GTS2012), the working definition had been the proposed lowest occurrence of ammonite *Cardioceras redcliffense* in a coastal section in Dorset, England (Page et al., 2010), but that event is difficult to correlate and might be a morphotype of the index species of the uppermost Callovian subzone (Pellenard et al., 2014a). The current Oxfordian working group is now favoring a slightly higher level with a brief, marked faunal turnover of the disappearance of several Callovian genera and the appearance of new species, especially *Hecticoceras (Brightia) thuouxensis* followed by the lowest *Cardioceras scarburgense*. In the candidate GSSP at Thuoux in southeastern France (Pellenard et al., 2014a), this level also coincides with the first appearance of dinocyst *Wanaea fimbriata*. The Dubki section candidate in European Russia is excellent for the basal Oxfordian, but is poor for the uppermost Callovian (Kiselev et al., 2013). Even though the candidate GSSP in France has been unsuitable for paleomagnetism, the comparison of ammonite assemblages

Base of the Aalenian Stage of the Jurassic System at Fuentelsaz, Spain

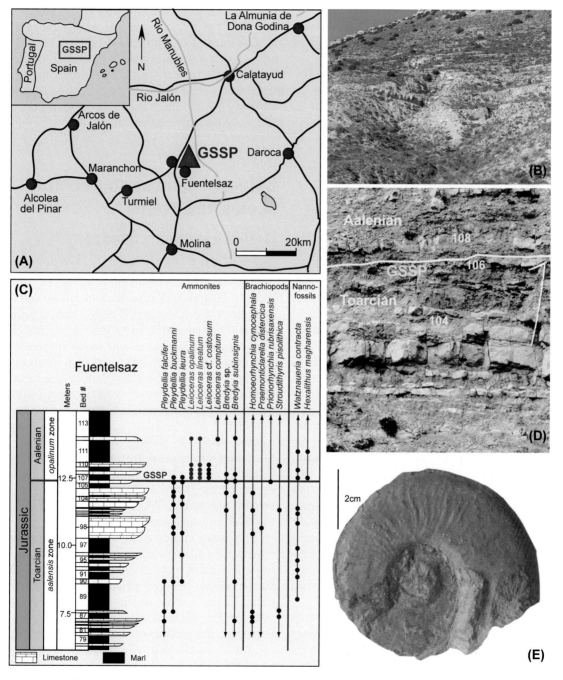

Figure 12.3 GSSP for base of the Middle Jurassic (base of Aalenian Stage) at Fuentelsaz section, Castilian Branch of the Iberian Range, Spain. The GSSP level coincides with the lowest occurrence of ammonite *Leioceras opalinum* (base of *Leioceras opalinum* Zone). (Photographs provided by Marie Soledad Ureta.)

enables approximate correlation to the magnetostratigraphy of the Dorset section at the base of the local *Cardioceras woodhamense* biohorizon (Kiselev et al., 2013), which would project this boundary as the lower part of Chron M36Br.

Kimmeridgian (mid-Upper Jurassic): The Colloque du Jurassique à Luxembourg in 1962 fixed the base of the Kimmeridgian as the base of the *Pictonia baylei* Zone (appearance of *Pictonia* genus) in the Sub-Boreal province, and, in 2007, the Kimmeridgian working group accepted the Flodigarry section (Wierzbowski et al., 2006) on the Isle of Skye, Scotland, to be the GSSP section pending a precise definition of the base of the *Pictonia baylei* Zone. However, the traditional placement of the Oxfordian–Kimmeridgian boundary in the Sub-Mediterranean province at the base of the ammonite *Sutneria platynota* Zone is over 1 myr younger than this level; indeed, the Sub-Boreal definition falls in the middle of the Tethyan ammonite *Epipeltoceras bimammatum* Zone. To allow a uniform Oxfordian–Kimmeridgian boundary that corresponds to an ammonite zonal boundary in the different provinces, which had been a major cause of the postponed decision, Wierzbowski and Matyja (2014) have used enhanced correlations to recommend the elevation of the *E. hypselum* subzone of the lower *E. bimammatum* Zone to zonal status (used in Fig. 12.4). In the candidate GSSP at the Isle of Skye, the lowest occurrence of a new *Pictonia flodigarriensis* taxon to define the base of the *P. baylei* Zone matches the base of the Boreal *Amoeboceras bauhini* Zone (first appearance of subgenus *Plasmatites*), and is essentially coeval with base of polarity Chron M26r (Przybylski et al., 2010).

Tithonian (uppermost Jurassic): The events near the Kimmeridgian–Tithonian boundary seem fairly well established in that the base of ammonite *Hybonoticeras hybonotum* Zone in Sub-Mediterranean province is coeval with the base of *Pectinatites (Virgatosphinctoides)* *elegans* Zone in the Sub-Boreal province and with the base of normal-polarity Chron M22An. However, the working group has not yet agreed on a GSSP section.

End-Jurassic (Tithonian–Berriasian boundary) is not yet defined, and is discussed in the Cretaceous chapter. Two alternative working definitions—the base of microfossil *Calpionella alpina* and the base of magnetic polarity Chron M18r—are shown in the stratigraphic figures.

Selected main stratigraphic scales and events

(1) Biostratigraphy (marine; terrestrial)

Ammonite workers in the Jurassic, in contrast to formalized practices in other periods, often use a "standard zone" that is only indirectly associated with the biotic range of the ammonite taxon that lends its name, and these are usually indicated by a nonitalicized name (e.g., as indicated above, the basal Oxfordian "Mariae Zone" does not begin with the lowest occurrence of *Quenstedtoceras mariae*). A synthesis of major evolutionary trends, regional paleogeographic distribution, and ammonite zones is compiled by Schweigert (2015). However, to show the taxonomic relationships of these ammonite zones, the full genus–species-name taxa are used in the stratigraphic charts in this book.

The other major marine biostratigraphic zonations utilize calcareous nannofossils and organic-walled cysts of dinoflagellates. These single-celled pelagic phytoplankton originated in the middle of the Triassic. Siliceous radiolarians, which had originated in the early Paleozoic, are a major component of deep-sea sediments, including "ribbon radiolarites" of the Tethyan basins. During the Tithonian, single-celled calcareous calpionellids appeared and are an important biostratigraphic tool in

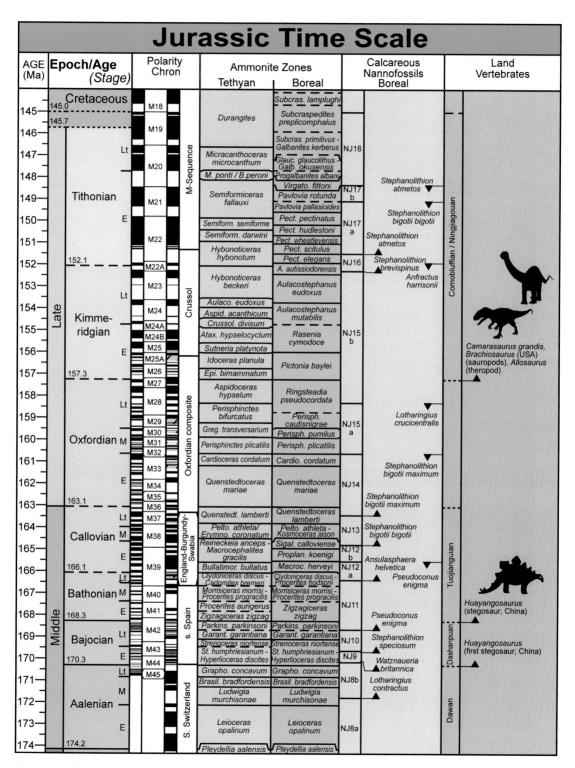

Figure 12.4 *(Continued)*

Jurassic Time Scale

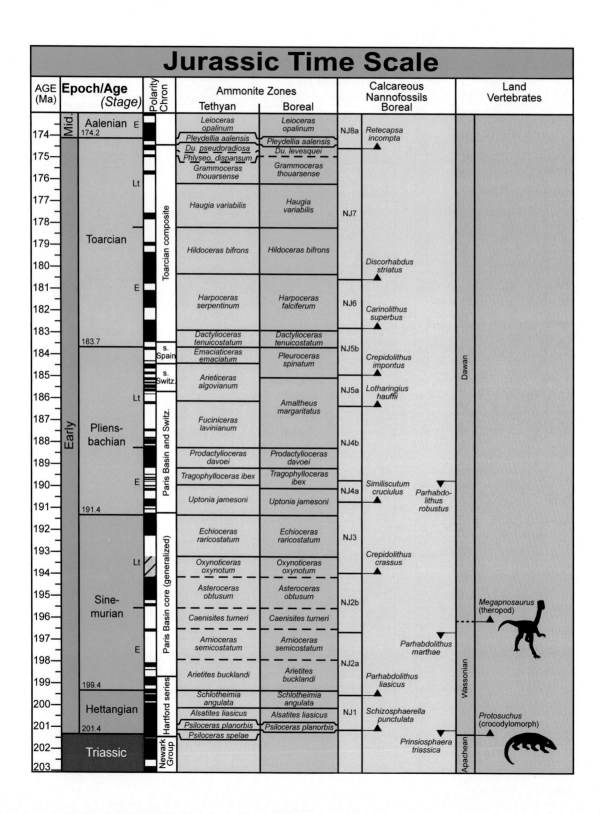

AGE (Ma)	Epoch/Age (Stage)	Polarity Chron	Ammonite Zones Tethyan	Ammonite Zones Boreal	Calcareous Nannofossils Boreal		Land Vertebrates
174	Mid. Aalenian 174.2	E	*Leioceras opalinum*	*Leioceras opalinum*	NJ8a	*Retecapsa incompta*	
175			*Pleydellia aalensis*	*Pleydellia aalensis*			
			Du. pseudoradiosa	*Du. levesquei*			
176		Lt	*Physeo. dispansum*				
			Grammoceras thouarsense	*Grammoceras thouarsense*			
177	Toarcian		*Haugia variabilis*	*Haugia variabilis*	NJ7		Dawan
178							
179			*Hildoceras bifrons*	*Hildoceras bifrons*		*Discorhabdus striatus*	
180							
181		E	*Harpoceras serpentinum*	*Harpoceras falciferum*	NJ6	*Carinolithus superbus*	
182							
183	183.7		*Dactylioceras tenuicostatum*	*Dactylioceras tenuicostatum*	NJ5b		
184		s. Spain	*Emaciaticeras emaciatum*	*Pleuroceras spinatum*		*Crepidolithus impontus*	
185		s. Switz.	*Arieticeras algovianum*		NJ5a	*Lotharingius hauffii*	
186		Lt		*Amaltheus margaritatus*			
187	Pliens- bachian		*Fuciniceras lavinianum*				
188			*Prodactylioceras davoei*	*Prodactylioceras davoei*	NJ4b		
189							
190		E	*Tragophylloceras ibex*	*Tragophylloceras ibex*	NJ4a	*Similiscutum cruciulus*	*Parhabdo- lithus robustus*
191	191.4		*Uptonia jamesoni*	*Uptonia jamesoni*			
192			*Echioceras raricostatum*	*Echioceras raricostatum*	NJ3		
193		Lt	*Oxynoticeras oxynotum*	*Oxynoticeras oxynotum*		*Crepidolithus crassus*	
194							
195	Sine- murian		*Asteroceras obtusum*	*Asteroceras obtusum*	NJ2b		*Megapnosaurus* (theropod)
196			*Caenisites turneri*	*Caenisites turneri*			
197		E	*Arnioceras semicostatum*	*Arnioceras semicostatum*		*Parhabdolithus marthae*	Wassonian
198	199.4		*Arietites bucklandi*	*Arietites bucklandi*	NJ2a	*Parhabdolithus liasicus*	
199							
200	Hettangian		*Schlotheimia angulata*	*Schlotheimia angulata*		*Schizosphaerella punctulata*	
201	201.4		*Alsatites liasicus*	*Alsatites liasicus*	NJ1		*Protosuchus* (crocodylomorph)
			Psiloceras planorbis	*Psiloceras planorbis*			
202	Triassic		*Psiloceras spelae*			*Prinsiosphaera triassica*	Apachean
203							

Polarity Chron labels (left to right): Toarcian composite; Paris Basin and Switz.; Paris Basin core (generalized); Hartford series; Newark Group

◄ **Figure 12.4 (A, B): Selected marine and terrestrial biostratigraphic zonations of the Jurassic.** ("*Age*" is the term for the time equivalent of the rock-record "*stage.*") Comparison of the M-sequence of marine polarity zones and its deep-tow extension to M45 (e.g., Tominaga et al., 2008) is plotted relative to its proposed calibration to a composite reference Middle and Late Jurassic magnetic polarity timescale from ammonite-zoned outcrops. Ammonite zone standards are shown for the Tethyan (Sub-Mediterranean province) and Boreal (Sub-Boreal for Late Jurassic) realms. An expanded listing of ammonoid zones with an explanation of each GTS2012 age calibration is in Ogg et al. (2012). Calcareous nannofossil zones with their Tethyan zone markers are a composite from Casellato (2010), Mattioli and Erba (1999) and Bown (1998). Land vertebrate zones and markers are from Lucas (2009). Additional zonations, biostratigraphic markers, geochemical trends, sea-level curves, and details on calibrations are compiled in the internal data sets within the *TimeScale Creator* visualization system (free at www.tscreator.org).

moderate-depth settings until their extinction in the mid-Valanginian of the early Cretaceous.

Dinosaurs are the famous Jurassic Park vertebrates, but the schematic Pangea-wide zonation of characteristic assemblages is only loosely calibrated to marine stages (e.g., Lucas, 2009, Fig. 12.4). Pollen-spore zonations and clam-shrimp conchostracans are useful in some terrestrial and marginal-marine settings.

(2) Magnetostratigraphy

The reference magnetic polarity scale for the Kimmeridgian to basal Aptian is the M-sequence of marine magnetic polarity anomalies from the Pacific spreading centers correlated to magnetostratigraphy of ammonite-zoned successions in the Sub-Mediterranean and Sub-Boreal realms. Deep-tow magnetic surveys have extended this "pre-M25" Pacific magnetic anomaly to "M45" on the relative fast-spreading "Japanese" lineations with a calibration to Ocean Drilling Site 801C (on anomaly/subchron "M42n.4r", dated near Bathonian–Bajocian boundary by biostratigraphy or $168.4 \pm 0.1.7$ Ma by $^{40}Ar/^{39}Ar$ ratio (Ar/Ar) radioisotopes recalibrated to GTS2012 standards) (e.g., Tivey et al., 2006; Tominaga et al., 2008). This pre-M25 pattern has been partly verified by deep-tow surveys on the older Hawaiian succession (Tominaga et al., 2015). The entire M45 through M0r marine magnetic anomaly pattern has been calibrated by numerous magnetostratigraphic

studies of ammonite-, nannofossil-, and/or calpionellid-zones outcrops (e.g., Channell et al., 2010; Przybylski et al., 2010; Ogg et al., 2010; Gipe, 2013; reviews in Ogg, 2012 and Ogg et al., 2012; etc.); and, with assumptions of a linearly very-slowly decreasing spreading rate, serves as the age model for the upper Aalenian through basal Aptian stages.

The Early Jurassic magnetic polarity time scale is currently a composite from magnetostratigraphy studies on ammonite-zoned outcrops in Europe [e.g., synthesis in Ogg et al., 2012; with partial verification of scaling of Kent and Olsen (2008) by Hüsing et al. (2014) using cyclostratigraphy for Hettangian to earliest Sinemurian].

(3) Stable-isotope stratigraphy and selected events

Major negative excursions in carbon isotopes occur in the Triassic–Jurassic boundary interval, basal Pliensbachian, a pair in the basal and lowest Toarcian, and in uppermost Callovian. The lower Toarcian event, one of the largest negative excursions in the Phanerozoic, is followed by widespread "anoxic event" enrichment in organic carbon in marine sections (Fig. 12.1). The main significant positive excursion in carbon isotopes is reported in the middle Oxfordian (e.g., Głowniak and Wierzbowski, 2007; Wierzbowski, 2015) within a broad elevated C-13 enrichment that is coeval with major carbon-rich sediment deposits that became the hydrocarbon source rocks of Saudi Arabia, North Sea, Siberia, and other regions.

Oxygen-isotope trends and other proxies suggest a general warming from Hettangian through Sinemurian, a cool Late Pliensbachian, a warm to very warm Toarcian, then a cooling Aalenian through Bajocian (e.g., Korte and Hesselbo, 2011; Korte et al., 2015). The generally warm climate of Callovian through Kimmeridgian was interrupted by an anomalous late Callovian cold spell (e.g., Dromart et al., 2003; Pellenard et al., 2014b). The Tithonian and Berriasian are generally interpreted as cooler and more arid in many regions.

However, most of these carbon-isotope, oxygen-isotope, and temperature trends are derived from individual basins in Western Europe, and a verified global synthesis has not yet been compiled (e.g., discussions in Wierzbowski, 2015).

Marine $^{87}Sr/^{86}Sr$ progressively decreased from an end-Norian peak to a low during basal Toarcian. Strontium isotope ratios rose during the Toarcian to crest with a sustained plateau through the Aalenian, and then decreased through Callovian to reach its lowest ratio of the entire Phanerozoic in early Oxfordian. After the middle Oxfordian, the strontium isotope ratio began a long-term increase that peaked in the Barremian of Early Cretaceous.

The Karoo–Ferrar volcanic province (southern Africa and preseparation Antarctica with extensive tuffs into preseparation Argentina) erupted during the basal and earliest Toarcian. As with many large igneous provinces, it coincides and was probably the cause of the carbon-isotope excursions, biotic turnover, organic-enrichment horizons, and elevated temperatures during the early Toarcian.

Numerical age model

GTS2012 age model; and potential future enhancements

The age of the base of the Jurassic is well constrained by U-Pb dates (see following). The Hettangian and lower Sinemurian are scaled by cyclostratigraphy in ammonite-zoned successions that includes the Sinemurian GSSP (Ruhl et al., 2010). However, due to lack of radiometric dating and verified cyclostratigraphy, the majority of the Sinemurian and Pliensbachian scaling in GTS2012 assumed a linearly decreasing $^{87}Sr/^{86}Sr$ ratio in ammonite-zoned successions in England tied to the negative excursion in carbon isotopes at the base-Toarcian in a reversed-polarity zone that was considered coeval with the initial reversed-polarity eruptive episode of the Karoo–Ferrar volcanic province.

There are essentially no direct radioisotope dates on ammonite-zoned deposits for the majority of the Jurassic that meets the standards used for most other geologic periods; therefore, the majority of the Jurassic age model is extrapolated using a combination of cycle stratigraphy and magnetostratigraphic correlations to a spreading-rate model for the M-sequence of Pacific marine magnetic anomalies. The middle Toarcian through Aalenian is scaled according to cyclostratigraphy interpretations of deposits in the Paris Basin (Huret et al., 2008; as revised by C. Huang in Ogg et al., 2012). This floating astronomical scale is constrained by the age for Aalenian–Bajocian boundary as calibrated by correlation of its magnetostratigraphy to the upper part of the relatively long reversed-polarity marine magnetic anomaly M44n.1r.8. The age model for this oldest part of the M-sequence incorporates a slow decrease in Pacific spreading rates derived from a linear fit to cyclostratigraphic-derived durations for groups of polarity zones/anomalies and constraints from selected radio-isotopic dates (e.g., Ocean Drilling Program [ODP] Site 801C) (detailed in Ogg, 2012).

The GTS2012 age model has been partially supported and partially questioned by later radioisotopic and cyclostratigraphic studies; but requires independent verification studies before making major revisions according to any single publication. Therefore, only the base-Toarcian in this GTS2016 was provisionally shifted ca. 1 myr older due to revised and enhanced dating by later studies. An example of a pending potential future revision possibility is a $^{40}Ar/^{39}Ar$ date of 156.1 ± 0.9 Ma from a volcanic ash layer that is considered to equate to Middle Oxfordian levels in another section (Pellenard et al., 2013). This report and one or two other published dates, if verified independently, would imply that much of the Late Jurassic age model for the M-sequence or a set of magnetostratigraphic correlations should be shifted about 3 myr younger than the GTS2012 age model.

Revised ages compared to GTS2012

Jurassic (Hettangian) base (minor shift): The dating of the bracketing ash beds by Schoene et al. (2010) was recalculated by them (in Wotzlaw et al., 2014) using updated EARTHTIME tracers. This changed the interpolated boundary age from 201.31 ± 0.18 Ma (as used in GTS2012) to a slightly older **201.36 ± 0.17 Ma.**

Sinemurian (199.4 ± 0.3 vs 199.3 ± 0.3 Ma in GTS2012): The duration of the Hettangian from cyclostratigraphy determines the assigned age for the Hettangian–Sinemurian boundary; therefore becomes slightly younger to be consistent with the revised base-Hettangian date.

Pliensbachian (191.36 ± 1.0 vs 190.8 ± 1.0 Ma in GTS2012): The projected age for the Sinemurian–Pliensbachian stage boundary is currently scaled relative to the assigned ages from radioisotopes and cyclostratigraphy for the base-Toarcian and base-Sinemurian; therefore, the age was adjusted when those other boundaries were revised.

Toarcian (183.7 ± 0.5 vs 182.7 ± 0.7 Ma in GTS2012): The most significant change from the GTS2012 age model is the projected age of the base Toarcian and its negative excursion in carbon isotopes during a reversed-polarity zone which was correlated to the onset of the Karoo–Ferrar volcanism. That onset and base-Toarcian boundary had been dated as 182.7 ± 0.7 Ma (e.g., Jourdan et al., 2007; Svensen et al., 2007); but those authors and others have revised their syntheses, especially in indicating an older age for the following normal-polarity most-voluminous phase of Karoo volcanism (ca. 183.0 Ma by Svensen et al., 2012) that is considered to initiate a second major negative excursion in the beginning of the second Toarcian ammonite zone. This conclusion is supported by dating an ash bed (183.2 ± 0.3 Ma; Sell et al., 2014) in the upper part of the lowest Toarcian ammonite zone that precedes this excursion. When incorporating an estimated 0.8-myr duration for this lowest Toarcian ammonite zone from cyclostratigraphy (Suan et al., 2008), then the base of the Toarcian is now projected as 183.7 Ma, or ca. 1-myr older than in the GTS2012 age model. However, because the upper Toarcian through Aalenian ammonite zones are scaled according to cyclostratigraphy in the French Basin with projected ages from the M-sequence spreading rate model, then the shift of base-Toarcian to this revised older date results in an anomalously long Middle Toarcian. Direct dating on the upper Toarcian through Bathonian stages are required to obtain a reliable age model that does not rely on these extrapolations.

Oxfordian (163.1 ± 1.1 vs 163.5 ± 1.1 Ma in GTS2012): The revised working definition of the basal Oxfordian ammonite zone shifted one ammonite horizon from the Oxfordian into the Callovian compared to that used on the GTS2012 scale, and this current ammonite marker is about 0.4 myr younger. This change in boundary definition does not significantly change the assigned dates to any other biostratigraphy datums, magnetic polarity, or other stratigraphic levels.

Cretaceous (Berriasian) base (**145.7 ± 0.8 vs 145.0 ± 0.8 Ma in GTS2012):** In GTS2012, a

GSSPs of the Jurassic Stages, with location and primary correlation criteria

Stage	GSSP Location	Latitude, Longitude	Boundary Level	Correlation Events	Reference
Tithonian	*candidates are Mt. Crussol or Canjuers (SE France) and Swabia, Germany*			*Near bases of* Hybonoticeras hybonotum *ammonite zone, of* Gravesia *genus, and of polarity chron M22An*	
Kimmeridgian	*candidate is Flodi-garry (Isle of Skye, NW Scotland)*			*Near base of* Pictonia flodigarriensis *ammonite zone of Boreal realm, and base of polarity chron M26r*	
Oxfordian	*Thuoux section, SE France*			*Ammonite, FAD of* Brightia thuouxensis	
Callovian	*candidates are Pfeffingen, Swabian Alb, SW Germany, and Novgorod region, Russia*			*Ammonite, FAD of Kepplerites (Kosmo-ceratidae); defines base of Macrocepha-lites herveyi Zone in UK and SW Germany*	
Bathonian	Ravin du Bès, Bas-Auran area, Alpes de Haute Provence, France	43°57'38"N 6°18'55"E*	base of limestone bed RB071	Ammonite, FAD of *Gonolkites convergens*	Episodes **32**/4, 2009
Bajocian	Murtinheira Section, Cabo Mondego, Portugal	40°11'57"N 8°54'15"W*	base of Bed AB11 of the Murtinheira Section	Ammonite, FAD of *Hyperlioceras mundum*	Episodes **21**/1, 1997
Aalenian	Fuentelsaz, Spain	41°10'15"N 1°50'W	base of Bed FZ 107 in Fuentelsaz Section	Ammonite, FAD of *Leioceras opalinum*	Episodes **24**/3, 2001
Toarcian	Ponta do Trovao, Peniche, Lusitanian Basin, Portugal	39°22'15"N 9°23'07"W	base of micritic lime-stone bed 15e	Ammonite, FAD of *Dactylioceras (Eodactylioceras) simplex*	
Pliensbachian	Robin Hood's Bay, Yorkshire Coast, England	41°10'15"N 1°50'W	base of Bed 73b at Wine Haven, Robin Hood's Bay	Ammonite, FAD of *Bifericeras donovani*	Episodes **29**/2, 2006
Sinemurian	East Quantoxhead, SW England	51°11'27.3"N 3°14'11.2"W*	0.90m above the base of Bed 145	Ammonite, FAD of *Vermiceras quantoxense*	Episodes **25**/1, 2002
Hettangian	Kuhjoch Section, Karwendel Mtns, Austria	47°29'02"N 11°31'50"E	5.80m above base of the Tiefengraben Mbr. of the Kendel-bach Formation	Ammonite, FAD of *Psiloceras spelae*	Episodes **36**/3, 2013

* according to Google Earth

Figure 12.5 Ratified GSSPs and potential primary markers under consideration for defining the Jurassic stages (status as of early 2016). Details of each GSSP are available at http://www.stratigraphy.org, https://engineering.purdue. edu/Stratigraphy/gssp/, and in the *Episodes* publications.

working definition of the base of magnetic Chron M18r was used to place the Jurassic–Cretaceous boundary. Now, as detailed at the beginning of the Cretaceous chapter, the current Jurassic–Cretaceous working group is favoring a slightly older level in the middle of Chron M19n at the "explosion" of small, globular *Calpionella alpina* (base of Calpionellid Zone B). Therefore, both options are shown for the base of the Cretaceous.

Estimated uncertainties on assigned ages on stage boundaries

The few radioisotopic dates with well-constrained biostratigraphic ages that constrain the Jurassic time scale typically have an uncertainty greater than 0.4 myr, when including external factors. In contrast, the durations for those stages that are scaled from cyclostratigraphy (e.g., Hettangian; Aalenian; Kimmeridgian–Tithonian; and portions of other stages) and for many events within those stages are considered to have an uncertainty of ca. 100-kyr (short-eccentricity cycle. The greatest uncertainty on the estimated numerical ages is for the "floating" timescale of the Aalenian through Tithonian stages, which is mainly derived from their magneto-stratigraphic correlations to a spreading-rate model for the Pacific M-sequence. Therefore, the GTS2012 age model projected an increasing ±0.8 to ±1.4 myr uncertainty for most of the Aalenian through base-Berriasian stage boundaries and "working definitions" for the yet-to-be-formalized stage boundaries of Callovian through Berriasian depending on their temporal "distance" from the anchoring radioisotopic dates at the lowermost Toarcian and the lowermost Aptian.

Acknowledgments

This brief Jurassic summary and update of the Jurassic chapter in GTS2012 was assisted by extensive discussions during the past four years on placement of GSSPs and improved zonations within stages with numerous colleagues. These include (*in alphabetical order*) Slah Boulila, Angela Coe, Bruno Galbrun, Andy Gale, Steve Hesselbo, Chunju Huang, Linda Hinnov, Mark Hounslow, Hugh Jenkyns, Pierre Pellenard, and Bill Wimbledon. All of them recognize that the age model and many of the calibrations require future research. Steve Hesselbo reviewed a draft of this chapter for general presentation.

Selected publications and websites

Cited publications

Only select publications were cited in this review with an emphasis on aspects of post-2011 updates. Pre-2011 literature is summarized in the synthesis by Ogg et al. (2012) and in some of the publications cited in the following.

Blackburn, T.J., Olsen, P.E., Bowring, S.A., McLean, N.M., Kent, D.V., Puffer, J., McHone, G., Rasbury, E.T., Et-Touhami, M., 2013. Zircon U-Pb geochronology links the end-Triassic extinction with the Central Atlantic Magmatic Province. *Science* **340**: 941–945. http://dx.doi.org/10.1126/science.1234204.

Bodin, S., Mattioli, E., Fröhlich, S., Marshall, J.D., Boutib, L., Lahsini, S., Redfern, J., 2010. Toarcian carbon isotope shifts and nutrient changes from the Northern margin of Gondwana (High Atlas, Morocco, Jurassic): palaeoenvironmental implications. *Palaeogeography, Palaeoclimatology, Palaeoecology* **297**: 377–390.

Bown, P.R. (Ed.), 1998. *Calcareous Nannofossil Biostratigraphy*. Chapman & Hall, London. 328 pp.

Callomon, J.H., Dietl, G., 2000. On the proposed basal boundary stratotype (GSSP) of the Middle Jurassic Callovian Stage. In: Hall, R.L., Smith, P.L. (Eds.), *Advances in Jurassic Research*, GeoResearch Forum, **6**, pp. 41–54.

Casellato, C.E., 2010. Calcareous nannofossil biostratigraphy of upper Callovian-lower Berriasian successions from the Southern Alps, North Italy. *Rivista Italiana di Paleontologia e Stratigrafia* **116**: 357–404.

Channell, J.E.T., Casellato, C.E., Muttoni, G., Erba, E., 2010. Magnetostratigraphy, nannofossil stratigraphy and apparent polar wander for Adria-Africa in the Jurassice-Cretaceous boundary interval. *Palaeogeography, Palaeoclimatology, Palaeoecology* **293**: 51–75.

Dromart, G., Garcia, J.P., Picard, S., Atrops, F., Lécuyer, C., Sheppard, S.M.F., 2003. Ice age at the Middle-Late Jurassic transition? *Earth and Planetary Science Letters* **213**: 205–220.

Gipe, R.A., 2013. Callovian (upper Middle Jurassic) magnetostratigraphy: a composite polarity pattern from France, Britain and Germany, and its correlation to the Pacific marine magnetic anomaly model. Purdue University Open Access Theses. Paper 36: 107 pp. http://docs.lib.purdue.edu/open_access_theses/36. Plus abstract at, Gipe, R.A., Ogg, J.G., and Coe, A.L., 2013. Magnetostratigraphy of the Callovian (upper Middle Jurassic) from outcrops in France, England and Germany and calibration of the Pacific M-sequence of magnetic anomalies. Geological Society of America Abstracts With Programs. 45, No. 7: 811, https://gsa.confex.com/gsa/2013AM/webprogram/Paper231332.html.

Głowniak, E., Wierzbowski, A., 2007. Comment on mid-Oxfordian (Late Jurassic) positive carbon-isotope excursion recognized from fossil wood in the British Isles by C.R. Pearce, S.P. Hesselbo, A.L. Coe, Palaeogeography, Palaeoclimatology. *Palaeoecology* **221**: 343–357 Palaeogeography, Palaeoclimatology, Palaeoecology, 248: 252–254.

Hardenbol, J., Thierry, J., Farley, M.B., Jacquin, Th, de Graciansky, P.-C., Vail, P.R., with numerous contributors, 1998. Mesozoic and Cenozoic sequence chronostratigraphic framework of European basins. In: de Graciansky, P.-C., Hardenbol, J., Jacquin, Th, Vail, P.R. (Eds.), *Mesozoic-Cenozoic Sequence Stratigraphy of European Basins*, **60**. SEPM Special Publication, pp. 763–781.

von Hillebrandt, A., Krystyn, L., 2009. On the oldest Jurassic ammonites of Europe (Northern Calcareous Alps, Austria) and their global significance. *Neues Jahrbuch für Geologie und Paläontologie Abhandlungen* **253**: 163–195.

von Hillebrandt, A., Krystyn, L., Kürschner, W.M., Bonis, N.R., Ruhl, M., Richoz, S., Schobben, M.A.N., Urlichs, M., Bown, P.R., Kment, K., McRoberts, C.A., Simms, M., Tomãsov́ych, A., 2013. Global Stratotype Sections and Point (GSSP) for the base of the Jurassic System at Kuhjoch (Karwendel Mountains, Northern Calcareous Alps, Tyrol, Austria). *Episodes* **36**: 162–198.

Huret, E., Hinnov, L.A., Galbrun, B., Collin, P.-Y., Gardin, S., Rouget, I., 2008. Astronomical calibration and correlation of the Lower Jurassic, Paris and Lombard basins (Tethys). In: *33rd International Geological Congress*. Norway, abstract, Oslo.

Hüsing, S.K., Beniest, A., van der Boon, A., Abels, H.A., Deenen, M.H.L., Krijgsman, W., 2014. Astronomically-calibrated magnetostratigraphy of the Lower Jurassic marine successions at St. Audrie's Bay and East Quantoxhead (Hettangian-Sinemurian; Somerset, UK). *Palaeogeography, Palaeoclimatology, Palaeoecology* **403**: 43–56.

Jenkyns, H.C., Jones, C.E., Gröcke, D.R., Hesselbo, S.P., Parkinson, D.N., 2002. Chemostratigraphy of the Jurassic System: applications, limitations and implications for palaeoceanography. *Journal of the Geological Society* **159**: 351–378.

Jourdan, F., Féraud, G., Bertrand, H., Watkeys, M.K., Renne, P.R., 2007. Distinct brief major events in the Karoo large igneous province clarified by new $^{40}Ar/^{39}Ar$ ages on the Lesotho basalts. *Lithos* **98**: 195–209.

Kemp, D.B., Coe, A.L., Cohen, A.S., Schwark, L., 2005. Astronomical pacing of methane release in the Early Jurassic Period. *Nature* **437**: 396–399.

Kent, D.V., Olsen, P.E., 2008. Early Jurassic magneto-stratigraphy and paleolatitudes from the Hartford continental rift basin (eastern North America): testing for polarity bias and abrupt polar wander in association with the central Atlantic magmatic province. *Journal of Geophysical Research* **113**(B6): 105. http://dx.doi.org/10.1029/2007JB005407.

Kiselev, D., Rogov, M., Glinskikh, L., Guzhikov, A., Pimenov, M., Mikhailov, A., Dzyuba, O., Matveev, A., Tesakova, E., 2013. Integrated stratigraphy of the reference sections for the Callovian-Oxfordian boundary in European Russia. *Volumina Jurassica* **11**: 59–96.

Korte, C., Hesselbo, S.P., 2011. Shallow marine carbon and oxygen isotope and elemental records indicate icehouse-greenhouse cycles during the Early Jurassic. *Paleoceanography* **26**: PA4219. http://dx.doi.org/10.1029/2011PA002160 (18 pp).

Korte, C., Hesselbo, S.P., Ullmann, C.V., Dietl, G., Ruhl, M., Scweigert, G., Thibault, N., 2015. Jurassic climate mode governed by ocean gateway (7 pp. plus 2 on-line suppliments) *Nature Communications* **6**: 10015. http://dx.doi.org/10.1038/ncomms10015.

Lucas, S.G., 2009. Global Jurassic tetrapod biochronology. *Volumina Jurassica* **6**: 99–108.

Mattioli, E., Erba, E., 1999. Synthesis of calcareous nannofossil events in Tethyan Lower and Middle Jurassic successions. *Rivista Italiana di Paleontologia e Stratigrafia* **105**: 343–376.

Mönnig, E., 2014. Report of the Callovian Stage Task Group, 2013. *Volumina Jurassica* **12**(1): 197–200.

Muttoni, G., Mazza, M., Mosher, D., Katz, M.E., Kent, D.V., Balini, M., 2014. A Middle-Late Triassic (Ladinian-Rhaetian) carbon and oxygen isotope record from the Tethyan Ocean. *Palaeogeography, Palaeoclimatology, Palaeoecology* **399**: 246–259 (and on-line data repository item 2015069).

Ogg, J.G., Coe, A.L., Przybylski, P.A., Wright, J.K., 2010. Oxfordian magnetostratigraphy of Britain and its correlation to Tethyan regions and Pacific marine magnetic anomalies. *Earth and Planetary Science*

Letters **289**: 433–448. http://dx.doi.org/10.1016/j.
epsl.2009.11.031.

Ogg, J.G., 2012. Geomagnetic polarity time scale. In:
Gradstein, F.M., Ogg, J.G., Schmitz, M., Ogg, G.,
(Coordinators), *The Geologic Time Scale 2012.*
Elsevier Publ, pp. 85–113.

Ogg, J.G., Hinnov, L.A., Huang, C., 2012. Jurassic. In:
Gradstein, F.M., Ogg, J.G., Schmitz, M., Ogg, G.,
(Coordinators), *The Geologic Time Scale 2012.*
Elsevier Publ., pp. 731–791 (An overview on all
aspects, including graphics on the ratified GSSPs
of the stages, diagrams and tables for the biostratigraphic scales, and discussion on the age
models.).

Page, K.N., Melendez, G., Hart, M.B., Price, G., Wright, J.K.,
Bown, P., Bello, J., 2010. Integrated stratigraphic study
of the candidate Oxfordian Global Stratotype Section
and Point (GSSP) at Redcliff Point, Weymouth, Dorset,
UK. *Volumina Jurassica* **7**: 101–111.

Pellenard, P., Nomade, S., Martire, L., De Oliveira
Ramalho, F., Monna, F., Guillou, H., 2013. The first
⁴⁰Ar-³⁹Ar date from Oxfordian ammonite-cali-
brated volcanic layers (bentonites) as a tie-point
for the Late Jurassic. *Geological Magazine* **150**:
1136–1142.

Pellenard, P., Fortwengler, D., Marchand, D., Thierry, J.,
Bartolini, A., Boulila, S., Collin, P.-Y., Enay, R.,
Galbrun, B., Gardin, S., Huault, V., Huret, E.,
Martinez, M., Chateau Smith, C., 2014a. Integrated
stratigraphy of the Oxfordian global stratotype
section and point (GSSP) candidate in the Subalpine
Basin (SE France). *Volumina Jurassica* **12**(1): 1–44.

Pellenard, P., Tramoy, R., Pucéat, E., Huret, E., Martinez,
M., Brueau, L., Thierry, J., 2014b. Carbon cycle and
sea-water palaeotemperature evolution at the
Middle-Late Jurassic transition, eastern Paris Basin
(France). *Marine and Petroleum Geology* **53**: 30–43.
http://dx.doi.org/10.1016/j.marpetgeo.2013.07.002.

Przybylski, P.A., Ogg, J.G., Wierzbowski, A., Coe, A.L.,
Hounslow, M.W., Wright, J.K., Atrops, F., Settles, E.,
2010. Magnetostratigraphic correlation of the
Oxfordian-Kimmeridgian boundary. *Earth Planet.
Sci. Lett.* **289**: 256–272.

Rocha, R.B., Mattioli, E., Duarte, L.V., Pittet, B., Elmi, S.,
Mouterde, R., Cabral, M.C., Comas-Rengifo, M.J.,
Gómez, J.J., Goy, A., Hesselbo, S.P., Jenkyns, H.C.,
Littler, K., Mailliot, S., Oliveira, L.C.V., Osete, M.L.,
Perilli, N., Pinto, S., Ruget, C., Suan, G., 2016. The
Global Stratotype Section and Point (GSSP) of the
Toarcian Stage at the base of the *polymorphum* zone
in the Peniche section (Portugal). *Episodes* in press.

Ruhl, M., Deenen, M.H.L., Abels, H.A., Bonis, N.R.,
Krijgsman, W., Kürschner, W.M., 2010. Astronomical
constraints on the duration of the early Jurassic

Hettangian stage and recovery rates following the
end-Triassic mass extinction (St Audrie's Bay/East
Quantoxhead, UK). *Earth and Planetary Science
Letters* **295**: 262–276.

Schoene, B., Guex, J., Bartolini, A., Schaltegger, U.,
Blackburn, T.J., 2010. Correlating the end-Triassic
mass extinction and flood basalt volcanism at the
100 ka level. *Geology* **38**: 387–390.

Schweigert, G., 2015. Chapter 14. Ammonoid biogeogra-
phy in the Jurassic. In: Klug, C., Korn, D., De Baets,
K., Kruta, I., Mapes, R.H. (Eds.), *Ammonoid Paleobiology: From Macroevolution to Paleogeography.*
Topics in Geobiology, **44**. Springer Publ, pp. 389–402.
http://dx.doi.org/10.1007/978-94-017-9633-0_13.

Scotese, C.R., 2014. *Atlas of Jurassic Paleogeographic
Maps, PALEOMAP Atlas for ArcGIS, Volume 4, the
Jurassic, Maps 32-42, Mollweide Projection.* PALEO-
MAP Project, Evanston, IL. https://www.academia.
edu/11245556/Atlas_of_Jurassic_
Paleogeographic_Maps.

Sell, B., Ovtcharova, M., Guex, J., Bartolini, A.,
Jourdan, F., Spangenberg, J.E., Vicente, J.-C., Schaltegger, U., 2014. Evaluating the temporal link between
the Karoo LIP and climatic–biologic events of the
Toarcian Stage with high-precision U–Pb geochronology. *Earth and Planetary Science* **408**: 48–56.

Suan, G., Pittet, B., Bour, I., Mattioli, E., Duarte, L.V.,
Mailliot, S., 2008. Duration of the Early Toarcian
carbon isotope excursion deduced from spectral
analysis – Consequence for its possible causes. *Earth
and Planetary Science Letters* **267**: 666–679.

Svensen, H., Planke, S., Chevallier, L., Malthe-Sørenssen,
A., Corfu, F., Jamtveit, B., 2007. Hydrothermal venting
of greenhouse gases triggering Early Jurassic global
warming. *Earth and Planetary Science Letters* **256**:
554–566.

Svensen, H., Corfu, F., Polteau, S., Hammer, Ø., Planke,
S., 2012. Rapid magma emplacement in the Karoo
large igneous province. *Earth and Planetary Science
Letters* **325–326**: 1–9.

Tivey, M.A., Sager, W.W., Lee, S.-M., Tominaga, M., 2006.
Origin of the Pacific Jurassic quiet zone. *Geology* **34**:
789–792.

Tominaga, M., Sager, W.W., Tivey, M.A., Lee, S.-M., 2008.
Deep-tow magnetic anomaly study of the Pacific
Jurassic Quiet Zone and implications for the
geomagnetic polarity reversal time scale and
geomagnetic field behavior. *Journal of Geophysical
Research* **B07110**(113). http://dx.doi.
org/10.1029/2007JB005527 20 pp.

Tominaga, M., Tivey, M.A., Sager, W.W., 2015. Nature of
the Jurassic magnetic quiet zone. *Geophysical
Research Letters* **42**: 8367–8372. http://dx.doi.org/10.
1002/2015GL065394.

Veizer, J., Prokoph, A., 2015. Temperatures and oxygen isotopic composition of Phanerozoic oceans. *Earth-Science Reviews* **146**: 92–104.

Wierzbowski, A., Coe, A.L., Hounslow, M.W., Matyja, B.A., Ogg, J.G., Page, K.N., Wierzbowski, H., Wright, J.K., 2006. A potential stratotype for the Oxfordian/ Kimmeridgian boundary: Staffin Bay, Isle of Skye, U.K. *Volumina Jurassica* **4**: 17–33.

Wierzbowski, A., Matyja, B.A., 2014. Ammonite biostratigraphy in the Polish Jura sections (central Poland) as a clue for recognition of the uniform base of the Kimmeridgian Stage. *Volumina Jurassica* **12**(1): 45–98.

Wierzbowski, H., 2015. Seawater temperatures and carbon isotope variations in Central European basins at the Middle-Late Jurassic transition (Late Callovian–Early Kimmeridgian). *Palaeogeography, Palaeoclimatology, Palaeoecology* **440**: 506–523.

Wotzlaw, J.-F., Guex, J., Bartolini, A., Gallet, Y., Krystyn, L., McRoberts, C.A., Taylor, D., Schoene, B., Schltegger, U., 2014. Towards accurate numerical calibration of the Late Triassic: high-precision U-Pb geochronology constraints on the duration of the Rhaetian. *Geology* **42**: 571–574.

Websites (selected)

International Subcommission on Jurassic Stratigraphy (of ICS)— *http://jurassic.earth.ox.ac.uk*—details on ratified GSSPs and status of future ones.

Jurassic of Russia and adjacent areas (maintained by Mikhaile Rogov and colleagues) —*http://www.jurassic.ru/eindex.htm*—extensive collection of Portable Document Formats (PDFs), including scans of difficult-to-find Russian papers; and hosting Callovian Working Group Website (*http://jurassic.ru/callovian.htm*)

The Jurassic Coast UNESCO World Heritage site; Dorset, United Kingdom) —*http://jurassiccoast.org*—

Volumina Jurassica (journal; published by Polish Geological Institute of the National Research Institute and the Faculty of Geology, University of Warsaw) —http://www.voluminajurassica.org.

Palaeos: Jurassic—*http://palaeos.com/mesozoic/jurassic/jurassic.htm*—A well-presented suite of diverse topics for a general science audience that was originally compiled by M. Alan Kazlev in 1998–2002.

13

CRETACEOUS

96.6 Ma Cretaceous

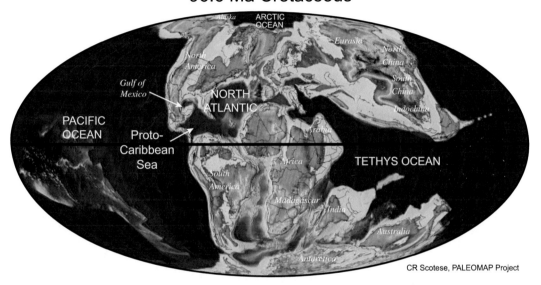

CR Scotese, PALEOMAP Project

Cenomanian paleogeographic reconstruction (Sea level + 80 m) from Scotese (2014).

Basal definition and international subdivisions

The Cretaceous, named from "*Creta*" (Latin for *chalk*), has 12 stages in 2 series (Fig. 13.1).

The Cretaceous is the only system/period of the Phanerozoic that has not yet been defined by a basal GSSP. Contributing factors include the lack of any significant evolutionary or physical/chemical event on either the regional or global scale within the commonly used boundary interval, the challenge to correlate any biostratigraphic datum among paleogeographic provinces, and the divergent traditional working definitions within each province (e.g., review in Wimbledon, 2014). The

current Jurassic–Cretaceous boundary working group has used the distinctive magnetic polarity pattern of the transitional interval to intercorrelate the main zonal schemes, paleontological markers, and regional reference sections (Fig. 13.2). Therefore, in GTS2012, the base of reversed-polarity Chron M18r was used as a temporary boundary placement for the base of the Berriasian stage (Wimbledon et al., 2011). The current boundary working group is favoring a slightly older level in the middle of Chron M19n at the "explosion" of small, globular *Calpionella alpina* (base of Calpionellid Zone B) in deep-shelf to pelagic limestones on either side of the former Tethys seaway from modern Mexico to Iraq. Both

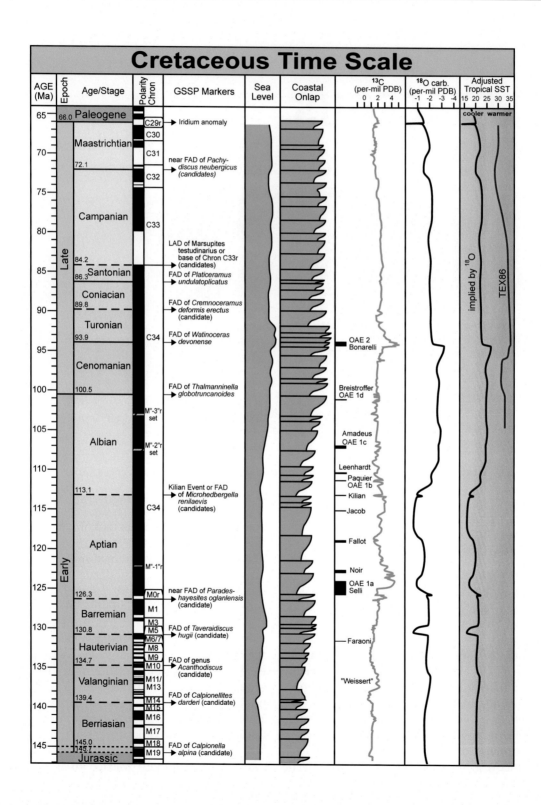

◀ **Figure 13.1 Cretaceous overview.** Main markers or candidate markers for Global Boundary Stratotype Sections and Points (GSSPs) of stages are detailed in text and in Fig. 13.5. ("Age" is the term for the time equivalent of the rock-record "stage".) See text and Fig. 13.2 for Jurassic/Cretaceous boundary options. Coastal onlap and schematic sea-level curve is modified from Haq (2014). The $\delta^{13}C$ curve (with widespread anoxic events and the "Weissert" positive excursion) is a synthesis of generalized trends with relative magnitudes: Berriasian through middle–late Aptian from Föllmi et al. (2006) with early Aptian modifications from Renard et al. (2005); late Aptian to early Albian generalized from Herrle et al. (2004) and Kennedy et al. (2014); middle and late Albian from Gale et al. (2011); Cenomanian through Campanian from Jarvis et al. (2006); early Maastrichtian from Voigt et al. (2010, 2012); and late Maastrichtian from Thibault et al. (2012); although values of the observed $\delta^{13}C$ measurements are usually systematically offset among different regions. The Late Cretaceous $\delta^{18}O$ curve is a schematic merger from several sources: Late Cretaceous from Norris et al. (2001) and Huber et al. (2002); early and middle Albian from Bottini et al. (2015); early Barremian from Mutterlose et al. (2014); and the other intervals and estimates of tropical sea-surface temperatures derived from those oxygen-18 values are modified from Veizer and Prokoph (2015). Estimates of tropical temperatures from the tetraether index of 86 carbon atoms (TEX86) technique (Forster et al., 2007; Linnert et al., 2014) are significantly hotter. *PDB*, PeeDee Belemnite ^{13}C and ^{18}O standard; *SST*, sea-surface temperature.

options for a future international Tithonian–Berriasian stage boundary are shown on the synthesis diagrams for the Jurassic and the Cretaceous in this handbook.

Lower Cretaceous

The Berriasian through Aptian stages of the Lower Cretaceous were originally derived from exposures in southeast France and adjacent northwest Switzerland. Unfortunately, the ranges for most of the fauna and microflora thriving in the tropical ocean of this Tethyan Realm during the Early Cretaceous did not extend into the colder Boreal Realm of northwest Europe and other northern regions. Nevertheless, the applications of strontium- and carbon-isotope trends, calcareous nannofossils, and other methods have constrained correlations and aided in the decisions for assigning these stage boundaries (e.g., Reboulet et al., 2014; Mutterlose et al., 2014). As of late 2015, only a GSSP for the Albian Stage has been submitted for formal recognition of a Lower Cretaceous stage boundary. The status of the probable biostratigraphic or other markers for the other stage boundaries are summarized in Figs. 13.1 and 13.5, and briefly below:

Valanginian: The proposed GSSP marker is the first appearance datum (FAD) of the calpionellid *Calpionellites darderi* (base of calpionellid Zone E), which is recorded from Mexico to Turkey, is nearly coincident with the base of the Tethyan ammonite *Thurmanniceras pertransiens* Zone, is just below the FAD of calcareous nannofossil *Calcicalathina oblongata*, and is used as the basal level in a detailed cycle-scaling of the ammonite and other events through the entire Valanginian stage in southeastern France (e.g., Charbonnier et al., 2013; Martinez et al., 2013). The leading GSSP candidate that also enables a direct correlation to the magnetic polarity scale (lower part of polarity Chron M14r) is near Caravaca in southern Spain (Aguado et al., 2000; Cretaceous Subcommission section of International Commission on Stratigraphy [ICS] annual report, 2014).

Hauterivian: The leading GSSP candidate is the base of the Tethyan ammonite *Acanthodiscus radiatus* Zone in the La Charce section of southeastern France (Cretaceous Subcommission section of ICS annual report, 2014). However, its precise correlation to the Boreal ammonite zonation is uncertain [e.g., compare Mutterlose et al. (2014) with Reboulet et al. (2014)], and the calibration of this event to the magnetic polarity scale is not yet determined (e.g., Martinez et al., 2015).

Barremian: The proposed GSSP and primary marker is the FAD of the Tethyan

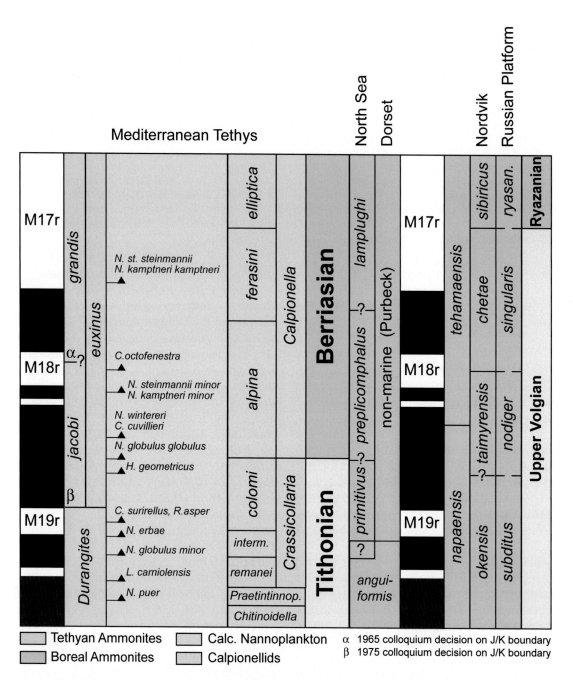

Figure 13.2 Relative placement of selected markers for defining the Jurassic–Cretaceous boundary among paleo-geographic regions of the Sub-Mediterranean Tethys, Sub-Boreal (North Sea, Dorset), and Boreal (Nordvik, Russian Platform) realms. Modified from Schnabl et al. (2015). In GTS2012, the preferred working definition was the base of Chron M18r (Wimbledon et al., 2011); but the working group is currently favoring the base of the calpionellid *C. alpina* Zone.

Base of the Cenomanian Stage of the Cretaceous System, Mont Risou, Hautes-Alpes, France.

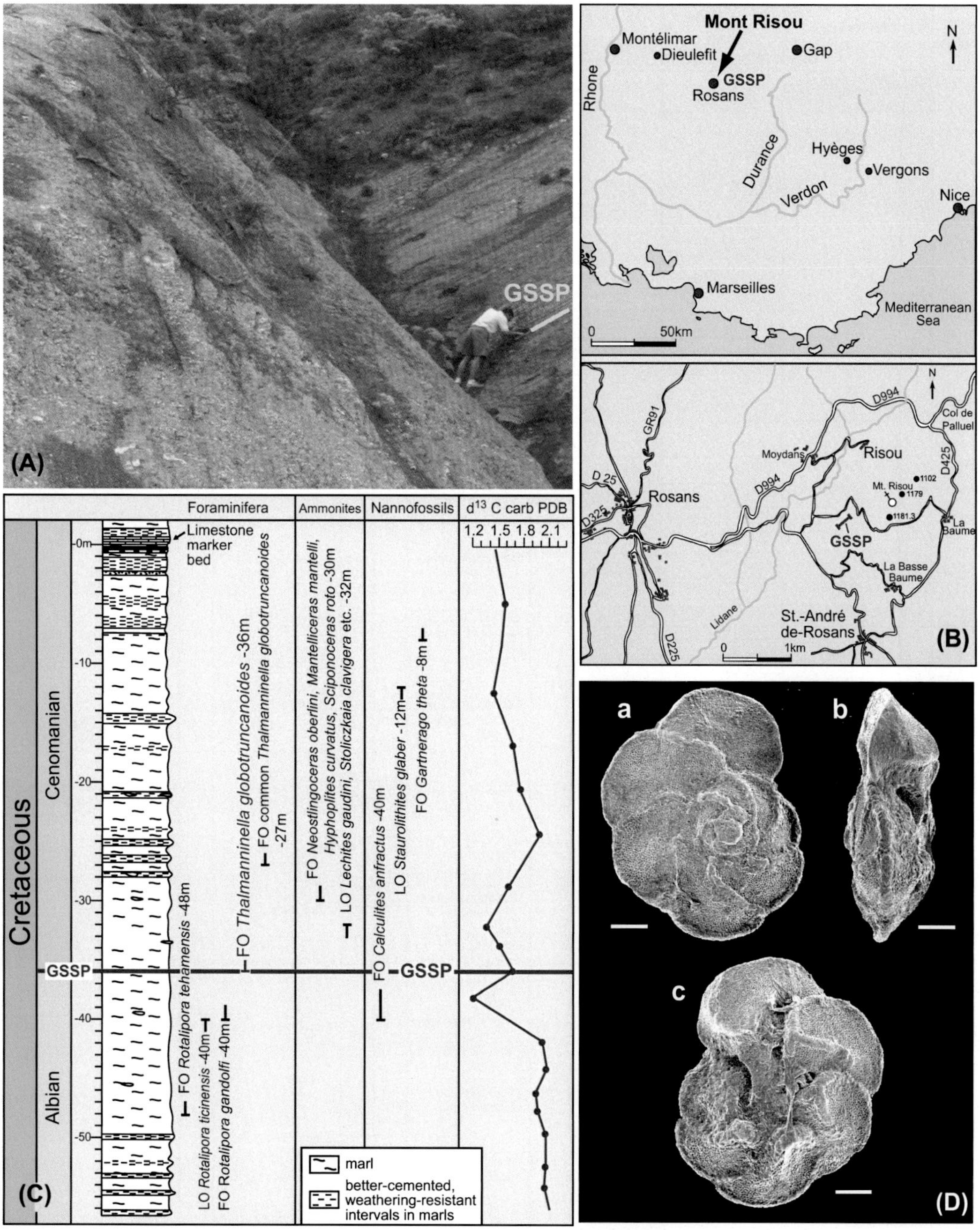

(A)

(B)

(C)

(D)

Figure 13.3 GSSP for base of the Upper Cretaceous (base of Cenomanian Stage) at Mont Risou, southeastern France. The GSSP level coincides with the lowest occurrence of the planktonic foraminifera *Thalmanninella globotruncanoides* (formerly *Rotalipora globotruncanoides*). *FO*, first occurrence; *LO*, last occurrence; *PDB*, PeeDee Belemnite ^{13}C standard. Photographs of the marker foraminifer were provided by Atsushi Ando.

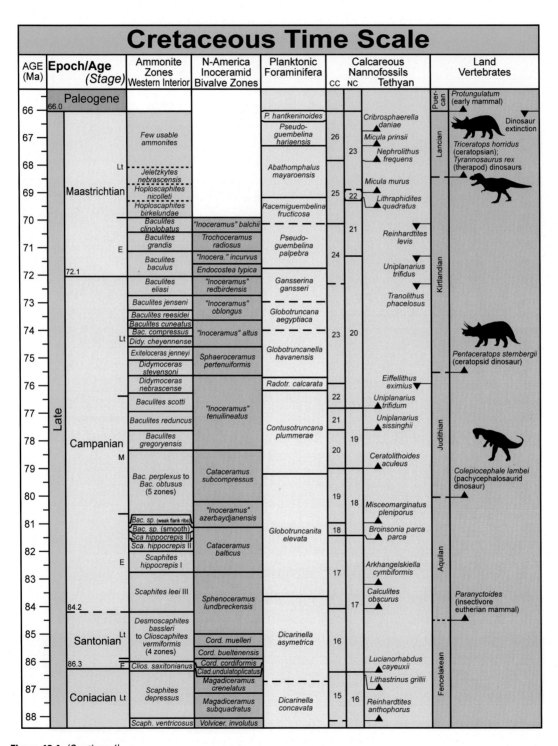

Cretaceous Time Scale

AGE (Ma)	Epoch/Age (Stage)	Ammonite Zones Western Interior	N-America Inoceramid Bivalve Zones	Planktonic Foraminifera	Calcareous Nannofossils		Calcareous Nannofossils Tethyan	Land Vertebrates	
					CC	NC			
66	Paleogene							Puer-can	Protungulatum (early mammal)
	66.0			P. hantkeninoides			Cribrosphaerella daniae		Dinosaur extinction
67		Few usable ammonites		Pseudo-guembelina hariaensis	26	23	Micula prinsii	Lancian	Triceratops horridus (ceratopsian); Tyrannosaurus rex (therapod) dinosaurs
68		Lt		Abathomphalus mayaroensis			Nephrolithus frequens		
69	Maastrichtian	Jeletzkytes nebrascensis			25	22	Micula murus		
		Hoploscaphites nicolleti					Lithraphidites quadratus		
70		Hoploscaphites birkelundae		Racemiguembelina fructicosa					
		Baculites clinolobatus	"Inoceramus" balchii			21			
71	E	Baculites grandis	Trochoceramus radiosus	Pseudo-guembelina palpebra			Reinhardtites levis	Kirtlandian	
72	72.1	Baculites baculus	"Inocera." incurvus		24		Uniplanarius trifidus		
			Endocostea typica						
73		Baculites eliasi	"Inoceramus" redbirdensis	Gansserina gansseri			Tranolithus phacelosus		
		Baculites jenseni	"Inoceramus" oblongus	Globotruncana aegyptiaca					
74	Lt	Baculites reesidei			23	20			Pentaceratops sternbergii (ceratopsid dinosaur)
		Baculites cuneatus							
		Bac. compressus	"Inoceramus" altus						
		Didy. cheyennense		Globotruncanella havanensis					
75		Exiteloceras jenneyi	Sphaeroceramus pertenuiformis						
		Didymoceras stevensoni							
76		Didymoceras nebrascense		Radotr. calcarata			Eiffellithus eximius		
77		Baculites scotti	"Inoceramus" tenuilineatus		22		Uniplanarius trifidum	Judithian	
		Baculites reduncus		Contusotruncana plummerae	21	19	Uniplanarius sissinghii		
78	M	Baculites gregoryensis			20				Colepiocephale lambei (pachycephalosaurid dinosaur)
79	Campanian						Ceratolithoides aculeus		
		Bac. perplexus to Bac. obtusus (5 zones)	Cataceramus subcompressus						
80					19	18	Misceomarginatus pleniporus		
81		Bac. sp. (weak flank ribs)	"Inoceramus" azerbaydjanensis		18		Broinsonia parca parca		
		Bac. sp. (smooth)		Globotruncanita elevata					
		Sca. hippocrepis III							
82	E	Sca. hippocrepis II	Cataceramus balticus				Arkhangelskiella cymbiformis	Aquilan	
		Scaphites hippocrepis I			17				
83		Scaphites leei III					Calculites obscurus		Paranyctoides (insectivore eutherian mammal)
84	84.2		Sphenoceramus lundbreckensis			17			
85	Santonian Lt	Desmoscaphites bassleri to Clioscaphites vermiformis (4 zones)	Cord. muelleri	Dicarinella asymetrica	16			Fencelakean	
86	86.3 E	Clios. saxitonianus	Cord. bueltenensis				Lucianorhabdus cayeuxii		
			Cord. cordiformis						
			Clad. undulatoplicatus				Lithastrinus grillii		
87			Magadiceramus crenelatus						
	Coniacian Lt	Scaphites depressus	Magadiceramus subquadratus	Dicarinella concavata	15	16	Reinhardtites anthophorus		
88		Scaph. ventricosus	Volvicer. involutus						

Figure 13.4 *(Continued)*

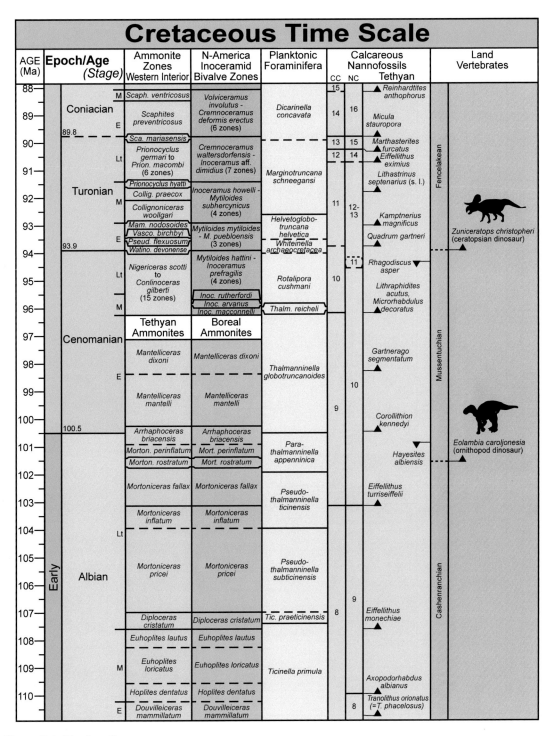

Figure 13.4 *(Continued)*

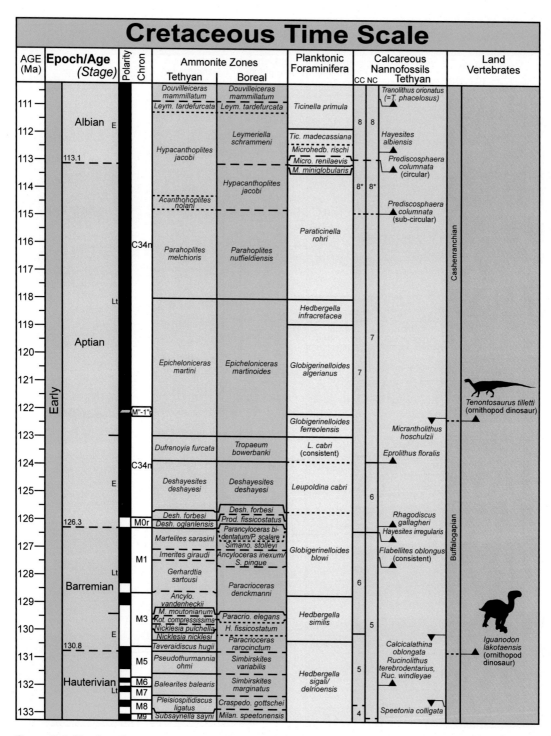

Figure 13.4 *(Continued)*

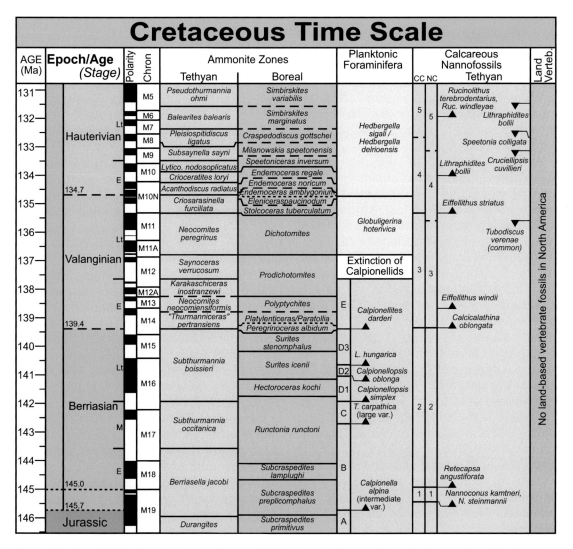

Figure 13.4 (A, B, C, D): Selected marine and terrestrial biostratigraphic zonations of the Cretaceous. (*"Age"* is the term for the time equivalent of the rock-record *"stage."*) Marine macrofossil zones for Late Cretaceous are ammonoids and inoceramids of the Western Interior of North America (e.g., Cobban et al., 2006; with partial modification by A. Gale, pers. comm., 2010; of North America). The Early Cretaceous ammonoid zones for the Tethyan realm (Sub-Mediterranean province) and Boreal realm (Sub-Boreal province) are mainly modified from Reboulet et al. (2014). An expanded listing of ammonoid zones with an explanation of each GTS2012 age calibration is in Ogg et al. (2012). Selected microfossil zones are planktonic foraminifers (e.g., Petrizzo et al., 2012; Huber and Petrizzo, 2014; Coccioni and Premoli Silva, 2015; among many other sources, including B. Huber and M.R. Petrizzo, pers. comm., 2011 and 2016) and Early Cretaceous calpionellid zones (numbered-zone system of Remane, 1998). Early Cretaceous calcareous nannofossil zones for the Tethyan (tropical) realm with selected zone/subzone markers are compiled from several sources, including J. Bergen, P. Bown, J. Lees, and D. Watkins, pers. comm., 2007–15). Land vertebrate zones and markers for North America are from Lucas et al. (2012); of which no fossil record exists for most of the lower Cretaceous. Additional zonations, biostratigraphic markers, geochemical trends, sea-level curves, and details on calibrations are compiled in the internal data sets within the *TimeScale Creator* visualization system (free at www.tscreator.org).

ammonite *Taveraidiscus hugii* at Bed 171 the Rio Argos section near Caravaca in southern Spain (Cretaceous Subcommission section of ICS annual report, 2014). This event is during the latter part of Chron M5n and ca. 0.7 myr after the onset of a regional "Faraoni" organic-enrichment episode that was accompanied by a small positive $\delta^{13}C_{carb}$ event (e.g., Martinez et al., 2012).

Aptian: The base of reversed-polarity Chron M0r at the Gorgo a Cerbara section in the Umbria–Marche Basin in central Italy is the proposed Aptian GSSP (e.g., Patruno et al., 2015). This level approximately coincides with the base of the Tethyan ammonite *Deshayesites oglanlensis* Zone, is above the FAD of the calcareous nannofossil *Rucinolithus irregularis*, and is below a nannoconid crisis.

Albian: A major turnover in planktonic foraminifers, during which long-ranging Aptian taxa are replaced by a brief episode of minute-sized *Microhedbergella*, occurs at the onset of a negative $\delta^{13}C_{carb}$ excursion and the base of a widespread "Kilian" organic-rich level in western Mediterranean sections (e.g., Huber and Leckie, 2011; Huber et al., 2011). The Albian GSSP that has been submitted for ratification (*status in January 2015*) is in the middle of this Niveau Kilian at the Col de Pré-Guittard section in southeastern France, where it coincides with the base of the planktonic foraminifer *Microhedbergella renilaevis* Zone (Petrizzo et al., 2012; Kennedy et al., 2014). This level is near the transition of the calcareous nannofossil *Prediscophaera columnata* from its subcircular to circular morphology; although this lowest occurrence may be diachronous among regions (e.g., review by Herrle et al., 2004).

Upper Cretaceous

Four of the six Upper Cretaceous stages have GSSPs. The primary marker for the Lower–Upper Cretaceous boundary, the GSSP for the **Cenomanian** Stage at Mont Risou,

coincides with the FAD of *Thalmanninella* (formerly *Rotalipora*) *globotruncanoides* (Fig. 13.3), which is slightly earlier than the base of the ammonite *Mantelliceras mantelli* Zone. The **Turonian** GSSP near Pueblo, Colorado, United States, coincides with the FAD of the ammonite *Watinoceras devonense*.

Coniacian: The working group had proposed a composite section for the GSSP candidate based on the FAD of the inoceramid *Cremnoceramus deformis erectus* as the primary correlation marker. This composite would have merged the succession at Salzgitter–Salder Quarry (Lower Saxony, Germany) where the boundary is condensed with a more expanded boundary interval (but limited to a narrow slice of stratigraphic time) at the Slupia Nadbrzena River cliff outcrop (central Poland). Other candidate sections in North America and Kazakhstan have condensed intervals with gaps, safety issues, or other problems, with the possible exception of Big Bend in southwest Texas (Cretaceous Subcommission section of ICS annual report, 2014).

Santonian: The GSSP for the Coniacian–Santonian boundary was ratified in 2013 to coincide with the local lowest occurrence of the inoceramid bivalve *Platyceramus undulatoplicatus* (called *Cladoceramus undulatoplicatus* in the US Western Interior zonation), above a significant inoceramid-barren interval at the Cantera de Margas quarry near Olazagutia in the Navarra province of northern Spain about 8 km from the Bay of Biscay (Lamolda et al., 2014). The lowest occurrence of the planktonic foraminifer *Globotruncana linneiana* within the *Dicarinella asymetrica* Zone is a secondary marker for the basal Santonian (Coccioni and Premoli Silva, 2015). In chalk successions of Britain, the FAD of *Platyceramus undulatoplicatus* is near a $\delta^{13}C$ stable carbon isotope event called "Michel Dean" (Jarvis et al., 2006), which may enable correlation, but there is not yet an unambiguous carbon-isotope signature from the Olazagutia

GSSP. The GSSP section lacks ammonites in this interval; but in the North American Western Interior, this inoceramid FAD marker is just below the base of the ammonite *Clioscaphites saxtonianus* Zone. If the inoceramid FAD in North America is coeval with its local FAD at the Olazagutia GSSP, then the radioisotopic-dated cyclostratigraphy of the Western Interior implies an age assignment of ca. 86.49±0.44 Ma (Sageman et al., 2014).

Campanian: There are two potential criteria as the main marker for the base of the Campanian: (1) The base of magnetic polarity Chron C33r, which is near the base of the ammonite *Scaphites leei III* Zone in the North American Western Interior, and is just above the last-appearance datum (LAD) of the planktonic foraminifer *Dicarinella asymetrica* (base of *Globotruncanita elevata* Zone), and near the FAD of the nannofossil *Aspidolithus parcus parcus* (also called *Broinsonia parca parca*) in Italian sections (e.g., review in Coccioni and Premoli Silva, 2015). (2) The extinction of the pelagic crinoid genus, *Marsupites*, as represented by the LAD of *Mar. testudinarius*, which has been the traditional placement in the chalk facies of the English–Paris–northern Germany basin and is observed worldwide (reviewed in Gale et al., 2008). But these two criteria may have a considerable offset. A high-resolution carbon-isotope stratigraphic comparison between the candidate GSSP section of Bottaccione near Gubbio in Italy and the main chalk reference section in northern Germany has been interpreted to imply that the option of the LAD of *Mar. testudinarius* occurs near the top of polarity Chron C33r, rather than near its base, which would imply a 3.5-myr offset in these two competing criteria (Razmjooei et al., 2014). However, other studies show nearly an exact equivalence between the base of Chron C33r and the extinction of *Marsupites* (e.g., A. Gale, written comm., Jan 2016). There is also a discrepancy in interpreted dating of the base of Chron C33r—the base of the ammonite S. *leei III* Zone in the Western Interior, which

is considered to coincide with this reversal, is constrained by radioisotopic and cyclostratigraphy at 84.19±0.38 Ma (Sageman et al., 2014); whereas the interpreted cyclostratigraphy of the Bottaccione section relative to the well-dated Cenomanian–Turonian boundary would imply that the reversal is at 83.06±0.4 Ma (Sprovieri et al., 2013)—over 1.1 myr higher. Until these reported discrepancies are resolved and the GSSP for the global Campanian is decided, we retain the working equivalence of the bases of Chron C33r and of the base of the ammonite S. *leei III* Zone with the revised age of 84.19±0.38 Ma (Sageman et al., 2014).

Maastrichtian: The level of the GSSP at Tercis, France, does not have a specific marker, but is constrained by several biostratigraphic datums and geochemical signatures. This level is close to the radioisotopic-dated base of the ammonite *Baculites baculus* Zone of the Western Interior, which is in the age model for the magnetic polarity time scale in GTS2012, projected to about 88% up in Subchron C32n.2n. This correlation to the upper part of Subchron C32n.2n was verified by detailed carbon-isotope and microfossil calibration to magnetostratigraphy in pelagic sections (e.g., Voigt et al., 2012; Thibault et al., 2012; and reviewed in Coccioni and Premoli Silva, 2015). Cyclostratigraphy of Maastrichtian sections and ocean drilling cores have enabled high-resolution age assignments among different regions (e.g., Thibault et al., 2012; Batenberg et al., 2014).

Selected main stratigraphic scales and events

As with the other chapters in this concise booklet, it was necessary to omit most details, interesting aspects, and important article citations on these topics. The GTS2012 book has a more extended discussion and bibliography, as do the selected later articles that are cited.

(1) Biostratigraphy (marine; terrestrial)

Ammonites, despite their ecological and facies restrictions, provide the primary reference scale for the majority of the Cretaceous in each paleogeographic region (e.g., syntheses in Lehmann, 2015; Lehmann et al., 2015; Ifrim et al., 2015). The Lower Cretaceous standard is the Tethyan ammonite succession in the western Mediterranean region with its revised and enhanced zonal schemes developed by the Kilian working group (e.g., Reboulet et al., 2014). The Upper Cretaceous ammonite and bottom-dwelling inoceramid bivalve successions of the Western Interior is in a carbonate-to-marl facies with abundant radioisotopic-dated volcanic ash beds and orbital-tuned cycles that are the basis for the Late Cretaceous age model (e.g., Sageman et al., 2014).

Calpionellid microfossils are important in pelagic carbonates of the upper Tithonian through Valanginian, followed by dominance by pelagic foraminifers, which are a major biostratigraphic tool from the Cretaceous through Cenozoic. Calcareous nannofossils are another main constituent of the Cretaceous chalk ("*creta*"). Dinoflagellate cysts are important in global correlation of organic-rich marine deposits.

Even though dinosaurs are the famous land-dwellers through the Cretaceous, there is only a generalized "land-vertebrate age" division for North America with poorly constrained correlations to marine-based stages (Lucas et al., 2012). Despite important fossil-rich beds that provide insights into the Cretaceous evolution of birds and mammals, most of these are only snapshots in time for a limited region. Only the uppermost Cretaceous deposits in North America have an adequate record of local mammals to enable a general zonation of land-mammal ages.

(2) Magnetic stratigraphy

The M-sequence of Early Cretaceous and Late Jurassic marine magnetic anomalies from the Hawaiian lineations has been the standard for relative durations of polarity chrons, and it has been calibrated to numerous ammonite, calpionellid, calcareous nannofossil, stable-isotope, and other datums. Cycle-calibrated durations of polarity zones in some of these sections indicate a fairly constant spreading rate for the Hawaiian magnetic lineations during the Valanginian through Barremian (e.g., Sprovieri et al., 2006). In contrast, methods that apply statistical techniques to partly normalize relative polarity widths and minimize spreading rate changes during the Early Cretaceous and Late Jurassic for all Pacific global spreading centers (Tominaga and Sager, 2010) or globally (Malinverno et al., 2012) would require a ca. 5-myr interval of relatively significant slower (73%) spreading for those Hawaiian magnetic lineations during much of the Hauterivian. If that hypothesis is supported by reevaluations of the published direct cycle-scaling of magnetostratigraphic sections, then it would imply a longer duration for the Hauterivian stage than is currently derived from its correlation to the quasiconstant-spreading-rate Hawaiian reference scale (e.g., Martinez et al., 2015).

A 40-myr "Cretaceous Quiet Zone" of constant normal-polarity extends from the early Aptian through the Santonian. The C-sequence of marine magnetic anomalies spans the Campanian to the Present.

(3) Stable-isotope stratigraphy and selected events

There are several major excursions, both positive and negative, in carbon-isotope ratios during the Cretaceous that are important for global correlation. Several of these are associated with organic-rich black shales or other low-oxygen-influenced horizons in oceanic to shelf settings (Fig. 13.1). The causes of these main "oceanic anoxic events" (OAEs) are thought to be massive releases of carbon dioxide by large igneous provinces. For example, the early Aptian OA1a, called the Selli level in Italian exposures, and its initial negative $\delta^{13}C$ excursion is synchronous with the massive outpouring of Ontong Java Plateau flood basalts in the Pacific at about 120 Ma (e.g., review by Erba et al., 2015). A relatively high level of carbon dioxide during the Aptian through middle Campanian, coupled with sequestering of calcium carbonate in the flooded continental interiors, resulted in enhanced carbonate dissolution in deeper waters and an elevated carbonate compensation depth (CCD) through much of this interval.

Even though the Cretaceous lacks documented glacial intervals, the trends in oxygen isotopes and localized suggestions of sea ice in high-latitude basins (e.g., possible ice-rafted dropstones in Australia during portions of the late Aptian–Albian) indicate that the sustained greenhouse was interrupted by several cool intervals. The oxygen-isotope and temperature trends in Fig. 13.1 are a schematic generalization from several sources. In contrast, the interpretation of tropical temperatures from the TEX86 ("*tetraether index of tetraethers consisting of 86 carbon atoms*") technique implies much higher temperatures (e.g., Forster et al., 2007; Linnert et al., 2014); indeed, at levels approaching those interpreted to have been lethal to most land animals during the similarly hot Early Triassic.

Strontium-isotope curves display major progressive rises during the Berriasian through Barremian and during the Coniacian through Maastrichtian. The comparisons of the high-resolution strontium curves among Tethyan, Boreal, and North American Interior Seaway successions has enabled intercorrelation of those regional biostratigraphic zonations (e.g., McArthur et al., 1994; Mutterlose et al., 2014; etc.).

Numerical age model

GTS2012 age model and potential future enhancements

The GTS2012 merged four main age models and primary scales: (1) Late Jurassic through Barremian was the M-sequence magnetic polarity time scale (Hawaiian pattern) with base of Chron M0r (base Aptian) as 126.3 ± 0.4 Ma (derived from the cycle-duration of Aptian–Albian relative to base-Cenomanian; and consistent with U-Pb and $^{40}Ar/^{39}Ar$ ages from Great Valley and Ocean Drilling sites) and a linear fit to spreading rates derived from five studies of cycle-durations on magnetostratigraphic intervals (e.g., Sprovieri et al., 2006); (2) Aptian–Albian cyclostratigraphy analysis of the Piobbico borehole in Italy and of sections in southeastern France that calibrated a set of planktonic foraminifer and calcareous nannofossil datums and of carbon-isotope events; in turn, constrained by a 100.5 ± 0.4 Ma age for the base of Cenomanian and 113.1 ± 0.3 Ma for the Aptian–Albian boundary (e.g., Selby et al., 2009; Huang et al., 2010; Gale et al., 2011; etc.); (3) spline fit for an array of radioisotopic dates and selected cycle-scaled durations of Cenomanian through early Campanian ammonite zones in the Western Interior (e.g., Siewert, 2011, and a preprint of Siewert et al. "*Geol. Soc. Amer. Bulletin*, 2012 in press"); and (4) cycle-scaled chrons of the C-sequence for the Campanian through Maastrichtian (and into Paleogene) (e.g., Husson et al., 2011).

Following the publication of GTS2012, some aspects of this age model required revision or have been questioned by other studies. As noted, the Late Cretaceous of GTS2012 had used "Siewert et al., *GSA Bulletin*, 2012 in press", which when eventually published in 2014 (Sageman et al., 2014) had incorporated enhanced radioisotopic dating and cyclostratigraphy for some of the ammonite and inoceramid zones and associated stage boundaries. The scaling of Aptian ammonite zones is also constrained by an $^{40}Ar/^{39}Ar$ date of 114.9 ± 0.4 Ma near the top of the *Parahoplites nutfieldiensis* Zone (Singer et al., 2015).

The derivation of the base-Aptian age of 126.3 Ma in GTS2012 has been examined by Erba et al. (2015) who prefer an age of 121.5 Ma, thereby implying a 5-myr shortening of the Aptian Stage relative to the cyclostratigraphy interpretations of the Piobbico reference core. However, the older age is currently supported by Ontong Java Plateau volcanism (124.3 ± 1.8 Ma; Chambers et al., 2004, recalculated using $^{40}Ar/^{39}Ar$ monitor standards in GTS2012) that is considered coeval with the lowermost Aptian carbon-13 excursion and by a U-Pb date of 124.1 ± 0.2 Ma from volcanic beds within the early Aptian calcareous nannofossil *Chiastozygus litterarius* subzone NC6b (Shimokawa, 2010). There is also an apparent discrepancy in the Berriasian through Barremian between $^{40}Ar/^{39}Ar$ dates from oceanic basalts drilled by the Ocean Drilling Program (ODP) and U-Pb-dated bentonites from the Great Valley (e.g., Mahoney et al., 2005; Shimokawa, 2010; Pringle and Duncan, 1995, as recalculated using $^{40}Ar/^{39}Ar$ monitor standards in GTS2012) and some new significantly younger U-Pb dates from volcanic ash beds in the Neuquén Basin of Argentina (e.g., Vennari et al., 2014; Aguirre-Urreta et al., 2015; Schwarz et al., 2016).

Until these incompatibilities are resolved by applying magnetostratigraphy and other verifications of the chronostratigraphic ages of these important Argentina and other sections and by additional radioisotopic dating of the array of ODP basalts and Great Valley bentonites, then the GTS2012 age model for the M-sequence polarity pattern and the associated magnetostratigraphic calibrations should be considered a working option.

Revised ages compared to GTS2012

The establishment of new GSSPs since 2011, changes in the preferred criteria from the working groups for some Cretaceous stages, and revision of some radioisotopic dates required some slight modification of the assigned ages from the GTS2012 age model. The revised-age scales used in these diagrams also have incorporated extensive revised biostratigraphic correlations within stages.

Cretaceous (Berriasian) base (**145.7 ± 0.8** vs 145.0 ± 0.8 Ma in GTS2012): In GTS2012, a working definition of the base of magnetic Chron M18r was used to place the Jurassic–Cretaceous boundary. Now, as detailed at the beginning of this chapter, the current Jurassic/Cretaceous working group is favoring a slightly older level in the middle of Chron M19n at the "explosion" of small, globular *Calpionella alpina* (base of Calpionellid Zone B). Therefore, both options are shown for the base of the Cretaceous.

Hauterivian base (**134.7 ± 0.7** vs 133.9 ± 0.6 Ma in GTS2012): The published estimate that the base of the Hauterivian is near the base of Chron M10n (e.g., Weissert et al., 1998) that was used in GTS2012 has been rejected by later ammonite and cyclostratigraphy studies, which have converged on a duration of the Valanginian stage at its candidate GSSP sections for its upper and lower boundaries of 4.7 myr, with only its base calibrated to magnetostratigraphy (lower Chron M14r) [e.g., Gréselle and Pittet, 2010; Gréselle, 2012; Martinez et al. (2013) as revised by Charbonnier et al., 2013]. Therefore, the base of the Hauterivian is placed 4.7 myr younger than the base of the Valanginian (currently 139.4 ± 0.7 Ma, as in GTS2012); and the scaling of ammonite zones within the Valanginian and much of the Hauterivian are according to those studies. [NOTE: The Early–Late Hauterivian boundary is the base of the ammonite *Subsaynella sayni* Zone; it was accidentally drawn one zone lower in one of the GTS2012 graphics; but text and table were correct.]

GSSPs of the Cretaceous Stages, with location and primary correlation criteria

Stage	GSSP Location	Latitude, Longitude	Boundary Level	Correlation Events	Reference
Maastrichtian	*Tercis les Bains, Landes, France*	43°40'46.1"N 1°06'47.9"W*	level 115.2 on platform IV of the geological site at Tercis les Bains	Mean of 12 biostrati-graphic criteria of equal importance. Near ammonite FAD of *Pachydiscus neubergicus*	Episodes **24/4**, 2001
Campanian	*candidates are in Italy and in Texas*			*Crinoid, LAD of Marsupites testudinarius or base of Chron C33r*	
Santonian	Olazagutia, Northern Spain	42°52'5.3"N 2°11'40"W	94.4m in the eastern border of the Cantera de Margas quarry	Inoceramid bivalve, FAD *Platyceramus undulatoplicatus*	Episodes **37/1**, 2014
Coniacian	*candidates are in Poland (Slupia Nadbrzena) and Germany (Salzgitter)*			*Inoceramid bivalve, FAD of Cremno-ceramus deformis erectus*	
Turonian	Pueblo, Colorado, USA	38°16'56"N 104°43'39"W*	base of Bed 86 of the Bridge Creek Limestone Member	Ammonite, FAD of *Watinoceras devonense*	Episodes **28/2**, 2005
Cenomanian	Mont Risou, Hautes-Alpes, France	44°23'33"N 5°30'43"E	36 meters below the top of the Marnes Bleues Formation on the south side of Mont Risou	Foraminifer, FAD of *Thalmanninella globotruncanoides*	Episodes **27/1**, 2004
Albian	*candidate is Pré-Guittard section, Southeastern France*			*Kilian Event or FAD of Microhedbergella renilaevis*	
Aptian	*candidate is Gorgo a Cerbara, Umbria-Marche, central Italy*			*Base of Chron M0r; near ammonite, FAD of Paradeshayesites oglanlensis*	
Barremian	*candidate is Río Argos near Caravaca, Murcia province, Spain*			*Ammonite, FAD of Taveraidiscus hugii*	
Hauterivian	*candidate is La Charce village, Drôme province, southeast France*			*Ammonite, FAD of genus Acanthodiscus*	
Valanginian	*candidate is near Caravaca (S. Spain)*			*Calpionellid, FAD of Calpionellites darderi*	
Berriasian				*Calpionellid, FAD of Calpionella alpina*	

* according to Google Earth

Figure 13.5 Ratified GSSPs and potential primary markers under consideration for defining the Cretaceous stages *(status as of early 2016).* Details of each GSSP are available at http://www.stratigraphy.org, https://engineering. purdue.edu/Stratigraphy/gssp/, and in the *Episodes* publications.

Albian base (**113.14**±0.4 vs 113.0±0.4 Ma in GTS2012): GTS2012 had used a working placement for the Aptian/Albian boundary as the FAD of calcareous nannofossil *P. columnata* (base of Nannofossil zone CC8 or NC8), which was the placement for the boundary in the Piobbico core used for cycle-scaling of calcareous nannofossil and planktonic foraminifer zones (Huang et al., 2010), and was assigned in that study above an organic-rich unit interpreted as the local manifestation of the Niveau Kilian event of southeastern France. The proposed Albian GSSP in France is in the middle of this Niveau Kilian bed (Kennedy et al., 2014). However, the correlation of this level to U-Pb-dated tuffs near a similar calcareous nannofossil datum in northern Germany (113.08±0.14 Ma; Selby et al., 2009) is uncertain, due to use of different taxonomic morphologies. [NOTE: The Early–Middle Albian is the base of the ammonite *Hoplites dentatus* zone; it was accidentally drawn one zone lower in GTS2012 graphic; but text and table were correct.]

Coniacian base (**89.75**±0.38 vs 89.8±0.4 Ma in GTS2012—no significant change), **Santonian** base (**86.49**±0.44 vs 86.3±0.5 Ma in GTS2012), and **Campanian** base (**84.19**±0.38 vs 83.6±0.3 Ma in GTS2012): As explained earlier, the age model for these Late Cretaceous stages was based on an array of "2012, in press" radioisotopic and cycle-stratigraphic calibrations from the North American Western Interior. However, when this publication was eventually published (Sageman et al., 2014), it included additional dates, revised biostratigraphic calibrations (and the use of the inoceramid marker for the base of the Santonian, rather than the ammonite marker), and other enhancements that also changed some of the scaling of the ammonite zones. The most significant change between the earlier 2012 draft and the 2014-published version is the derivation of an older age for the base of the Campanian; which now omits a relatively young ^{40}Ar/^{39}Ar date from the regional "base Campanian" ammonite zone. However, it must be emphasized that the Campanian stage has not yet been defined by an international GSSP.

Estimated uncertainties on assigned ages on stage boundaries

Radioisotopic dates that constrain the Cretaceous timescale typically have an uncertainty greater than 0.4 myr, when including external factors. In contrast, the durations are considered to have an uncertainty of ca. 100-kyr (short-eccentricity cycle for those stages that are scaled from cyclostratigraphy and for many events within those stages). The greatest uncertainty is on the floating timescale for the Early Cretaceous stages, for which there are not yet verified constraints from high-precision radioisotopic dates with reliable international biostratigraphic correlations. Therefore, the GTS2012 age model arbitrarily projected a ±0.5 to ±0.8 myr uncertainty for most of the "working definitions" for those yet-to-be-formalized stage boundaries.

Acknowledgments

This brief summary and update of the Cretaceous chapter in GTS2012 was assisted by extensive discussions during the past four years on placement of GSSPs and improved zonations within stages with numerous colleagues, including (*in alphabetical order*) Elisabetta Erba, Andy Gale, Chunju Huang, Brian Huber, Linda Hinnov, Steve Meyers, Brad Singer, and Bill Wimbledon. Andrew Gale and W. James Kennedy provided advice on correlation of the Aptian–Albian boundary GSSP between Sub-Mediterranean and Sub-Boreal realms. Brian Huber and Andy Gale reviewed an early draft and recommended many enhancements and updates.

Selected publications and websites

Cited publications

NOTE: To keep the length of this chapter as short as possible, only select publications were cited in this review with an emphasis on aspects of post-2011 updates. Pre-2011 literature is summarized in the synthesis by Ogg et al. (2012) and in some of the publications cited below.

Aguado, R., Company, M., Tavera, J., 2000. The Berriasian/Valanginian boundary in the Mediterranean region: new data from the Caravaca and Cehegín sections, SE Spain. *Cretaceous Research* **21**: 1–21.

Aguirre-Urreta, B., Lescano, M., Schmitz, M.D., Tunik, M., Concheyro, A., Rawson, P.F., Ramos, V.A., 2015. Filling the gap: new precise Early Cretaceous radioisotopic ages from the Andes. *Geological Magazine* **152**(3): 557–564. http://dx.doi.org/10.1017/S001675681400082X.

Battenburg, S.J., Gale, A.S., Sprovieri, M., Hilgen, F.J., Thibault, N., Boussaha, M., Orue-Etxebarria, X., 2014. An astronomical time scale for the Maastrichtian based on the Zumaia and Sopelana sections (Basque country, northern Spain). *Journal of the Geological Society* **171**: 165–180. http://dx.doi.org/10.1144/jgs2013-015.

Bottini, C., Erba, E., Tiraboschi, D., Jenkyns, H.C., Schouten, S., Sinninghe Damsté, J.E., 2015. Climate variability and ocean fertility during the Aptian Stage. *Climates of the Past* **11**: 383–402. http://dx.doi.org/10.5194/cp-11-383-2015.

Chambers, L.M., Pringle, M.S., Fitton, J.G., 2004. Phreatomagmatic eruptions on the Ontong Java Plateau: an Aptian $^{40}Ar/^{39}Ar$ age for volcaniclastic rocks at ODP Site 1184. In: Fitton, J.G., Mahoney, J.J., Wallace, P.J., Sanders, A.D. (Eds.), *Origin and Evolution of the Ontong Java Plateau*, **vol. 229**, pp. 325–331. Geological Society Special Publications.

Charbonnier, G., Boulila, S., Gardin, S., Duchamp-Alphonse, S., Adatte, T., Spangenberg, J.E., Föllmi, K.B., Colin, C., Galbrun, B., 2013. Astronomical calibration of the Valanginian "Weissert" episode: the Orpierre marl-limestone succession (Vocontian Basin, southeastern France). *Cretaceous Research* **45**: 25–42.

Coccioni, R., Premoli Silva, I., 2015. Revised Upper Albian–Maastrichtian planktonic foraminiferal biostratigraphy and magnetostratigraphy of the classical Tethyan Gubbio section (Italy). *Newsletters on Stratigraphy* **48/1**: 47–90. http://dx.doi.org/10.1127/nos/2015/0055.

Cobban, W.A., Walaszczyk, I., Obradovich, J.D., McKinney, K.C., 2006. *A USGS Zonal Table for the Upper Cretaceous Middle Cenomanian-Maastrichtian of the Western Interior of the United States Based on Ammonites, Inoceramids, and Radiometric Ages*. U.S. Geological Survey Open-File Report, 2006-1250: 45 pp.

Erba, E., Duncan, R.A., Bottini, C., Daniele, T., Weissert, H., Jenkyns, H.C., Malinverno, A., 2015. Environmental consequences of Ontong Java Plateau and Kerguelen Plateau volcanism. In: Neel, C.R., Sager, W.W., Sano, T., Erba, E. (Eds.), The Origin, Evolution and Environmental Impact of Oceanic Large Igneous Provinces. *Geological Society of America Special Paper* **511**: 271–303. http://dx.doi.org/10.1130/2015.2511(15).

Föllmi, K.B., Godet, A., Bodin, S., Linder, P., 2006. Interactions between environmental change and shallow water carbonate buildup along the northern Tethyan margin and their impact on the Early Cretaceous carbon isotope record. *Paleoceanography* **21**: PA4211. http://dx.doi.org/10.1029/2006PA001313.

Forster, A., Schouten, S., Moriya, K., Wilson, P.A., Sinninghe Damsté, J.S., 2007. Tropical warming and intermittent cooling during the Cenomanian/Turonian Oceanic Anoxic Event 2: sea surface temperature from the equatorial Atlantic. *Paleoceanography* **22**: 1–14.

Gale, A.S., Hancock, J.M., Kennedy, W.J., Petrizzo, M.R., Lees, J.A., Walaszczyk, I., Wray, D.S., 2008. An integrated study (geochemistry, stable oxygen and carbon isotopes, nannofossils, planktonic foraminifera, inoceramid bivalves, ammonites and crinoids) of the Waxahachie Dam Spillway section, north Texas, a possible boundary stratotype for the base of the Campanian Stage. *Cretaceous Research* **29**: 131–167.

Gale, A.S., Bown, P., Caron, M., Crampton, J., Crowhurst, S.J., Kennedy, W.J., Petrizzo, M.R., Wray, D.S., 2011. The uppermost Middle and Upper Albian succession at the Col de Palluel, Hautes-Alpes, France: an integrated study (ammonites, inoceramid bivalves, planktonic foraminifera, nannofossils, geochemistry, stable oxygen and carbon isotopes, cyclostratigraphy). *Cretaceous Research* **32**: 59–130.

Gréselle, B., Pittet, B., 2010. Sea-level reconstructions from the Peri-Vocontian zone (South-east France) point to Valanginian glacio-eustasy. *Sedimentology* **57**: 1640–1684.

Gréselle, B., May 2012. A high-resolution multi-stratigraphic approach to global stratigraphy, biostratigraphy and paleoclimatology: the mid-Valanginian crisis. In: Stephenson, M., Simmons, M., ad Molyneux, S., conveners (Eds.), *High Fidelity: The Quest for Precision in Stratigraphy and its Applications*, pp. 16–17 The Geologic Society, abstract book: p. 16.

Haq, B.U., 2014. Cretaceous eustasy revisited. *Global and Planetary Change* **113**: 44–58.

Herrle, J.O., Kössler, P., Friedrich, O., Erlenkeuser, H., Hemleben, C., 2004. High-resolution carbon isotope records of the Aptian to Lower Albian from SE France and the Mazagan Plateau (DSDP Site 545): a stratigraphic tool for paleoceanographic and paleobiologic reconstruction. *Earth and Planetary Science Letters* **218**: 149–161.

Huang, C., Hinnov, L.A., Fischer, A.G., Grippo, A., Herbert, T., 2010. Astronomical tuning of the Aptian Stage from Italian reference sections. *Geology* **38**: 899–902.

Huber, B.T., Norris, R.D., Macleod, K.G., 2002. Deep sea paleotemperature record of extreme warmth during the Cretaceous. *Geology* **30**: 123–126.

Huber, B.T., Leckie, R.M., 2011. Planktic foraminiferal species turnover across deep-sea Aptian/Albian boundary sections. *Journal of Foraminiferal Research* **41**: 53–95.

Huber, B.T., Macleod, K.G., Gröcke, D., Kucera, M., 2011. Paleotemperature and paleosalinity inferences and chemostratigraphy across the Aptian/Albian boundary in the subtropical North Atlantic. *Paleoceanography* **26**: PA4221 http://dx.doi.org/4210.1029/2011PA002178.

Huber, B.T., Petrizzo, M.R., 2014. Evolution and taxonomic study of the Cretaceous planktonic foraminifer genus *Helvetoglobotruncana* Reiss, 1957. *Journal of Foraminiferal Research* **44**: 40–57.

Husson, D., Galbrun, B., Laskar, J., Hinnov, L.A., Thibault, N., Gardin, S., Locklair, R.E., 2011. Astronomical calibration of the Maastrichtian. *Earth and Planetary Science Letters* **305**: 328–340.

ICS (International Commission on Stratigraphy), Annual Report 2014. Submitted to International Union of Geological Sciences (IUGS) at: http://iugs.org/uploads/ICS%202014.pdf.

Ifrim, C., Lehmann, J., Ward, P., 2015. Chapter 10. Paleobiogeography of Late Cretaceous ammonoids. In: Klug, C., Korn, D., De Baets, K., Kruta, I., Mapes, R.H. (Eds.), Ammonoid Paleobiology: From Macroevolution to Paleogeography. *Topics in Geobiology* **44**: 259–275. http://dx.doi.org/10.1007/978-94-017-9633-0_13 Springer Publication.

Jarvis, I., Gale, A.S., Jenkyns, H.C., Pearce, M.A., 2006. Secular variation in Late Cretaceous carbon isotopes: a new δ13C carbonate reference curve for the Cenomanian-Campanian (99.6-70.6 Ma). *Geological Magazine* **143**: 561–608.

Kennedy, W.J., Gale, A.S., Huber, B.T., Petrizzo, M.R., Bown, P., Barchetta, A., Jenkyns, H.C., 2014. Integrated stratigraphy across the Aptian/Albian boundary at Col de Pré-Guittard (southeast France): a candidate Global Boundary Stratotype Section. *Cretaceous Research* **51**: 248–259.

Lamolda, M.A., Paul, C.R.C., Peryt, D., Pons, J.M., 2014. The Global Boundary Stratotype and Section Point (GSSP) for the base of the Santonian Stage, "Cantera de Margas", Olazagutia, northern Spain. *Episodes* **37**: 2–13.

Lehmann, J., 2015. Chapter 15. Ammonite biostratigraphy of the Cretaceous – an overview. In: Klug, C., Korn, D., De Baets, K., Kruta, I., Mapes, R.H. (Eds.), Ammonoid Paleobiology: From Macroevolution to Paleogeography. *Topics in Geobiology* **44**: 403–423. http://dx.doi.org/10.1007/978-94-017-9633-0_13 Springer Publication.

Lehmann, J., Ifrim, C., Bulot, L., Frau, C., 2015. Chapter 9. Paleobiogeography of Early Cretaceous ammonoids. In: Klug, C., Korn, D., De Baets, K., Kruta, I., Mapes, R.H. (Eds.), Ammonoid Paleobiology: From Macroevolution to Paleogeography. *Topics in Geobiology* **44**: 229–258. http://dx.doi.org/10.1007/978-94-017-9633-0_13 Springer Publication.

Linnert, C., Robinson, S.A., Lees, J.A., Bown, P.R., Pérez-Rodriguez, I., Petrizzo, M.R., Falzoni, F., Littler, K., Arz, J.A., Russell, E.E., 2014. Evidence for global cooling in the Late Cretaceous. *Nature Communications* **5**: 4194. http://dx.doi.org/10.1038/ncomms5194.

Lucas, S.G., Sulliva, R.M., Spielmann, 2012. Cretaceous vertebrate biochronology, North American Western Interior. *Journal of Stratigraphy* **36**(2): 426–461.

Mahoney, J.J., Duncan, R.A., Tejada, M.L.G., Sager, W.W., Bralower, T.J., 2005. Jurassic–Cretaceous boundary age and mid-ocean-ridge-type mantle source for Shatsky Rise. *Geology* **33**: 185–188.

Malinverno, A., Hildebrandt, J., Tominaga, M., Channell, J.E.T., 2012. M-sequence geomagnetic polarity time scale (MHTC12) that steadies global spreading rates and incorporates astrochronology constraints. *Journal of Geophysical Research* **117**: B06104. http://dx.doi.org/10.1029/2012JB009260 (17 pages).

Martinez, M., Pellenard, P., Deconinck, J.F., Monna, F., Riquier, L., Boulila, S., Moirou, M., Company, M., 2012. An orbital floating time scale of the Hauterivian/Barremian GSSP from a magnetic susceptibility signal (Río Argos, Spain). *Cretaceous Research* **36**: 106–115.

Martinez, M., Deconinck, J.F., Pellenard, P., Reboulet, S., Riquier, L., 2013. Astrochronology of the Valanginian stage from reference sections (Vocontian Basin, France) and palaeoenvironmental implications for the Weissert Event. *Palaeogeography, Palaeoclimatology, Palaeoecology* **376**: 91–102.

Martinez, M., Deconinck, J.F., Pellenard, P., Riquier, L., Company, M., Reboulet, S., Moiroud, M., 2015. Astrochronology of the Valanginian–Hauterivian stages (Early Cretaceous): Chronological relationships between the Paraná–Etendeka large igneous province and the Weissert and the Faraoni events. *Cretaceous Research* **131**: 158–173.

McArthur, J.M., Kennedy, W.J., Chen, M., Thirlwall, M.F., Gale, A.S., 1994. Strontium isotope stratigraphy for the Late Cretaceous: direct numerical age calibration of the Sr-isotope curve for the U.S. Western Interior Seaway. *Palaeogeography, Palaeoclimatology, Palaeoecology* **108**: 95–119.

Mutterlose, J., Bodin, S., Fähnrich, L., 2014. Strontium-isotope stratigraphy of the Early Cretaceous (Valanginian–Barremian): implications for Boreal–Tethys correlation and paleoclimate. *Cretaceous Research* **50**: 252–263.

Norris, R.D., Kroon, D., Huber, B.T., Erbacher, J., 2001. Cretaceous–Palaeogene ocean and climate change in the subtropical North Atlantic. In: Kroon, D., Norris, R.D., Klaus, A. (Eds.), *Western North Atlantic Palaeogene and Cretaceous Palaeoceanography*, **vol. 183**, pp. 1–22. Geological Society, London, Special Publications.

Ogg, J.G., Hinnov, L.A., Huang, C., 2012. Cretaceous. In: Gradstein, F.M., Ogg, J.G., Schmitz, M., Ogg, G., (Coordinators), *The Geologic Time Scale 2012*, Elsevier Publication, pp. 793–853. (Overview with extensive details and graphics on the GSSPs of the stages, the different marine macrofossil and microfossil biostratigraphic scales, synthesis of radio-isotopic age and cyclo-stratigraphy constraint as of 2011, and the astronomical tuning of the magnetic polarity scale.).

Patruno, S., Triantaphyllou, M.V., Erba, E., Dimiza, M.D., Bottini, C., Kaminski, M.A., 2015. The Barremian and Aptian stepwise development of the 'oceanic anoxic event 1a' (OAE 1a) crisis: integrated benthic and planktic high-resolution palaeoecology along the Gorgo a Cerbara stratotype section (Umbria–Marche Basin, Italy). *Palaeogeography, Palaeoclimatology, Palaeoecology* **424**: 147–182.

Petrizzo, M.R., Huber, B.T., Gale, A.S., Barchetta, A., Jenkyns, H.C., 2012. Abrupt planktic foraminiferal turnover across the Niveau Kilian at Col de Pré-Guittard (Vocontian Basin, southeast France): new criteria for defining the Aptian/Albian boundary. *Newsletters on Stratigraphy* **45/1**: 55–74.

Pringle, M.S., Duncan, R.A., 1995. Radiometric ages of basaltic lavas recovered at Lo-En, Wodejebato, MIT, and Takuyo-Daisan Guyots, northwestern Pacific Ocean. Proceedings of the Ocean Drilling Program, Scientific Results **144**: 547–557.

Razmjooei, M.J., Thibault, N., Kani, A., Mahanipour, A., Boussaha, M., Korte, C., 2014. Coniacian–Maastrichtian calcareous nannofossil biostratigraphy and carbon-isotope stratigraphy in the Zagros Basin (Iran): consequences for the correlation of Late Cretaceous stage boundaries between the Tethyan and Boreal realms. *Newsletters on Stratigraphy* **47/2**: 183–209.

Reboulet, S., Szives, O., Aguirre-Urreta, B., Barragán, R., Company, M., Idakieva, V., Ivanov, M., Kakabadze, M.V., Moreno-Bedmar, J.A., Sandoval, J., Baraboshkin,

E.J., Çaglar, M.K., Fözy, I., González-Arreola, C., Kenjo, S., Lukeneder, A., Raisossadat, S.N., Rawson, P.F., Tavera, J.M., 2014. Report on the 5th international meeting of the IUGS Lower Cretaceous Working Group, the Kilian Group (Ankara, Turkey, 31st August 2013). *Cretaceous Research* **50**: 126–137.

Renard, M., de Raféllis, M., Emmanuel, L., Moullade, M., Masse, J.-P., Kuhnt, W., Bergen, J.A., Tronchetti, G., 2005. Early Aptian $\delta^{13}C$ and manganese anomalies from the historical Cassis-La Bédoule stratotype sections (S.E. France): relationship with a methane hydrate dissociation event and stratigraphic implications. *Carnets de Géologie/ Notebooks on Geology*. Article 2005/04: 18p. Available at: http://paleopolis.rediris.es/cg/ CG2005_A04/index.html.

Remane, J., 1998. Calpionellids (Part of columns for Jurassic chart of Mesozoic and Cenozoic sequence chronostratigraphic framework of European basins, by Hardenbol, J., Thierry, J., Farley, M.B., Jacquin, Th., de Graciansky, P.-C., and Vail, P.R. (coordinators)). In: de Graciansky, P.-C., Hardenbol, J., Jacquin, Th., Vail, P.R. (Eds.), *Mesozoic-Cenozoic Sequence Stratigraphy of European Basins*, **vol. 60**, pp. 763–781 SEPM Special Publication.

Sageman, B.B., Singer, B.S., Meyers, S.R., Siewert, S.E., Walaszczyk, I., Condon, D.J., Jicha, B.R., Obradovich, J.D., Sawyer, D.A., 2014. Integrating $^{40}Ar/^{39}Ar$, U-Pb, and astronomical clocks in the Cretaceous Niobrara Formation, Western Interior Basin, USA. *Geological Society of America Bulletin* **126**: 956–973. http:// dx.doi.org/10.1130/B30929.1.

Schnabl, P., Pruner, P., Wimbledon, W.A.P., 2015. A review of magnetostratigraphic results in the Tithonian-Berriasian of Nordvik. (Siberia) and possible biostratigraphic constraints. *Geologica Carpathica* **66**: 489–498. [and W.A.P. Wimbledon, written commun., July 2015].

Scotese, C.R., 2014. *Atlas of Late Cretaceous Paleogeographic Maps, PALEOMAP Atlas for ArcGIS, Volume 2, the Cretaceous, Maps 16–22, Mollweide Projection.* PALEOMAP Project, Evanston, IL. https://www. academia.edu/11193387/Atlas_of_Late_Cretaceous_ Paleogeographic_Maps. https://www.academia. edu/11245383/Atlas_of_Early_Cretaceous_ Paleogeographic_Maps.

Schwarz, E., Spalletti, L.A., Veiga, G.D., Fanning, C.M., 2016. First U–Pb SHRIMP age for the Pilmatué Member (Agrio Formation) of the Neuquén Basin, Argentina: implications for the Hauterivian lower boundary. *Cretaceous Research* **58**: 223–233.

Selby, D., Mutterlose, J., Condon, D.J., 2009. U/Pb and Re/Os geochronology of the Aptian/Albian and Cenomanian/Turonian stage boundaries; implications for timescale calibration, osmium isotope sea water composition and Re-Os systematics in organic-rich sediments. *Chemical Geology* **265**(3–4): 394–409.

Shimokawa, A., 2010. *Zircon U-Pb Geochronology of the Great Valley Group: Recalibrating the Lower Cretaceous Time Scale* (M.S. thesis) University of North Carolina at Chapel Hill, p. 46. Also: Shimokawa, A., Coleman, D.S., and Bralower, T.J., 2010. Recalibrating the Lower Cretaceous time scale with U-Pb zircon ages from the Great Valley Group. Geological Society of America Annual Meeting, Denver, 31 Oct–3 Nov 2010, Abstract, 160–7 https://gsa.confex.com/gsa/2010AM/finalprogram/abstract_182413.htm. and T.J. Bralower, pers. commun., Nov. 2015.

Siewert, S.E., 2011. *Integrating $^{40}Ar/^{39}Ar$, U-Pb and Astronomical Clocks in the Cretaceous Niobrara Formation.* University of Wisconsin at Madison, p. 74 (M.S. thesis).

Singer, B.S., Meyers, S.R., Sageman, B.B., Jicha, B.R., Condon, D., 2015. Improving Cretaceous time scale uncertainties via multi-collector $^{40}Ar/^{39}Ar$ dating. *Geological Society of America Abstracts With Programs.* **47**(7): 172. https://gsa.confex.com/gsa/2015AM/webprogram/Paper264460.html.

Sprovieri, M., Coccioni, R., Lirer, F., Pelosi, N., Lozar, F., 2006. Orbital tuning of a lower Cretaceous composite record (Maiolica Formation, central Italy). *Paleoceanography* **21**: PA4212. http://dx.doi.org/10.1029/2005PA001224.

Sprovieri, M., Sabatino, N., Pelosi, N., Batenburg, S.J., Coccioni, R., Iavarone, M., Mazzola, S., 2013. Late Cretaceous orbitally-paced carbon isotope stratigraphy from the Bottaccione Gorge (Italy). *Palaeogeography Palaeoclimatology Palaeoecology* **379–380**: 81–94.

Thibault, N., Husson, D., Harlou, R., Gardin, S., Galbrun, B., Huret, E., Minoletti, F., 2012. Astronomical calibration of upper Campanian–Maastrichtian carbon isotope events and calcareous plankton biostratigraphy in the Indian Ocean (ODP Hole 762C): implications for the age of the Campanian–Maastrichtian boundary. *Palaeogeography, Palaeoclimatology, Palaeoecology* **337–338**: 52–71.

Tominaga, M., Sager, W.W., 2010. Revised Pacific M-anomaly geomagnetic polarity time scale. *Geophysical Journal International* **182**: 203–232.

Veizer, J., Prokoph, A., 2015. Temperatures and oxygen isotopic composition of Phanerozoic oceans. *Earth-Science Reviews* **146**: 92–104.

Vennari, V.V., Lescano, M., Naipauer, M., Aguirre-Urreta, B., Concheyro, A., Schaltegger, U., Armstrong, R., Pimentel, M., Ramos, V.A., 2014. New constraints on the Jurassic–Cretaceous boundary in the High Andes using high-precision U-Pb data. *Gondwana Research* **26**: 374–385.

Voigt, S., Friedrich, O., Norris, R.D., Schönfeld, J., 2010. Campanian-Maastrichtian carbon isotope stratigraphy: shelf-ocean correlation between the European shelf sea and the tropical Pacific Ocean. *Newsletters on Stratigraphy* **44**: 57–72. http://dx.doi.org/10.1127/0078-0421/2012/0016.

Voigt, S., Gale, A.S., Jung, C., Jenkyns, H.C., 2012. Global correlation of Upper Campanian-Maastrichtian successions using carbon-isotope stratigraphy: development of a new Maastrichian timescale. *Newsletters on Stratigraphy* **45**: 25–53.

Weissert, H., Lini, A., Föllmi, K.B., Kuhn, O., 1998. Correlation of Early Cretaceous carbon isotope stratigraphy and platform drowning events: a possible link? *Palaeogeography, Palaeoclimatology, Palaeoecology* **137**: 189–203.

Wimbledon, W.A.P., 2014. Warsaw remarks – Berriasian progress. *Volumina Jurassica* **XII**(1): 107–112.

Wimbledon, W.A.P., Casellato, C.E., Reháková, D., Bulot, L.G., Erba, E., Gardin, S., Verreussel, R.M.C.H., Munsterman, D.K., Hunt, C.O., 2011. Fixing a basal Berriasian and Jurassic–Cretaceous (J-K) boundary – is there perhaps some light at the end of the tunnel? *Rivista Italiana di Paleontologia e Stratigrafia* **117**: 295–307.

Websites (selected)

All Things Cretaceous (by Jen Aschoff, Montana State University, as part of the DLESE Community Services Project on Integrating Research in Education; ca.2009)—*http://serc.carleton.edu/research_education/cretaceous/index.html*—an assortment of digital resources for teaching.

Palaeos: Cretaceous—*http://palaeos.com/mesozoic/cretaceous/cretaceous.htm*—A well-written compendium of many aspects of Cretaceous life, climate, tectonics, and other aspects for a general science audience that was originally compiled by M. Alan Kazlev in 1998–2002.

Cretaceous Publications (maintained by Mikhail Rogov and colleagues)—*http://www.jurassic.ru/cretaceous.eng.htm*—extensive collection of Portable Document Formats (PDFs), including scans of difficult-to-find Russian papers.

Cretaceous Research (journal; published by Elsevier)—*http://www.journals.elsevier.com/cretaceous-research/*.

There are also extensive Cretaceous pages and links at Smithsonian Institution, Wikipedia, and other organizations.

14

PALEOGENE

52.2 Ma Paleogene

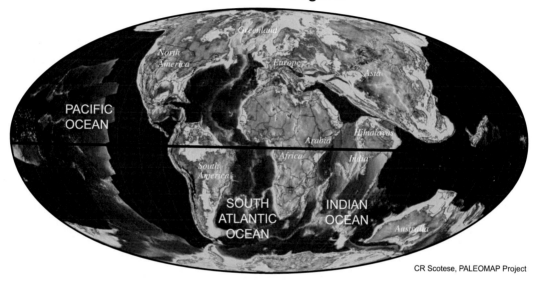

CR Scotese, PALEOMAP Project

Early Eocene (Ypresian) paleogeographic reconstruction (Sea level + 120 m) from Scotese (2014).

Basal definition and international subdivisions

The Cenozoic Era had been traditionally divided either into the Tertiary and Quaternary ("third" and "fourth") periods or into the Paleogene and Neogene ("old"- and "new"-"birth") periods. In 2009, the International Union of Geological Sciences (IUGS) formally adopted a three-fold division—Paleogene, Neogene, and Quaternary.

The Paleogene [*palaios*=old, *genes*=born] of the Cenozoic [*kainos*=new, *zoic*=animal life] is subdivided into the Paleocene, Eocene, and Oligocene epochs [respectively "old," "dawn"

and "few" + "new"]. Their component stages are named after the locations of deposits in Denmark, Germany, England, Belgium, and France, except for the Priabonian stage of uppermost Eocene that was named after a location in northern Italy. However, because most of those deposits were delimited by unconformities at sea-level lowstands, all of the ratified Global Boundary Stratotype Sections and Points (GSSPs) are within uplifted marine deposits in Tunisia, northern Spain, Egypt, and central Italy. The GSSP levels were selected to coincide with global oxygen- or carbon-isotope excursions, magnetic polarity reversals, or calcareous nannofossil appearances (Figs. 14.1 and 14.6).

Cenozoic Time Scale

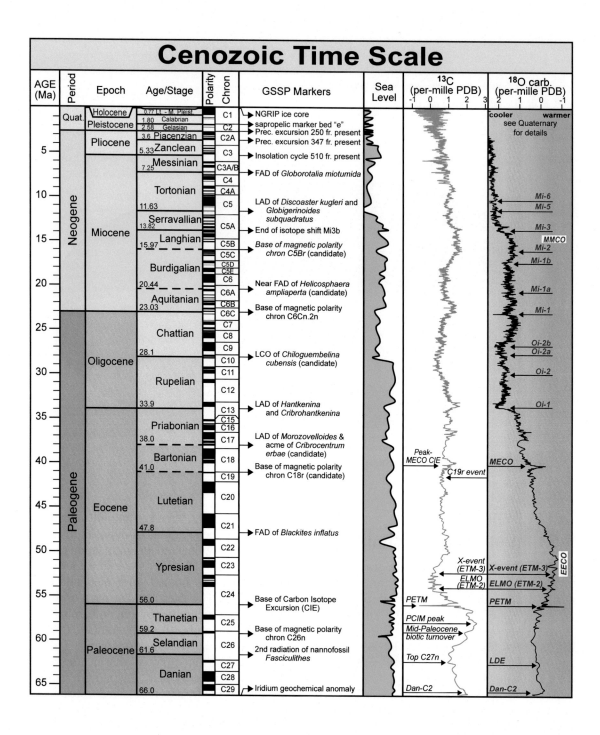

AGE (Ma)	Period	Epoch	Age/Stage	Polarity	Chron	GSSP Markers	Sea Level	13C (per-mille PDB)	18O carb. (per-mille PDB)

Quat. — Holocene — 0.77 Lt. - M. Pleist.; Pleistocene — 1.80 Calabrian, 2.58 Gelasian

- NGRIP ice core — C1
- sapropelic marker bed "e" — C2
- Prec. excursion 250 fr. present
- Pliocene: 3.6 Piacenzian (C2A) — Prec. excursion 347 fr. present
- 5.33 Zanclean (C3) — Insolation cycle 510 fr. present
- 7.25 Messinian (C3A/B) — FAD of *Globorotalia miotumida*
- C4
- Tortonian (C4A)
- 11.63 (C5) — LAD of *Discoaster kugleri* and *Globigerinoides subquadratus*
- Serravallian (C5A)
- 13.82 — End of isotope shift Mi3b
- Langhian (C5B) — Base of magnetic polarity chron C5Br (candidate)
- 15.97 (C5C)
- Burdigalian (C5D, C5E)
- 20.44 (C6) — Near FAD of *Helicosphaera ampliaperta* (candidate)
- Aquitanian (C6A)
- 23.03 (C6B)
- (C6C) — Base of magnetic polarity chron C6Cn.2n
- Oligocene: Chattian (C7, C8, C9)
- 28.1 (C10) — LCO of *Chiloguembelina cubensis* (candidate)
- Rupelian (C11, C12)
- 33.9 (C13) — LAD of *Hantkenina* and *Cribrohantkenina*
- Priabonian (C15, C16, C17) — LAD of *Morozovelloides* & acme of *Cribrocentrum erbae* (candidate)
- 38.0 (C18)
- Bartonian — Base of magnetic polarity chron C18r (candidate)
- 41.0 (C19)
- Lutetian (C20)
- Eocene (C21) — FAD of *Blackites inflatus*
- 47.8 (C22)
- Ypresian (C23)
- (C24) — Base of Carbon Isotope Excursion (CIE)
- 56.0
- Thanetian (C25) — Base of magnetic polarity chron C26n
- 59.2
- Selandian 61.6 (C26) — 2nd radiation of nannofossil *Fasciculithes*
- (C27)
- Paleocene: Danian (C28)
- 65 — 66.0 (C29) — Iridium geochemical anomaly

13C markers: Peak-MECO CIE; C19r event; X-event (ETM-3); ELMO (ETM-2); PETM; PCIM peak; Mid-Paleocene biotic turnover; Top C27n; Dan-C2

18O carb. markers (cooler ← → warmer, see Quaternary for details): Mi-6, Mi-5, Mi-3, MMCO, Mi-2, Mi-1b, Mi-1a, Mi-1, Oi-2b, Oi-2a, Oi-2, Oi-1, MECO, EECO, X-event (ETM-3), ELMO (ETM-2), PETM, LDE, Dan-C2

◀ **Figure 14.1 Cenozoic overview.** Main markers or candidate markers for GSSPs of Paleogene, Neogene, and Quaternary stages are detailed in text and in Fig. 14.6, 15.4, and 16.6, respectively. ("Age" is the term for the time equivalent of the rock-record "stage." Abbreviation "*Prec. Excursion 250 fr. Present*" = orbital precession cycle excursion #250 before present). Mean sea-level curve (schematic between ca. 0 and 250 m) is modified from Hardenbol et al. (1998) and does not show high-frequency oscillations during the Quaternary. Oxygen-isotope curve from benthic foraminifers and the carbon-isotope curve are derived from a nine-point moving window of recalibrated data from Cramer et al. (2009). Subsets of named Oligocene–Miocene events from Boulila et al. (2011) based on definitions by Miller et al. (1991, 1998) and additional calibrations by Pekar et al. (2002); and of named Paleocene–Eocene events after Zachos et al. (2010), Westerhold et al (2008, 2014, 2015), and Dinarès-Turell et al. (2014) [*CIE*, carbon-isotope excursion; *EECO*, early Eocene climatic optimum; *ETM*, Eocene thermal maximum; *LDE*, late Danian event; *MECO*, mid-Eocene climatic optimum; *PCIM*, Paleocene carbon-isotope excursion maximum; *PETM*, Paleocene–Eocene thermal maximum; *PDB*, PeeDee Belemnite ^{13}C and ^{18}O standards]. The vertical scale of this diagram is standardized to match the vertical scales of the first stratigraphic summary figure in all other Phanerozoic chapters.

The base of the Cenozoic (base of Paleocene) is placed at the catastrophic asteroid impact that terminated the majority of Mesozoic species in the oceans and on land. The GSSP of the Paleogene System/Period (Paleocene Series/Epoch, **Danian** Stage/Age) at an exposure in Tunisia of that impact ejecta layer and the mass extinction of planktonic microorganisms was ratified by IUGS in 1991, but not formally published until 2006 (Molina et al., 2006, 2009) (Fig. 14.2). The age of this impact has been dated from ash beds in Montana as 66.043 ± 0.043 Ma (Renne et al., 2013) and occurs at a minimum in a short-eccentricity 100-kyr orbital–climate cycle (e.g., Husson et al., 2014). The global ecosystems may have been already stressed by several short cold episodes from the ongoing eruptions of the Deccan Traps volcanic province in India (Renne et al., 2013; Schoene et al., 2014).

The base of the Eocene (Fig. 14.3) is placed at the onset of a pronounced negative excursion in carbon-13 and a PETM (Fig. 14.3). In contrast, the base of the Oligocene was selected close to the onset of a major cooling event (Fig. 14.4).

As of 2015, three stages in the upper Eocene and Oligocene await international definition:

Bartonian (upper Eocene): The provisional base for the Lutetian–Bartonian boundary is base of reversed-polarity Chron C18r (top of brief Chron C19n), for which the Gubbio section in central Italy is a possible candidate. A former possible candidate at the Cape of Oyambre in Cantabria, Spain, failed to yield a magnetostratigraphy for that interval, but also indicated a questionable reliability of planktonic foraminifer Zone E11, that had been presumed to span the boundary interval (Payros et al., 2015). The status is continuously updated at the working group website (http://www.paleogene.org/working-group/bartonian/).

Priabonian (uppermost Eocene): The candidate levels in central and northern Italy for the Middle/Late Eocene subepoch boundary focus on the double extinction in planktonic foraminifera of the distinctive "muricate" genus *Morozovelloides* and the large acarininids (Zone E13/E14 boundary) which occurred within 11 kyr in the middle (ca. 40% up) of magnetic polarity Subchron C17n.3n (Wade et al., 2012) and the acme of nannofossil *Cribrocentrum erbae*. The leading candidate for the GSSP is at the Alano di Piave section, located ~80 km northeast of the historical Priabonian Stratotype in northeastern Italy. The position of this foraminifer-zone boundary is lower than the provisional definition used in GTS2012 of the base of polarity Chron C17n.1n. The status is continuously updated at the working group website (http://www.paleogene.org/working-group/priabonian/).

Base of the Danian Stage of the Paleogene System at El Kef, Tunisia

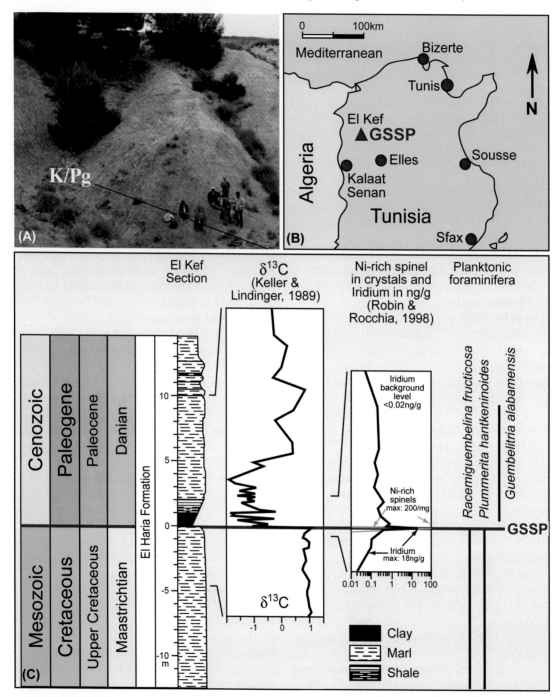

Figure 14.2 GSSP for base of the Paleogene (base of Cenozoic Era, Paleocene Epoch, and Danian Stage) at the El Kef section, Tunisia. The GSSP level coincides with the mass extinction of Cretaceous microfossils, with brief enrichment in iridium (note that scale is logarithmic) and with a sudden negative shift in $\delta^{13}C_{carb}$. Photograph (A) is from Ogg et al. (2008) and stratigraphic data (C) are from Molina et al. (2006).

Base of the Ypresian Stage of the Paleogene System at Dababiya, Egypt

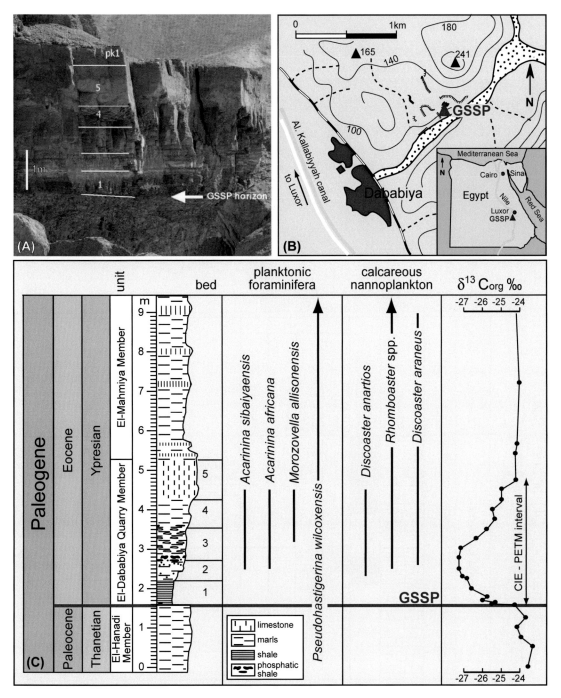

Figure 14.3 GSSP for base of the Eocene (base of Ypresian Stage) at the Dababiya section, Egypt. The GSSP level coincides with the onset of a major negative excursion in carbon isotopes. "CIE—PETM interval" refers to this carbon-isotope excursion and associated Paleocene/Eocene boundary thermal maximum. Photograph and stratigraphic data are modified from Aubry et al. (2007).

Base of the Rupelian Stage of the Oligocene System in the Massignano section near Ancona, Italy.

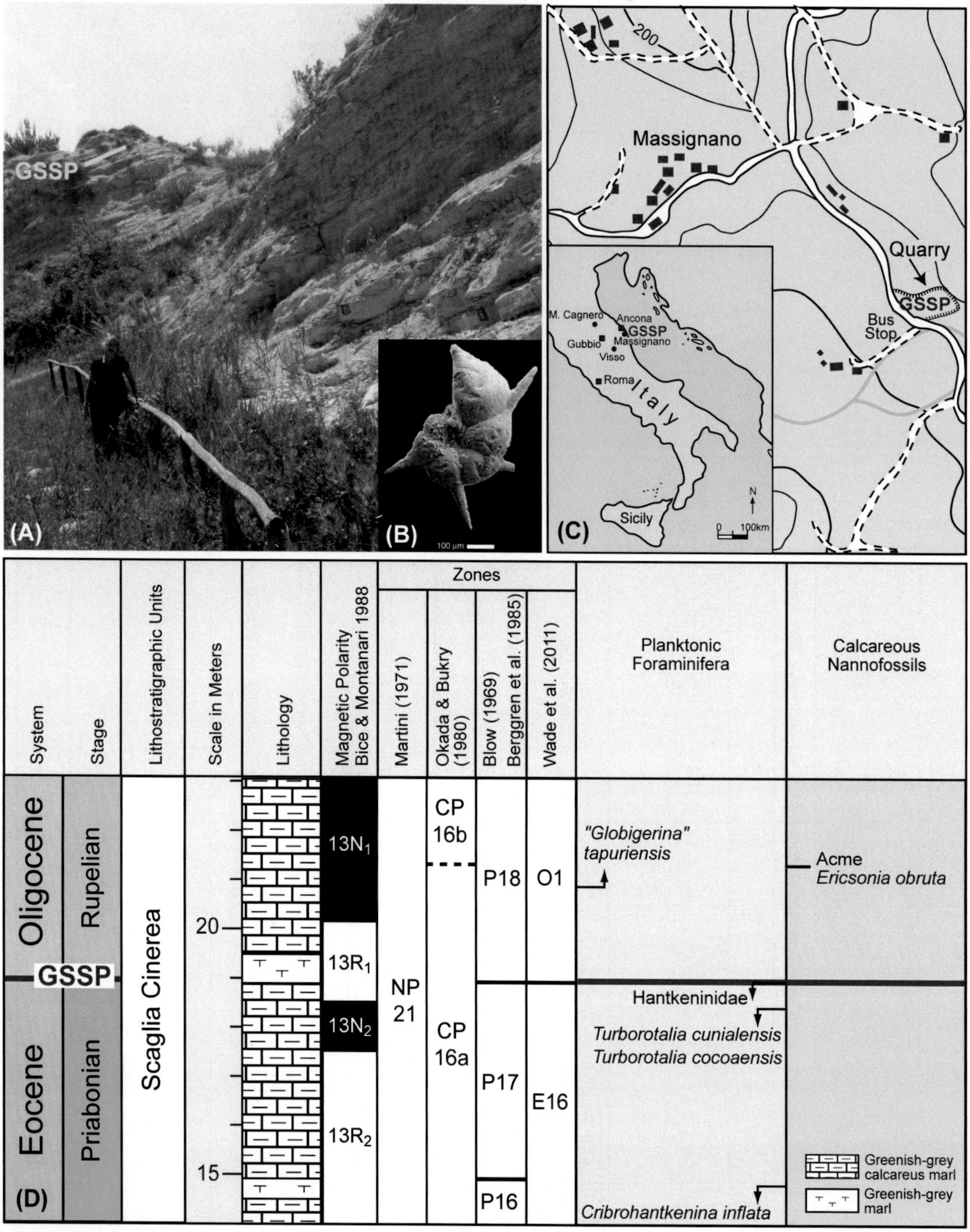

GSSP
(A)
(B)
100 µm
(C)
Massignano
200
Quarry
GSSP
Bus Stop
M. Cagnero
Ancona
GSSP
Gubbio
Massignano
Visso
Roma
Italy
Sicily
N
100km
System
Stage
Lithostratigraphic Units
Scale in Meters
Lithology
Magnetic Polarity Bice & Montanari 1988
Zones
Martini (1971)
Okada & Bukry (1980)
Blow (1969) Berggren et al. (1985)
Wade et al. (2011)
Planktonic Foraminifera
Calcareous Nannofossils
Oligocene
Rupelian
GSSP
Eocene
Priabonian
Scaglia Cinerea
20
15
13N1
13R1
13N2
13R2
NP 21
CP 16b
CP 16a
P18
P17
P16
O1
E16
"Globigerina" tapuriensis
Hantkeninidae
Turborotalia cunialensis
Turborotalia cocoaensis
Cribrohantkenina inflata
Acme
Ericsonia obruta
Greenish-grey calcareus marl
Greenish-grey marl
(D)

Figure 14.4 GSSP for base of the Oligocene (base of Rupelian Stage) at the Massignano section, central Italy. The GSSP level coincides with extinction of Hankeninidae foraminifers. (CP and NP are Paleogene calcareous nannoplankton zones, P and O/E are planktonic foraminifera zones). Photograph of the outcrop provided by E. Molina; photograph of *Hantkenina alabamensis* provided by Brian Huber, and stratigraphic data based on Premoli Silva and Jenkins (1993).

Chattian (mid-Oligocene): The section at Monte Cagnero in central Italy had been proposed as a candidate with the GSSP assigned at meter level 188 to coincide with the highest (last) common occurrence (HCO) of planktonic foraminifer *Chiloguembelina cubensis* (base of Zone O5) (Coccioni et al., 2008, 2013). That placement of the HCO level was in the upper half of magnetic polarity Chron C10n at Monte Cagnero, and below an oxygen-isotope cooling event Oi2a (Fig. 14.1) that is associated with a widespread sea-level lowstand (sequence boundary "Ch1"; Fig. 14.5). In GTS2012, the age of the Rupelian–Chattian boundary used the base of magnetic polarity Subchron C10n.1n (near middle of Chron C10n; 28.1 Ma); which is close to the interpolated ages of 27.99 Ma from astronomical-cycle interpolation and of 28.27 ± 0.1 Ma from dating bracketing volcanic ash beds in that Monte Cagnero section. [*The paleomagnetic sampling within magnetic polarity zone C10n at Monte Cagnero does not yet have the resolution to resolve the brief C10n.1r subzone* (Coccioni et al., 2008).] The HCO of *Ch. cubensis* is commonly reported from a level within polarity zone C10n in the ocean basins. However, in some sites, the HCO is observed as high as the lower part of polarity zone C9n, approximately 1 myr later, which led King and Wade (2015) to suggest that these occurrences "question the legitimacy of the biostratigraphic utility of *Chiloguembelina cubensis* as a reliable boundary marker for the early/late Oligocene." Nevertheless, following additional extensive sampling of foraminifer abundances at Monte Cagnero, the proposed GSSP level using a revised "HCO" concept was shifted 9 m upward to meter level 197 (in lower part of polarity zone C9n) which is above the oxygen-isotope cooling event Oi2a and would precede the cooling event Oi2b (Oligocene Glacial Maximum of Van Simaeys, 2004) by ca. 0.5 myr (Coccioni et al., 2016, unpublished resubmission to Subcommission on Paleogene Stratigraphy). Dating of volcanic ash beds bracketing this level, coupled with an interpretation of potential climatic cycles, yield an estimated age of 27.8 Ma for this revised GSSP candidate in lower Chron C9n (Coccioni et al., 2016; unpublished); however, that age is ca. 0.5 myr older than on the Oligocene cycle-magnetic timescale of Pälike et al. (2006) which had been used in GTS2012 (Fig. 14.5A). Pending formal decision on the Chattian GSSP and reexamination of the astronomical tuning of the Oligocene time scale by cycle stratigraphers, the placement of the Rupelian–Chattian boundary by Vandenberghe et al. (2012) is retained here (Fig. 14.5A). The current status of the Chattian GSSP candidate is at the working group website (http://www.paleogene.org/working-group/chattian/).

Selected main stratigraphic scales and events

(1) Biostratigraphy (marine; terrestrial)

Detailed intercalibrated datums and zones for calcareous, siliceous, and organic-walled microfossils have been compiled from micropaleontology studies of Cenozoic deposits drilled from the ocean basins, continental shelves, and interior basins. The correlation of these events to orbital–climate cycles and magnetic polarity chrons has been progressively enhanced (e.g., Wade et al., 2011; for planktonic foraminifers). For calcareous nannofossils in low- to midlatitude settings, many of the traditional markers were found diachronous or difficult to apply between basins; therefore, Agnini et al. (2014) assembled a new zonation which is diagrammed in Fig. 14.5. For terrestrial deposits, the evolution and migratory exchanges of mammals on the distributed continents are used for different regional scales.

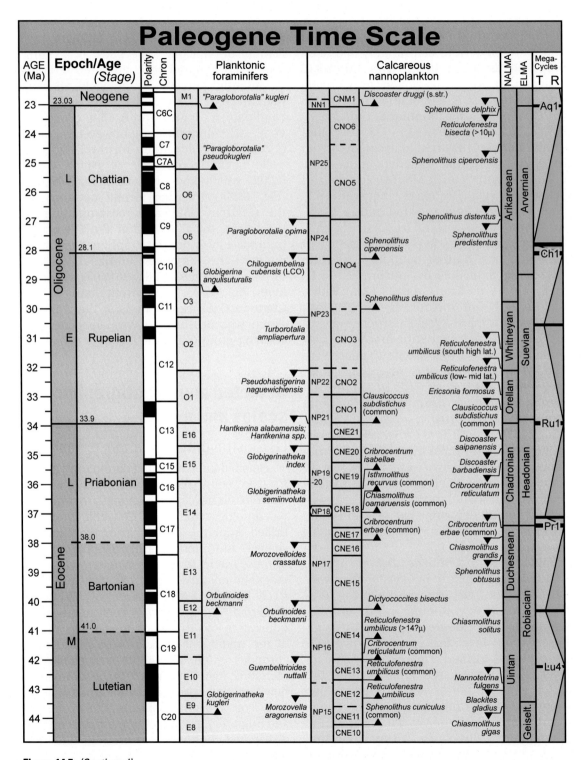

Figure 14.5 *(Continued)*

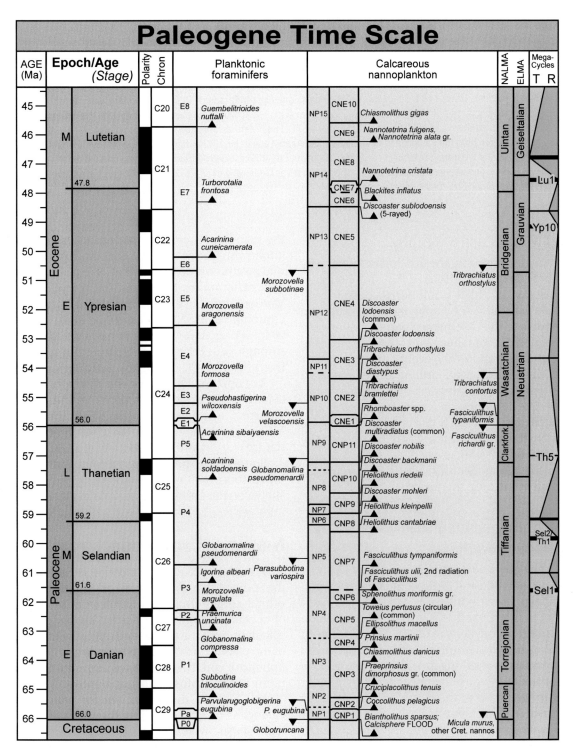

Figure 14.5 *(Continued)*

◀ **Figure 14.5 (A, B): Selected marine and terrestrial biostratigraphic zonations of the Paleogene.** ("*Age*" is the term for the time equivalent of the rock-record "*stage*"; and subepoch groupings of these "ages" into "late," "early" are informal common usage.) Magnetic polarity zones are scaled to astronomical cycles (e.g., Vandenberghe et al., 2012; Westerhold et al., 2015). Planktonic foraminifer zones and main markers are modified from Wade et al (2011; and pers. comm.). Placement of proposed base-Chattian marker of last common occurrence (LCO) of *Chiloguembelina cubensis* relative to magnetic polarity chrons of the Oligocene, hence the assigned age to Rupelian–Chattian boundary, is uncertain (see text). Calcareous nannofossil "CN" zones and markers from Agnini et al. (2014) are shown together with the commonly used Paleogene nannofossil ("NP") zonation of Martini (1971). Land Mammal Ages of North America (NALMA) and Europe (ELMA) are from the compilation by Hooker (in Vandenberghe et al., 2012). Major sea-level sequence boundaries and highstands are from Hardenbol et al. (1998). Additional zonations, biostratigraphic markers, geochemical trends, sea-level curves, and details on calibrations are compiled in Vandenberghe et al. (2012) and in the internal data sets within the *TimeScale Creator* visualization system (free at www.tscreator.org).

(2) Magnetostratigraphy

Beginning with the landmark papers by Cande and Kent (1992, 1995) and Berggren et al. (1995), a high-resolution Paleogene time scale was derived by applying spreading-rate models to the C-sequence patterns of marine magnetic anomalies, and then correlating microfossil and other events to those polarity chrons in ocean drilling cores and land outcrops. In the past decade, a suite of more precise durations and ages for these polarity chrons have been derived from extensive cyclostratigraphy studies; and this philosophy begun by Berggren et al. (1995) is still the basis for the age model for most Paleogene microfossil events (e.g., Agnini et al., 2014).

(3) Stable-isotope stratigraphy and selected events

The negative excursion in carbon-13 and a PETM at the base of the Eocene is also recorded in deep-sea carbonates by a dissolution event that is interpreted as a rise in the carbonate compensation depth (CCD). This release of ^{12}C-enriched carbon dioxide and/or methane and the resulting greenhouse warming is considered to have been triggered by a main eruptive phase of the North Atlantic Igneous Province. The identification of these PETM signatures in oceanic sediment deposits led to

the recognition of precursor episodes and of a following series of early Eocene hyperthermal events composing the EECO (e.g., reviews and calibrations to astronomical cycles in Vandenberghe et al., 2012; Littler et al., 2014; Lauretano et al., 2015; Westerhold et al., 2014, 2015). The general cooling trend of the Middle through Late Eocene is interrupted by a relatively brief MECO during the early Bartonian.

In contrast, the Oligocene through Miocene benthic oxygen-isotope curve indicates recurrent cooling episodes (Oi-1 through Mi-6), which seem to be of a semiperiodic nature (e.g., synthesis by Boulila et al., 2011).

Some of the major hyperthermal and cooling episodes and carbon-isotope excursions are shown on the Cenozoic time scale in Fig. 14.1.

Numerical age model

The entire Cenozoic timescale is based on the progressive recognition of Milankovitch orbital–climate cycles to tune reference sections of oceanic sediments to precise astronomical calculations of the orbital and obliquity history of the planet Earth. Those reference sections of uplifted or cored deposits have magnetostratigraphy and microfossil biostratigraphy.

In GTS2012, the placement for nearly all Paleogene events, including the bases of geologic stages, were derived from the direct or indirect placement of biostratigraphic datums and zones relative to magnetic polarity chrons. The age model for those polarity chrons (Vandenberghe et al., 2012) was in three segments—(1) astronomical cycle tuning from the base of the Miocene (set as 23.0 Ma) "downward" to the base of the Oligocene (polarity Chron C13r) by Pälike et al. (2006), (2) cycle tuning of durations "upward" from the base-Cenozoic (set as 66.0 Ma) through early Eocene to polarity Chron C23n, and (3) a bridge by interpolating a smoothed spreading-rate spline applied to the seafloor marine magnetic anomaly record for middle through late Eocene Chrons C22n through C13r.

That temporary spline bridge for interpolating within the Eocene has now been replaced by astronomical-cycle tuning of the middle and late Eocene from a suite of ocean drilling cores (Westerhold et al., 2014, 2015). This astronomical-tuned time scale for Chrons C22n through C13r (used here) implied that the assigned ages for events (including the provisional base of the Bartonian stage) are shifted slightly from the GTS2012 extrapolations (Fig. 14.5).

Revised ages compared to GTS2012; and potential future enhancements

Danian base (**66.04** Ma): The mean age derived by Renne et al. (2013) of 66.043 ± 0.43 Ma for the Cretaceous–Paleogene boundary from a suite of U-Pb and ^{40}Ar/^{39}Ar dates has verified the GTS2012 estimate of 66.04 ± 0.05 Ma derived from the placement of the boundary within the astronomical-tuned age for magnetic polarity Chron C29r.

Bartonian base (**41.03** vs 41.15 Ma in GTS2012): The revised astronomical tuning for the provisional definition of the base of the Bartonian as the base of Chron C18r has shifted this chron age slightly younger by 0.1 myr.

Priabonian base (**37.97** vs 37.7 Ma in GTS2012): In GTS2012, a provisional definition of the

base of magnetic polarity chron C17n.1n was employed; and the revised astronomical tuning would shift this to 37.4 Ma. However, the current working group favors use of the last occurrence of the *Morozovelloides* genus, which occurs in the middle of Chron C17n.3n, therefore at 37.97 Ma in the current age model.

The age assigned to the base of the Chattian, provisionally set as the Chron 10n.1n, may change when a GSSP definition has been decided (see previous discussion of the potential shift to Chron 9n). In addition, as explained in the next chapter on the Neogene, the assigned 23.0 Ma for the Oligocene–Miocene boundary has not yet been verified by detailed astronomical tuning; and, if it is revised, this may have minor impacts on the assigned ages for many of late Oligocene events.

Future enhancements for the Paleogene time scale include extending correlations to the high-latitude oceanic realms and improving the correlations to terrestrial events and deposits on the different continents. The recognition of the suites of hyperthermal and cooling events from the deep-sea records, when coupled with magnetostratigraphy, is one important tool for these goals.

Estimated uncertainties on assigned ages on stage boundaries

All Paleogene stage boundaries and magnetic polarity zones, plus some of the marine microfossil and calcareous nannofossil markers, have age assignments according to astronomical tuning to calculated long-eccentricity (405-kyr frequency) and short-eccentricity (ca. 100-kyr) cycles with additional precision in the Oligocene through tuning to a full Milankovitch target curve that incorporates obliquity (ca. 40-kyr) and precession (ca. 20-kyr). Most published calibrations do not explicitly give uncertainties. None of the Paleogene stage boundary GSSPs is in a cycle-tuned reference section, but are correlated through stable-isotope or biostratigraphic markers. Therefore, ignoring the additional factors of the not-yet-formalized stage boundaries and possible Oligocene–Miocene boundary recalibration, it is probable that the uncertainty for stage boundaries in the current compilation is ca. 50 kyr (one-half of a 100-kyr short-eccentricity cycle).

Acknowledgments

The compilation of this Paleogene summary relied heavily on the GTS2012 chapter by Noel Vandenberghe and his colleagues and on microfossil/nannofossil calibrations from Bridget Wade, Paul Bown, and Erik Anthonissen. Advice on improving the GTS2012 age model was from Thomas Westerhold, Frits Hilgen, and Linda Hinnov. Nöel Vandenberghe and Heiko Pälike reviewed an early draft and provided updates on the status of the Chattian GSSP candidate.

GSSPs of the Paleogene Stages, with location and primary correlation criteria

Stage	GSSP Location	Latitude, Longitude	Boundary Level	Correlation Events	Reference
Chattian	*Candidate section at Monte Cagnero, Urbania, central Italy*		at meter level 188	*Foraminifer, LCO of Chiloguembelina cubensis*	
Rupelian	Massignano, near Ancona, Italy	43°31'58.2"N 13°36'03.8"E*	Base of a 0.5m thick, greenish-grey marl bed, 19 m above base of section	Foraminifer, LAD of *Hantkenina* and *Cribrohantkenina*	Episodes **16**/3, 1993
Priabonian	*Candidate section at Alano di Piave, NE Italy*			*Foraminifer, LAD of Morozovelloides and acme of nannofossil Cribrocentrum erbae*	
Bartonian	*Candidate section at Gubbio, central Italy*			*provisional: base of magnetic polarity chronozone C18r*	
Lutetian	Gorrondatxe beach section near Getxo village, W Pyrenees, Spain	43°22'46.47"N 3°00'51.61"W	167.85m above the base of the Gorrondatxe section in a dark marly level	Calcareous nanno-fossil, FAD of *Blackites inflatus*	Episodes **34**/2, 2011
Ypresian	Dababiya, near Luxor, Egypt	25°30'04.70"N* 32°31'19.14"E*	Base of Bed 1 in DBH subsection	Base of Carbon Isotope Excursion (CIE)	Episodes **30**/4, 2007
Thanetian	Zumaia section, northern Spain	43°17'58.4"N 02°15'39.1"W*	6.5m above the base of Member B of the Itzurun Formation	Base of magnetic polarity chronozone C26n	Episodes **34**/4, 2011
Selandian	Zumaia section, northern Spain	43°17'57.1"N 02°15'39.6"W*	Base of the red marls of the Itzurun Formation	2nd radiation of nannofossil *Fasciculithes*	Episodes **34**/4, 2011
Danian	Oued Djerfane, west of El Kef, Tunisia	36°09'13.2"N 8°38'54.8"E	Reddish layer at the base of the 50cm thick, dark boundary clay	Iridium geochemical anomaly	Episodes **29**/4, 2006

* according to Google Earth

Figure 14.6 Ratified GSSPs and potential primary markers under consideration for defining the Paleogene stages (status as of early 2016). Details of each GSSP are available at http://www.stratigraphy.org, https://engineering. purdue.edu/Stratigraphy/gssp/, and in the *Episodes* publications.

Selected publications and websites

Cited publications

*Only select publications were cited in this review
with an emphasis on aspects of post-2011 updates.
Pre-2011 literature is well summarized in the
synthesis by Vandenberghe et al. (2012) and in some
of the publications cited in the following.*

Agnini, C., Fornaciari, E., Raffi, I., Catazariti, R., Pälike,
H., Backman, J., Rio, D., 2014. Biozonation and
biochronology of Paleogene calcareous nannofossils
from low and middle latitudes. *Newsletters on
Stratigraphy* **47**: 131–181.

Aubry, M.-P., Ouda, K., Dupuis, C., Berggren, W.A., Van
Couvering, J.A., the Members of the Working Group
on the Paleocene/Eocene Boundary, 2007. The
Global Standard Stratotype-section and Point (GSSP)
for the base of the Eocene Series in the Dababiya
section (Egypt). *Episodes* **30**: 271–286.

Berggren, W.A., Kent, D.V., Flynn, J.J., 1985. Paleogene
geochronology and chronostratigraphy. In: Snelling,
N.J. (Ed.), *The Chronology of the Geological Record.
Geological Society Memoir* **10**, pp. 141–195.

Berggren, W.A., Kent, D.V., Swisher III, C.C., Aubry, M.-P.,
1995. A revised Cenozoic geochronology and
chronostratigraphy. In: Berggren, W.A., Kent, D.V.,
Aubry, M.-P., Hardenbol, J. (Eds.), *Geochronology,
Time Scales, and Global Stratigraphic Correlation*, **54**,
pp. 129–212 SEPM Special Publication.

Bice, D.M., Montanari, A., 1988. Magnetic stratigraphy of
the Massignano section across the Eocene/
Oligocene boundary. In: Premoli Silva, I., Coccioni,
R., Montanari, A. (Eds.), *The Eocene/Oligocene
Boundary in the Marche-Umbria Basin (Italy)*.
Aniballi, Ancona, pp. 111–217.

Blow, W.H., 1969. Late middle Eocene to Recent
planktonic foraminiferal biostratigraphy. In:
Bronnimann, P., Renz, H.H. (Eds.), *Proceedings 1st
International Conference on Planktonic Microfossils,
Geneva, 1967*. **1**. E.J. Brill, Leiden, pp. 199–422.

Boulila, S., Galbrun, B., Miller, K.G., Pekar, S.F., Browning,
J.V., Laskar, J., Wright, J.D., 2011. On the origin of
Cenozoic and Mesozoic "third-order" eustatic
sequences. *Earth-Science Reviews* **109**: 94–112.

Cande, S.C., Kent, D.V., 1992. A new geomagnetic
polarity time scale for the Late Cretaceous and
Cenozoic. *Journal of Geophysical Research* **97**:
13917–13951.

Cande, S.C., Kent, D.V., 1995. Revised calibration of the
geomagnetic polarity timescale for the Late
Cretaceous and Cenozoic. *Journal of Geophysical
Research* **100**: 6093–6095.

Coccioni, R., Bellanca, A., Bice, D.M., Brinkhuis, H.,
Church, N., Deino, A., Lirer, F., Macalady, A.,
Maiorano, P., Marsili, A., Mcdaniel, A., Monechi, S.,
Neri, R., Nini, C., Nocchi, M., Pross, J., Rochette, P.,
Sagnotti, L., Sprovieri, M., Tateo, F., Touchard, Y.,
Simaeys, S.V., Williams, G.L., 2008. Integrated
stratigraphy of the Oligocene pelagic sequence in the
Umbria-Marche basin (northeastern Apennines,
Italy): a potential Global Stratotype Section and Point
(GSSP) for the Rupelian/Chattian boundary.
Geological Society of America Bulletin **120**: 487–511.
http://dx.doi.org/10.1130/B25988.1.

Coccioni, R., Montanari, A., Bellanca, A., Bice, D.M.,
Brinkhuis, H., Church, N., Deino, A., Frontalini, F.,
Lirer, F., Macalady, A., Maiorano, P., McDaniel, A.,
Marsili, A., Monechi, S., Neri, R., Nini, C., Nocchi, M.,
Pross, J., Rochette, P., Sagnotti, L., Sideri, M.,
Sprovieri, M., Tateo, F., Touchard, Y., Van Simaeys, S.,
Williams, G.L., July 1–7, 2013. Integrated stratigraphy
of the Monte Cagnero pelagic sequence in the
Umbria-Marche basin (northeastern Apennines,
Italy): A potential candidate for defining the Global
Stratotype Section and Point (GSSP) for the Rupe-
lian/Chattian boundary. *STRATI 2013*, Lisbon, p. 15.
Ciencias da Terra, Numero Especial VII.

Coccioni, R., Montanari, A., Bice, D.M., Brinkhuis, H.,
Deino, A., Frontalini, F., Lirer, F., Maiorano, P.,
McDaniel, A., Monechi, S., Pross, J., Rochette, P.,
Sagnotti, L., Sideri, M., Sprovieri, M., Tateo, F.,
Touchard, Y., Van Simaeys, S., Williams, G.L., 2016.
*Proposal for Establishing the GSSP for the Base of the
Chattian Stage in the Monte Cagnero Section, Italy*
(revised GSSP proposal submitted 16 Jan 2016 to
International Subcommission on Paleogene
Stratigraphy; 31 pp plus figures/tables).
(unpublished).

Cramer, B.S., Toggweiler, J.R., Wright, J.D., Katz, M.E.,
Miller, K.G., 2009. Ocean overturning since the Late
Cretaceous: inferences from a new benthic forami-
niferal isotope compilation. *Paleoceanography* **24**:
PA4216. http://dx.doi.org/10.1029/2008PA001683.

Dinarès-Turell, J., Westerhold, T., Pujalte, V., Röhl, U.,
Kroon, D., 2014. Astronomical calibration of the
Danian Stage (Early Paleocene) revisited: settling
chronologies of sedimentary records across the
Atlantic and Pacific Oceans. *Earth and Planetary
Science Letters* **405**: 119–131.

Hardenbol, J., Thierry, J., Farley, M.B., Jacquin, Th., de
Graciansky, P.-C., Vail, P.R., with numerous contribu-
tors, 1998. Mesozoic and Cenozoic sequence
chronostratigraphic framework of European basins.
In: de Graciansky, P.-C., Hardenbol, J., Jacquin, Th.,
Vail, P.R. (Eds.), Mesozoic-cenozoic Sequence
Stratigraphy of European Basins. *SEPM Special
Publication* **60**: 763–781.

Husson, D., Galbrun, B., Gardin, S., Thibault, N., 2014. Tempo and duration of short-term environmental perturbations across the Cretaceous-Paleogene boundary. *Stratigraphy* **11**: 159–171.

Keller, G., Lindinger, M., 1989. Stable isotope, TOC and CaCO₃ record across the Cretaceous/Tertiary boundary at El Kef, Tunisia. *Palaeogeography, Palaeoclimatology, Palaeoecology* **73**: 243–265.

King, D.J., Wade, B.S., 2015. The extinction of *Chiloguembelina cubensis* in the Pacific Ocean (Sites U1334 and 1237): implications for defining the base of the Chattian. In: *STRATI 2015* (2nd International Congress on Stratigraphy; 19–23 July 2015, Graz Austria) *Abstracts* (*Berichte des Institutes für Erdwissenschaften Karl-Franzens-Universität Graz*, Band 21): 202.

Lauretano, V., Littler, K., Polling, M., Zachos, J.C., Lourens, L.J., 2015. Frequency, magnitude and character of hyperthermal events at the onset of the Early Eocene Climatic Optimum. *Climate of the Past* **11**: 1795–1820.

Littler, K., Röhl, U., Westerhold, T., Zachos, J.C., 2014. A high-resolution benthic stable-isotope record for the South Atlantic: implications for orbital-scale changes in Late Paleocene–Early Eocene climate and carbon cycling. *Earth and Planetary Science Letters* **401**: 18–30. http://dx.doi.org/10.1016/j.epsl.2014.05.054.

Martini, E., 1971. Standard Tertiary and Quaternary calcareous nannoplankton zonation. In: Farinacci, A. (Ed.), *Proceedings of the II Planktonic Conference, Roma, 1969*. Tecnoscienza, Rome, pp. 739–785.

Miller, K.G., Wright, J.D., Fairbanks, R.G., 1991. Unlocking the ice house: Oligocene-Miocene oxygen isotopes, eustasy, and margin erosion. *Journal of Geophysical Research* **96**: 6829–6848.

Miller, K.G., Mountain, G.S., Browning, J.V., Kominz, M.A., Sugarman, P.J., Christie-Blick, N., Katz, M.E., Wright, J.D., 1998. Cenozoic global sea level, sequences, and the New Jersey Transect: results from coastal plain and continental slope drilling. *Reviews of Geophysics* **36**: 569–601.

Molina, E., Alegret, L., Arenillas, I., Arz, J.A., Gallala, N., Hardenbol, J., Von Salis, K., Steurbaut, E., Vandenbeghe, N., Zaghbib-Turki, D., 2006. The Global Boundary Stratotype Section and Point for the base of the Danian Stage (Paleocene, Paleogene, "Tertiary," Cenozoic) at El Kef, Tunisia: original definition and revision. *Episodes* **29**(4): 263–273.

Molina, E., Alegret, L., Arenillas, I., Arz, J.A., Gallala, N., Grajales, M., Murillo-Muñetón, G., Zaghbib, D., 2009. The Global Boundary Stratotype Section and Point for the base of the Danian Stage (Paleocene, Paleogene, "Tertiary", Cenozoic): auxiliary sections and correlation. *Episodes* **32**(2): 84–95.

Ogg, J.G., Ogg, G.M., Gradstein, F.M., 2008. *Concise Geologic Time Scale*. Cambridge University Press (177 pages).

Okada, H., Bukry, D., 1980. Supplementary modification and introduction of code number to the low latitude coccolith biostratigraphic zonation (Bukry, 1973, 1975). *Marine Micropaleontology* **5**: 321–325.

Pälike, H., Norris, R.D., Herrie, J.O., Wilson, P.A., Coxall, H.K., Lear, C.H., Shackleton, N.J., Tripati, A.K., Wade, B.S., 2006. The heartbeat of the Oligocene climate system. *Science* **414**: 1894–1898.

Payros, A., Dinarès-Turell, J., Monechi, S., Orue-Etxebarria, Ortiz, S., Apellaniz, E., Martínez-Braceras, 2015. The Lutetian/Bartonian transition (Middle Eocene) at the Oyambre section (northern Spain): implications for standard chronostratigraphy. *Palaeogeography, Palaeoclimatology, Palaeoecology* **440**: 234–248.

Pekar, S.F., Christie-Blick, N., Kominz, M.A., Miller, K.G., 2002. Calibrating eustasy to oxygen isotopes for the early icehouse world of the Oligocene. *Geology* **30**: 903–906.

Premoli Silva, I., Jenkins, D.G., 1993. Decision on the Eocene-Oligocene boundary stratotype. *Episodes* **16**: 379–382.

Renne, P.R., Deino, A.L., Hilgen, F.J., Kuiper, K.F., Mark, D.F., Mitchell III, W.S., Morgan, L.E., Mundil, R., Smit, J., 2013. Time scales of critical events around the Cretaceous-Paleogene boundary. *Science* **339**: 684–687.

Robin, E., Rocchia, R., 1998. Ni-rich spinel at the Cretaceous-Tertiary boundary of El Kef, Tunisia. *Bulletin de la Sociètè géologique de France* **169**(3): 365–372.

Schoene, B., Samperton, K.M., Eddy, M.P., Keller, G., Adatte, T., Bowring, S.A., Khadri, S.F.R., Gertsch, B., 2014. U-Pb geochronology of the Deccan Traps and relation to the end-Cretaceous mass extinction. *Science* **347**: 182–184. http://dx.doi.org/10.1126/science.aaa0118.

Scotese, C.R., 2014. *Atlas of Paleogene Paleogeographic Maps (Mollweide Projection), Maps 8-15, PALEOMAP Atlas for ArcGIS, Version 2, Volume 1, the Cenozoic*. PALEOMAP Project, Evanston, IL. https://www.academia.edu/11099001/Atlas_of_Paleogene_Paleogeographic_Maps.

Van Simaeys, S., 2004. The Rupelian–Chattian boundary in the North Sea Basin and its calibration to the international time scale. *Netherlands Journal of Geosciences* **83**: 241–248.

Vandenberghe, N., Hilgen, F.J., Speijer, R.P., with contributions by, Ogg, J.G., Gradstein, F.M., Hammer, O., Hollis, C.J., Hooker, J.J., 2012. The Paleogene Period. In: Gradstein, F.M., Ogg, J.G., Schmitz, M., Ogg, G., (Coordinators), *The Geologic Time Scale 2012*. Elsevier Publication, pp. 855–921 (A comprehensive Paleogene overview, especially with extensive details on marine microfossil and mammal biostratigraphy and on the age model compiled from cycle stratigraphy, plus graphics on the GSSPs of the stages and the different biostratigraphic scales).

Wade, B.S., Pearson, P.N., Berggren, W.A., Pälike, H., 2011. Review and revision of Cenozoic tropical planktonic foraminiferal biostratigraphy and calibration to the geomagnetic polarity and astronomical time scale. *Earth-Science Reviews* **104**: 111–142.

Wade, B.S., Fucek, V.P., Kamikuri, S.-I., Bartol, M., Luciani, V., Pearson, P.N., 2012. Successive extinctions of muricate planktonic foraminifera (*Morzovelloides* and *Acarinina*) as a candidate for marking the base Priabonian. *Newsletters on Stratigraphy* **45**: 245–262. http://dx.doi. org/10.1127/0078-0421/2012/0023.

Westerhold, T., Röhl, U., Raffi, I., Fornaciari, E., Monechi, S., Reale, V., Bowles, J., Evans, H.F., 2008. Astronomical calibration of the Paleocene time. *Palaeogeography, Palaeoclimatology, Palaeoecology* **257**: 377–403.

Westerhold, T., Röhl, U., Pälicke, H., Wilkens, R., Wilson, P.A., Acton, G., 2014. Orbitally tuned timescale and astronomical forcing in the middle Eocene to early Oligocene. *Climate of the Past* **10**: 955–973. http://dx.doi.org/10.5194/cp-10-955-2014.

Westerhold, T., Röhl, U., Frederichs, T., Bohaty, S.M., Zachos, J.C., 2015. Astronomical calibration of the geological timescale: closing the Middle Eocene gap. *Climate of the Past.* **11**: 1181–1195. http://dx.doi. org/10.5194/cp-11-1181-2015. www.clim-past. net/11/1181/2015.

Zachos, J.C., McCarren, H., Murphy, B., Röhl, U., Westerhold, T., 2010. Tempo and scale of late Paleocene and early Eocene carbon isotope cycles: implications for the origin of hyperthermals. *Earth and Planetary Science Letters* **299**: 242–249.

Websites (selected)

Subcommission on Paleogene Stratigraphy (ICS)— *http://www.paleogene.org* or *http://wzar.unizar.es/ isps/*—well-illustrated Website with extensive details of each stage, its GSSPs, including Portable Document Formats (PDFs).

Palaeos: The Paleogene—*http://palaeos.com/cenozoic/ paleogene.html*—A well-presented suite of diverse topics for a general science audience that was originally compiled by M. Alan Kazlev in 1998–2002.

NEOGENE

14.9 Ma Neogene

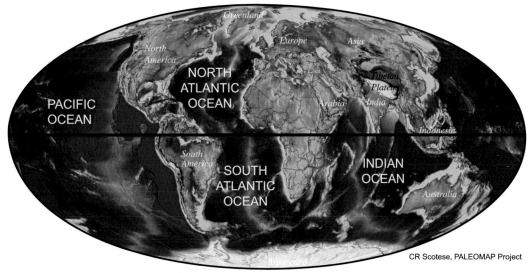

CR Scotese, PALEOMAP Project

Middle Miocene (Langhian) paleogeographic reconstruction (Sea level + 80 m) from Scotese (2014).

Basal definition and international subdivisions

A Neogénique period was recommended in 1894 to the International Geological Congress to incorporate the Miocene, Pliocene, Pleistocene epochs [respectively, "*less*," "*more*," and "*most*" + "*new*"], and the "*Recent*" [now called *Holocene*, meaning "*entire*" + "*new*"].

The base of the Neogene Period and the Miocene Epoch (Aquitanian Stage) is defined by a GSSP in Italy (Fig. 15.1) that corresponds to the base of magnetic polarity chron C6Cn.2n. At this level, the positive shift Mi-1 in oxygen isotopes marks the onset of a global cooling event and an associated major sequence boundary on continental margins (Fig. 14.1).

The top of the Neogene was revised in 2009 when the International Union of Geological Sciences (IUGS) ratified a Quaternary Period that encompassed a redefined Pleistocene Epoch beginning with the first major episode of continental glaciation (base of the Gelasian Stage) at ~2.6 Ma (see Chapter 16 on Quaternary).

The historical bases of the stages of the Miocene and Pliocene were placed at major sequence boundaries on the European and Mediterranean shelves. For global correlation and dating, the GSSP levels for these stages were placed in uplifted deep-marine carbonate

Base of the Aquitanian Stage of the Miocene Series at Lemme-Carrosio, Italy.

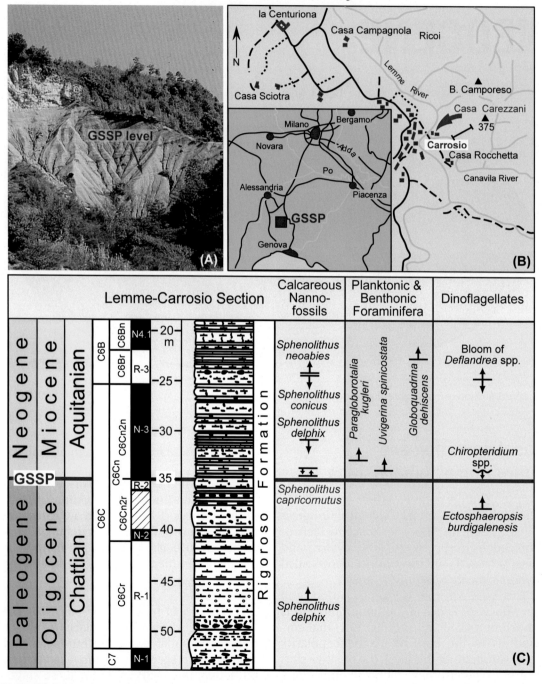

Figure 15.1 GSSP for base of the Neogene (base of Miocene Epoch, base of Aquitanian Stage) at Lemme–Carrosio, northern Italy. The GSSP level coincides with the lowest occurrence of calcareous nannofossil *Sphenolithus capricornutus* and the base of magnetic polarity Chron C6Cn.2n. From Hilgen et al. (2012).

successions in the Mediterranean region that exhibited cyclic alternations of clay and/or organic contents and were close to magnetic polarity reversals, oxygen-isotopic shifts, and/or microfossil datums (Hilgen et al., 2012; see Cenozoic summary Fig. 14.1). For example, the GSSP for the Pliocene was placed at the flooding of the Mediterranean following the Messinian salinity crisis, but its precise age was determined from the orbital–climate cycles and magnetic stratigraphy above that GSSP (Fig. 15.2). Those lithologic cycles could be correlated and tuned to Milankovitch insolation curves for precise age control.

As of 2015, the bases of the Burdigalian and Langhian stages of lower Miocene have not yet been formally defined, partly because the examined land exposures do not have adequate simultaneous preservation of reliable orbital–climate cycles, magnetic stratigraphy, and micropaleontology. Therefore, one alternative under discussion is to use the detailed records in ocean drilling cores spanning the potential boundary intervals.

Burdigalian (lower Miocene): The Aquitanian–Burdigalian boundary in GTS2004 and GTS2012 was provisionally placed at the lowest occurrence of calcareous nannofossil *Helicosphaera ampliaperta*, which was dated by astronomical cycles at Ceara Rise as 20.43 Ma (e.g., Hilgen et al., 2012) and is near the base of magnetic polarity Chron C6An.1r (20.44 Ma; used here). An older option for this boundary is a level coinciding with the last occurrence of planktonic foraminifer *Paragloborotalia kugleri* (base of zone M2; formerly called N4) at 21.12 Ma; and younger options are levels to coincide with either the base of Chron C6r (20.04 Ma) or the lowest occurrence of calcareous nannofossil *Sphenolithus belemnos* at 19.03 Ma. No suitable land outcrops for this boundary interval have been identified in the Mediterranean region; therefore, the working group is considering a definition that uses Ocean Drilling Program/Integrated Ocean Drilling Program (ODP/IODP) cores.

Langhian (base of middle Miocene): The working definition for the Burdigalian–Langhian boundary in GTS2012 is the base of magnetic polarity Chron C5Br (15.97 Ma), which is near the lowest occurrence of the planktonic foraminifer *Praeorbulina* genus. Potential GSSP sections are at La Vedova beach (northern Italy) and St. Peter's Pool (Malta).

Selected main stratigraphic scales and events

(1) Biostratigraphy (marine; terrestrial)

The extensive micropaleontology studies of Cenozoic deposits drilled from the ocean basins, continental shelves, and interior basins have enabled the compilation of very detailed intercalibrated datums and zones for calcareous, siliceous, and organic-walled microfossils that are correlated to orbital–climate cycles and to magnetic polarity chrons. The main reference scales for calcareous microfossils are summarized in Fig. 15.3, including the low- to midlatitude calcareous nannofossil biozonation by Backman et al. (2012). For terrestrial deposits, the evolution and migratory exchanges of mammals on the distributed continents are used for different regional scales.

(2) Magnetostratigraphy

Prior to the success of progressive high-resolution astronomical tuning for dating of marine deposits, the Neogene time scale was largely based on applying spreading-rate models to the C-sequence patterns of marine magnetic anomalies (e.g., Cande and Kent, 1992, 1995; Berggren et al., 1995). In the present Neogene scale, the ages and durations of most polarity chrons are adjusted according to these cyclostratigraphy calibrations (e.g., synthesis by Hilgen et al., 2012).

Base of the Zanclean Stage of the Pliocene Series at Eraclea Minoa, Italy

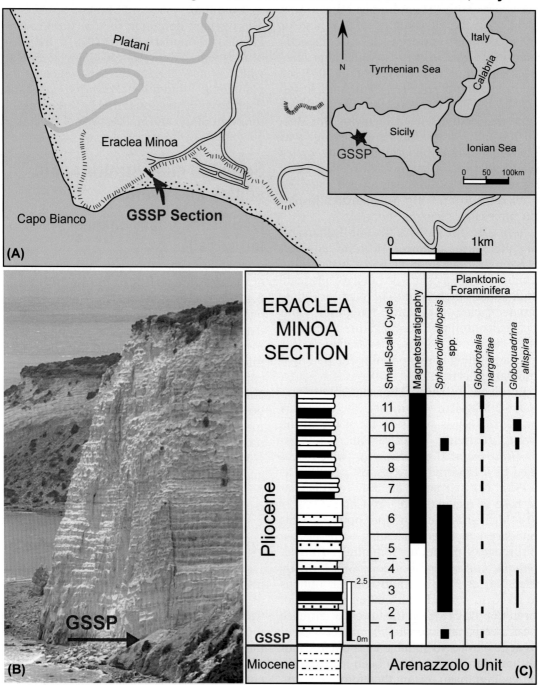

Figure 15.2 GSSP for base of the Pliocene (base of Zanclean Stage) at the Eraclea Minoa section in southwestern Sicily, Italy. The GSSP level at the base of the Trubi Marl Formation coincides with the flooding of the Mediterranean following the Messinian salinity crisis at the end of the Miocene and is positioned five precession-related cycles below the base of magnetic polarity Chron C3n.1n. From Hilgen et al. (2012).

(3) Stable-isotope stratigraphy and selected events

The past 34 myr of the Oligocene Epoch through Neogene and Quaternary periods were a relatively cold "ice-house" interval. The volume of the ice sheets at high latitudes responded to orbital climate cycles. It appears that the 1.2-myr modulation of the amplitude of obliquity was a major control on the medium-term oscillations of these ice sheets. Intervals of reduced obliquity amplitudes correspond to cooler climates ("Oi" and "Mi" oxygen-isotope events during the Oligocene and Miocene, respectively) and to low-stands in global sea level (e.g., Boulila et al., 2011). These "Oi" and "Mi" isotope events and simultaneous sequence boundaries on continental shelves were superimposed on the long-term trends toward a mid-Miocene warming followed by the pronounced global cooling from the Serravallian Stage through early Pliocene (Cenozoic summary in Fig. 14.1). Most stage boundaries occur near one of these events.

Numerical age model

A major international collaboration has enabled the development of an astronomical time scale for the variations of Earth's orbital parameters that affect climate and the correlation of variations in pelagic sedimentary successions to these orbital–climate cycles. Indeed, this astronomical tuning of the Neogene time scale has also improved the ages assigned to the monitor standard used for argon–argon ($^{40}Ar/^{39}Ar$) dating methods. The GSSP levels for Neogene stages had been selected to correspond to a particular orbital–climate cycle, therefore the uncertainty in their assigned ages is generally less than one-half–precession cycle, hence less than 10,000 years. An approximate correlation of marine Neogene events, isotopic excursions, and magnetic polarity chrons to the astronomical time scale has been accomplished for most of the major marine microfossil markers, and has enabled partial quantification of the diachroneity of taxa appearances and extinctions among ocean basins and climatic latitudes (Fig. 15.4) (e.g., Backman et al., 2012).

Revised ages compared to GTS2012; and potential future enhancements

Only a portion of the Early Miocene has not yet been fully tuned; and this presents an uncertainty on the estimated astronomical age for the base of the Miocene. The current age assignment for the Oligocene–Miocene boundary of 23.03 Ma was based mainly on interpretations of obliquity-dominated pelagic sediments from ODP Leg 154 on the Ceara Rise in the equatorial Atlantic (e.g., Shackleton et al., 2000; Pälike et al., 2006a,b), but these deposits did not preserve a magnetostratigraphy. Until the full astronomical tuning is accomplished in continuous overlapping sections to the base of the Miocene and into the upper Oligocene, including the polarity zones, there is the possibility of a small adjustment in the age of the Oligocene–Miocene boundary (Hilgen et al., 2012; p. 963). This will slightly impact the age model for the Early Miocene and Late Oligocene events.

The main frontiers in Neogene stratigraphy, other than fixing the two remaining Miocene stage GSSPs, include compiling more precise time scales for terrestrial events and deposits on the different continents and extending correlations to the Arctic and high southern-latitude oceans.

Neogene & Quaternary Time Scale

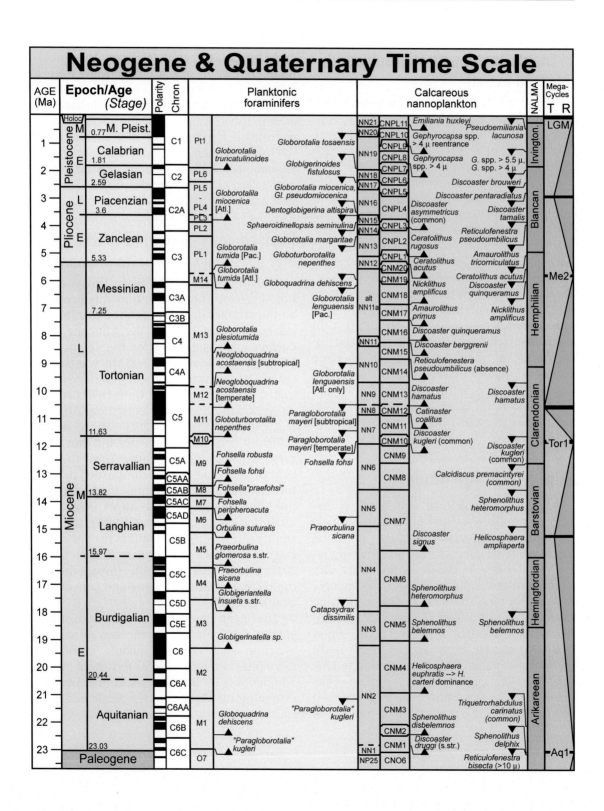

Figure 15.3 Selected marine and terrestrial biostratigraphic zonations of the Neogene and Quaternary. ("*Age*" is the term for the time equivalent of the rock-record "*stage*"; and subepoch groupings of these "ages" into "late," "early" are informal common usage.) Magnetic polarity zones are scaled to astronomical cycles (e.g., Hilgen et al., 2012). Planktonic foraminifer zones and main markers are from GTS2012 (Hilgen et al., 2012; Anthonissen and Ogg, 2012); but the late Pliocene details for PL4–PL5 between the Atlantic and Pacific basins have been omitted. Calcareous nannofossil "CN" zones and markers from Backman et al. (2012) are shown together with the commonly used "NN" zonation of Martini (1971). Land Mammal Ages of North America (NALMA) are from the compilation by Van Dam (in Hilgen et al., 2012). Major sea-level sequence boundaries and highstands are from Hardenbol et al. (1998). Additional zonations, biostratigraphic markers, geochemical trends, sea-level curves, and details on calibrations are compiled in Hilgen et al. (2012) and in the internal data sets within the *TimeScale Creator* visualization system (free at *www.tscreator.org*).

GSSPs of the Neogene Stages, with location and primary correlation criteria

Stage	GSSP Location	Latitude, Longitude	Boundary Level	Correlation Events	Reference
Piacenzian	Punta Piccola, Sicily, Italy	37°17'20"N 13°29'36"E*	base of the beige marl bed of carbonate cycle an age of 3.6 Ma	*precessional excursion 347 from the present with an astrochronological age estimate of 3.600 Ma*	Episodes **21/2**, 1998
Zanclean	Eraclea Minoa, Sicily, Italy	37°23'30"N 13°16'50"E	base of the Trubi Formation	Insolation cycle 510 counted from the present with an age of 5.33 Ma	Episodes **23/3**, 2000
Messinian	Oued Akrech, Morocco	33°56'13"N 6°48'45"W	base of reddish layer of sedimentary cycle number 15	Foraminifer, regular FAD of *Globorotalia miotumida* and calc. nannofossil FAD of *Amaurolithus delicatus*	Episodes **23/3**, 2000
Tortonian	Monte dei Corvi Beach, near Ancona, Italy	43°35'12"N 13°34'10"E	mid-point of sapropel layer of basic cycle number 76	Common LAD of calc nannofossil *Discoaster kugleri* and foraminifer *Globigerinoides subquadratus*	Episodes **28/1**, 2005
Serravallian	Ras il Pellegrin section, Fomm Ir-Rih Bay, west coast of Malta	35°54'50"N 14°20'10"E	formation boundary between the Globigerina Limestone and Blue Clay	younger end of Mi3b oxygen-isotopic event (global cooling episode)	Episodes **32/3**, 2009
Langhian	candidates are La Vedova (Italy) and St. Peter's Pool (Malta)			*Base of magnetic polarity chron C5Br*	
Burdigalian	Potentially in astronomically-tuned ODP core			*Near FAD of calc. nannofossil Helicosphaera ampliaperta*	
Aquitanian	Lemme-Carrioso Section, Allessandria Province, Italy	44°39'32"N 08°50'11"E	35m from the top of the section	Base of magnetic polarity chron C6Cn.2n; FAD of calc. nannofossil *Sphenolithus capricornutus*	Episodes **20/1**, 1997

* according to Google Earth

Figure 15.4 Ratified GSSPs and potential primary markers under consideration for defining the Neogene stages *(status as of early 2016)*. Details of each GSSP are available at http://www.stratigraphy.org, https://engineering.purdue.edu/Stratigraphy/gssp/, and in the *Episodes* publications.

Acknowledgments

The compilation of this Neogene summary relied extensively on discussions and the GTS2012 chapter on Neogene by Frits Hilgen and his colleagues.

Selected publications and websites

Cited publications

Only select publications were cited in this review with an emphasis on aspects of post-2011 updates. Pre-2011 literature is well summarized in the synthesis by Hilgen et al. (2012) and in some of the publications cited in the following.

Anthonissen, D.E., Ogg, J.G., compilers, 2012. Cenozoic and Cretaceous biochronology of planktonic foraminifera and calcareous nannofossils. In: Gradstein, F.M., Ogg, J.G., Schmitz, M., Ogg, G., (Coordinators), *The Geologic Time Scale 2012*. Elsevier Publ., pp. 1083–1127.

Backman, J., Raffi, I., Rio, D., Fornaciari, E., Pälike, H., 2012. Biozonation and biochronology of Miocene through Pleistocene calcareous nannofossils from low and middle latitudes. *Newsletters on Stratigraphy* **45**: 221–244.

Berggren, W.A., Kent, D.V., Swisher III, C.C., Aubry, M.-P., 1995. A revised Cenozoic geochronology and chronostratigraphy. In: Berggren, W.A., Kent, D.V., Aubry, M.-P., Hardenbol, J. (Eds.), *Geochronology, Time Scales, and Global Stratigraphic Correlation*, SEPM Special Publication 54, pp. 129–212.

Boulila, S., Galbrun, B., Miller, K.G., Pekar, S.F., Browning, J.V., Laskar, J., Wright, J.D., 2011. On the origin of Cenozoic and Mesozoic "third-order" eustatic sequences. *Earth-Science Reviews* **109**: 94–112.

Cande, S.C., Kent, D.V., 1992. A new geomagnetic polarity time scale for the Late Cretaceous and Cenozoic. *Journal of Geophysical Research* **97**: 13917–13951.

Cande, S.C., Kent, D.V., 1995. Revised calibration of the geomagnetic polarity timescale for the Late Cretaceous and Cenozoic. *Journal of Geophysical Research* **100**: 6093–6095.

Hardenbol, J., Thierry, J., Farley, M.B., Jacquin, Th., de Graciansky, P.-C., Vail, P.R., (with numerous contributors), 1998. Mesozoic and Cenozoic sequence chronostratigraphic framework of European basins. In: de Graciansky, P.-C., Hardenbol, J., Jacquin, Th., Vail, P.R. (Eds.), *Mesozoic-Cenozoic Sequence Stratigraphy of European Basins*, SEPM Special Publication 60, pp. 763–781.

Hilgen, F., Lourens, L.J., Van Dam, J.A., with contributions by Beu, A.G., Boyes, A.F., Cooper, R.A., Krigsman, W., Ogg, J.G., Piller, W.E., Wilson, D.S., 2012. The Neogene Period. In: Gradstein, F.M., Ogg, J.G., Schmitz, M., Ogg, G., (Coordinators), *The Geologic Time Scale 2012*, Elsevier Publ., pp. 923–978. [A comprehensive Neogene overview, especially extensive details and graphics on the astronomical tuning and GSSPs of the stages, and on the different marine microfossil and terrestrial mammal biostratigraphic scales.]

Martini, E., 1971. Standard Tertiary and Quaternary calcareous nannoplankton zonation. In: Farinacci, A. (Ed.), *Proceedings of the II Planktonic Conference, Roma, 1969*. Tecnoscienza, Rome, pp. 739–785.

Pälike, H., Norris, R.D., Herrle, J.O., Wilson, P.A., Coxall, H.K., Lear, C.H., Shackleton, N.J., Tripati, A.K., Wade, B.S., 2006a. The heartbeat of the Oligocene climate system. *Science* **314**: 1894–1898.

Pälike, H., Frazier, J., Zachos, J.C., 2006b. Extended orbitally forced palaeoclimatic records from the equatorial Atlantic Ceara Rise. *Quaternary Science Reviews* **25**: 3138–3149.

Scotese, C.R., 2014. *Atlas of Neogene Paleogeographic Maps (Mollweide Projection), Maps 1-7, PALEOMAP Atlas for ArcGIS, Volume 1, the Cenozoic*, PALEOMAP Project, Evanston, IL. https://www.academia.edu/11082185/Atlas_of_Neogene_Paleogeographic_Maps.

Shackleton, N.J., Hall, M.A., Raffi, I., Tauxe, L., Zachos, J.C., 2000. Astronomical calibration age for the Oligocene-Miocene boundary. *Geology* **28**: 447–450.

Websites (selected)

Subcommission on Neogene Stratigraphy (International Commission on Stratigraphy [ICS])—*http://www.sns.unipr.it/*—excellent details of GSSPs, including Portable Document Formats (PDFs) (although, as of Mar 2016, the site had not been enhanced since late 2013).

Palaeos: The Neogene—*http://palaeos.com/cenozoic/neogene.html*—A well-presented suite of diverse topics for a general science audience that was originally compiled by M. Alan Kazlev in 1998–2002. Includes the Pleistocene and Holocene epochs that now constitute the Quaternary Period.

QUATERNARY

16

21000 Years Quaternary

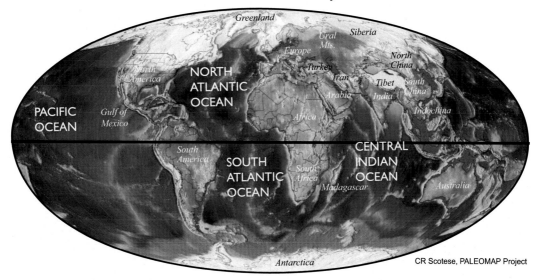

CR Scotese, PALEOMAP Project

Late Pleistocene (Last Glacial Maximum; ca. 21 ka) paleogeographic reconstruction (Sea level–120 m) from Scotese (2014).

Basal definition and international subdivisions

The Quaternary is characterized by glacial–interglacial cycles and by the evolution, migration, and increasing dominance of hominids. In 2009, the International Union of Geological Sciences (IUGS) ratified the Quaternary Period to begin with the first major episode of continental glaciation near the base of the Gelasian Stage, dated as ca. 2.6 Ma. Prior to this formalization of the Quaternary and required redefinition of its Pleistocene Series, the Gelasian had been the uppermost stage in the Pliocene Series.

The Pleistocene will have four stages, of which the lower two (Gelasian and Calabrian) are formally defined by Global Boundary Stratotype Sections and Points (GSSPs). The Holocene Epoch begins with the termination of the Younger Dryas cold episode at ca. 11.7 ka and has three potential stages (Walker et al., 2012). The *Anthropocene* has been proposed as an additional unit of epoch or stage status that is characterized by the major role of humans in geologic, climatic, and evolutionary processes on Earth.

The 2015 status of the Quaternary series and stages are compiled in a special volume of

Quaternary International (Head et al., 2015), especially the summary by Head and Gibbard (2015a). The evolution of this Quaternary time scale is detailed in Pillans and Gibbard (2012).

Pleistocene

Gelasian: The base of the Quaternary System and Pleistocene Series was assigned to the previously ratified GSSP of Gelasian Stage in Sicily that is just above "warm" Marine oxygen-Isotope Stage (MIS) 103 and is 1 m above the base of reversed-polarity Chron C2r (Gauss–Matuyama polarity boundary) (Rio et al., 1998; Gibbard et al., 2010). This level is slightly younger than the 2.6–2.7 Ma (MIS 104 and 110) onsets of the first major influx of ice-rafted debris into midlatitudes of the North Atlantic, deposition of glacial till into midcontinent North America, and the major glacial-caused global sea-level sequence boundary and lowstand "Ge1" (e.g., Haug et al., 2005; reviews in Head et al., 2008 and Ogg and Pillans, 2008). However, the Gelasian GSSP with its near coincidence with the onset of the Matuyama Chron (Chron C2r) enabled an unambiguous and precise global marker between continental and oceanic deposits. Therefore, the Quaternary System was defined by the ICS/IUGS at this established Gelasian GSSP (Figs. 16.1 and 16.6). This level has a recommended age from astronomical calibration of 2.588 Ma (Rio et al., 1998); but other estimates are 2.58 Ma based on its relative position within that precession cycle (Gibbard and Head, 2009) or 2.60 Ma based on revised $^{40}Ar/^{39}Ar$ dating of the underlying Gauss–Matuyama polarity boundary as 2.61 Ma (Singer, 2014).

Calabrian (2nd stage): The base of the Calabrian Stage at its GSSP at the Vrica section in Calabria of southern Italy is also at the top of a sapropel (MPRS 176) that corresponds to Marine Isotope Stage (MIS) 65/64 transition with an astronomical age of 1.80 Ma (Cita et al., 2012) (Fig. 16.2). This level is within the upper part of the normal-polarity *Olduvai* Chron C2n. This Calabrian Stage was redefined by IUGS in 2011 to be the second stage of the Pleistocene Series (Cita et al., 2012).

Third Stage of Pleistocene (unnamed; "*Ionian*" is a regional marine stage name used in the Mediterranean region, e.g., Cita et al., 2006, and "*Chiban*" is a potential name if the GSSP is in Japan): There is a major change in the oscillations of the Earth's climate and glacial systems between ca. 1.4 and 0.4 Ma as mirrored in the $\delta^{18}O$ values of benthic foraminifera (reviewed by Head and Gibbard, 2015b) (Fig. 16.2). The Early Pleistocene is characterized by frequent (41-kyr) and relatively low-amplitude climate cycles responding primarily to regular oscillations in Earth's tilt (obliquity). In contrast, the Middle and Late Pleistocene is characterized by quasi-100-kyr sawtooth climatic cycles. Within each of these 100-kyr cycles, the cooling and glaciation progressively increases to a maximum, then suddenly warms into a brief interglacial. Within this transition, the Calabrian–Middle Pleistocene boundary" will have the base of normal-polarity *Brunhes* Chron C1n as the primary marker for precise global correlation. The middle of this polarity transition has an age of 773 ka (Singer, 2014). Even though the main land–sea correlation criteria and its dating have been decided, it has been a challenge to select a physical GSSP location that has both preserved the polarity reversal and has other secondary biological and geochemical markers (reviewed in Head and Gibbard, 2015a). There are two candidate sections in Italy and one in Japan. The two candidates in southern Italy are well correlated with cycle stratigraphy. However, the Valle di Manche section (Capraro et al., 2015), which has magnetostratigraphy, spans only a narrow stratigraphic interval compared to the expanded Montalbano Jonico section (Marino et al., 2015), which does not preserve magnetostratigraphy. The Chiba section near Tokyo in

Base of the Gelasian Stage of the Pleistocene Series of the Quaternary System at Monte San Nicola, Italy

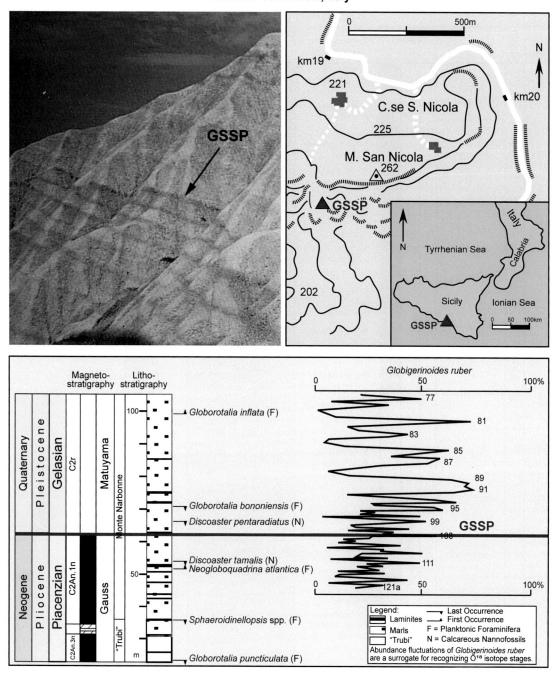

Figure 16.1 GSSP for base of the Quaternary (base of Pleistocene Series, base of Gelasian Stage) at Monte San Nicola, Sicily, Italy. The GSSP level at the base of the marly layer overlying organic-rich "Mediterranean Precession-Related Sapropel" (MPRS) 250 is just above the base of the Matuyama reversed-polarity magnetic chron (Chron C2r). (Graphics modified from Rio et al., 1998; and from Pillans and Gibbard, 2012.)

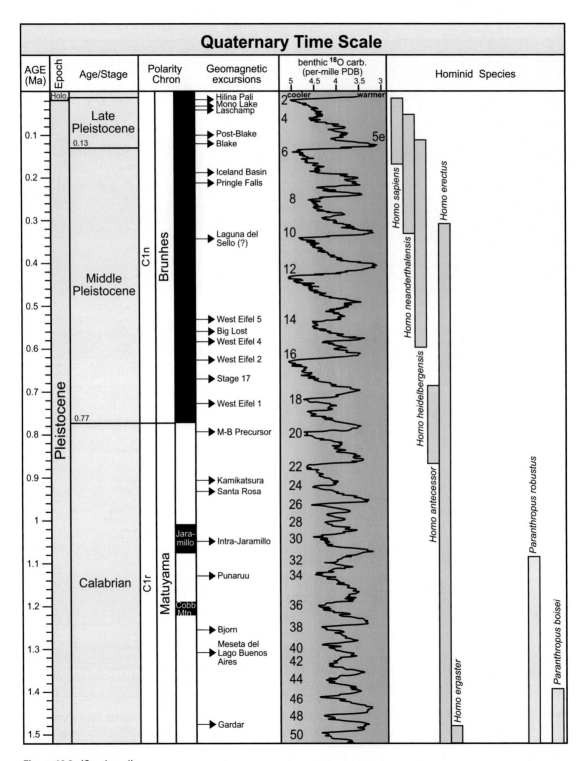

Figure 16.2 *(Continued)*

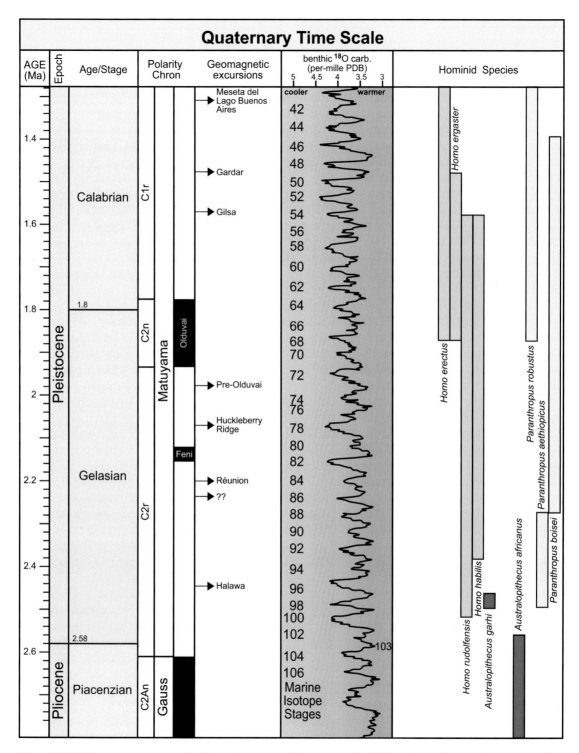

Figure 16.2 Pleistocene stages, marine isotope stages, and geomagnetic polarity chrons. Oxygen-isotope curve is the benthic foraminifer δ¹⁸O stack of Lisiecki and Raymo (2005) with selected labeling of Marine Isotope Stages (MIS). MIS with even numbers are cold intervals. Geomagnetic chrons and nomenclature of excursions are from Singer (2014). The approximate ranges of major hominid species are modified from Catt and Maslin (2012). See Fig. 15.3 in the Neogene chapter for selected marine microfossil zones and datums. *PDB*, PeeDee Belemnite ¹³C and ¹⁸O standard.

Japan (Kazaoka et al., 2015) does not appear to have either of these drawbacks.

Fourth Stage of Pleistocene (unnamed; "*Tarantian/Tarentian*" is the regional Italian marine stage, and "*Eemian*" is the regional nonmarine interglacial stage name used in northern Europe): The beginning of the warm Last Interglacial interval (marine oxygen-isotope substage 5e) at ca. 130 ka will be the placement of the "Middle/Upper" Pleistocene boundary (Figs. 16.2 and 16.3). The IUGS had rejected a proposed GSSP level in a drillcore documenting the beginning of the regional "Eemian" stage in pollen-rich freshwater facies near Amsterdam (Litt and Gibbard, 2008) because it was not in a marine outcrop and other concerns about correlations (summarized in Head and Gibbard, 2015a). The Fronte section of marine facies near Taranto (or Tarentum in ancient Roman) in southern Italy records the global deglaciation as the onset of a maximum flooding zone (Negri et al., 2015). Unfortunately, the facies of the transition interval does not preserve the oxygen-isotope record of the base of MIS 5e. Other possible GSSP placements for the onset of this interglacial could be within Antarctic ice cores that record the sharp rise in atmospheric methane or within speleothems from Asia (Head and Gibbard, 2015a). This fourth stage of the Pleistocene spans the Last Interglacial and the progressive cooling through the maximum of the Last Glacial (Fig. 16.3).

Holocene

Early: The termination of the Last Glacial was a multistep process, during which a warming trend was interrupted by a brief Younger Dryas cold episode, which was originally named after the anomalous abundance of pollen in Europe from the *Dryas octopetala* flower that grows in cold climates. The Younger Dryas ended in a very rapid warming, which is used as the definition for the base of the Holocene Epoch. The GSSP of the Holocene was assigned to the level in Greenland ice cores that records the beginning of this temperature rise as indicated by an abrupt decline in deuterium-excess values and other geochemical and physical parameters (Fig. 16.4). The banding from annual snow accumulation enables dating of the base of the Holocene as $11,700 \pm 99$ years before AD 2000; which is close to an uncalibrated ^{14}C age of 10,000 years BP. This GSSP placement within an ice core was ratified by IUGS in 2008.

Middle: Standardized subdivisions of the Holocene are important for geomorphology and geological hazard maps and for archeology and climatic history publications. The Holocene Epoch has three proposed stages, which are separated by the widely recorded major climatic excursions at 8.2 and 4.2 ka preserved in ice and lake cores, cave speleothems, and arid facies (Walker et al., 2012; Head and Gibbard, 2015a). The Lower–Middle Holocene stage boundary at the 8.2 ka cooling event is documented in Greenland ice cores by a pronounced excursion toward lighter $\delta^{18}O$ (Figs. 16.3 and 16.5); therefore, a GSSP might be placed in one of these cored intervals. The increased aridification associated with cooler North Atlantic waters at 8.2 ka may have triggered the spread of early farmers and the Mesolithic–Neolithic transition in the Near East and Mediterranean Europe (Walker et al., 2012).

Early: A widespread cultural disturbance at ca. 4.2 ka from southern Asia through North Africa appears to have been caused by a major arid episode. The collapsing civilizations include the pyramid-building Old Kingdom of Egypt, the Akkadian empire of Mesopotamia, and the Harappan culture of the Indus valley (Fig. 16.5). Aridity is recorded by geologic and cave speleothem deposits from North America to India (Walker et al., 2012). In contrast, it seems that Europe became wetter, and China experienced severe flooding. A GSSP might be potentially placed within the speleothem record in the Mawmluh Cave of northeastern

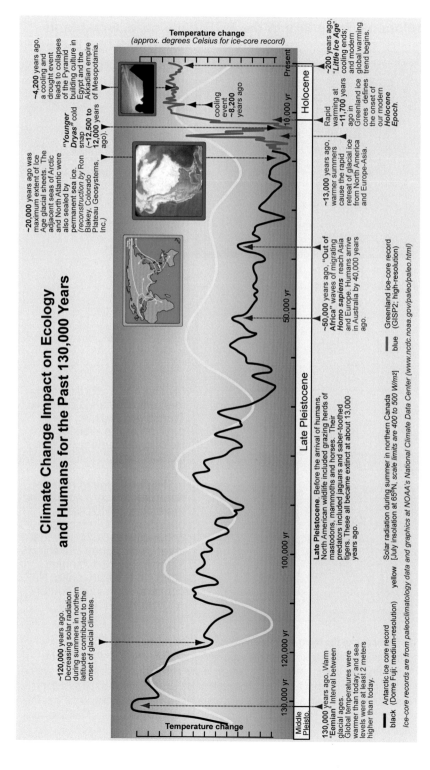

Figure 16.3 Late Pleistocene through Holocene ice core temperature trends and selected events. Temperature curves are smoothed degree-change trends from the interpretation of ice-core records as archived at NOAA's National Climate Data Center, rather than absolute temperatures. The high-resolution ice-core record from the past 20,000 years of Greenland has been adjusted to smoothly overlap and continue the medium-resolution record from Antarctica to the present. The Holocene Epoch begins at the rapid warming recorded in Greenland ice cores at 11,700 calendar years ago; the proposed Early–Middle and Middle–Late Holocene stage boundaries are cooling excursions recorded in those ice-core records at 8200 and 4200 years ago. Until the past ca. 4000 years, the averaged trends in global temperature also mirror most of the main oscillations in solar radiation received during summer months in the high latitudes of the North Hemisphere; but then the predicted cooling from decreasing summer radiation has been partly offset by rising atmospheric greenhouse gases.

Pleistocene - Holocene Boundary in the NGRIP ice core, Greenland

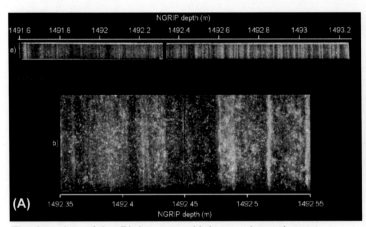

The location of the Pleistocene–Holocene boundary at 1492.45m is shown in the enlarged lower image.

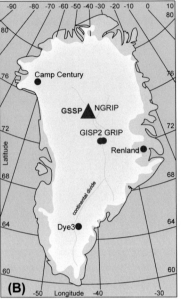

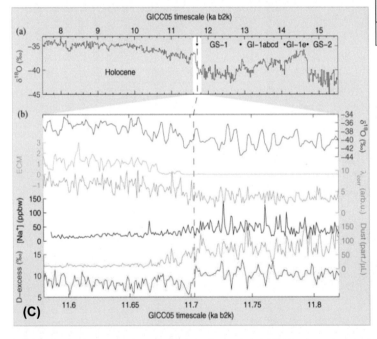

(a) The $\delta^{18}O$ record through the Last Glacial–Interglacial Transition, showing the position of the Pleistocene Holocene boundary in the NGRIP core.

(c) High-resolution multi-parameter record across the Pleistocene–Holocene boundary: $\delta^{18}O$, electrical conductivity (ECM), annual layer thicknesses corrected for flow-induced thinning (Icorr) in arbitrary units, Na^+ concentration, dust content, and deuterium excess

Figure 16.4 GSSP for base of the Holocene in the NorthGRIP ice core, central Greenland. The GSSP level at 1492.45 m depth corresponds to the end of the Younger Dryas cold spell. Visual stratigraphy of the NGRIP core between 1491.6 and 1493.25 m depth obtained using a digital line scanner. In this photograph, the image is "reversed" so that clear ice shows up black, and cloudy ice (which contains impurities such as micrometer-sized dust particles) shows up white. The visual stratigraphy is essentially a seasonal signal and reveals annual banding in the ice. The position of the Pleistocene–Holocene boundary at 1492.45 m is shown in the enlarged lower image. (Modified from Walker et al., 2009; and Pillans and Gibbard, 2012.)

India, where the broad 4.2 ka event consists of two abrupt excursions at 4.3 and 4.1 ka (Walker et al., 2012).

"Anthropocene"

Humans have become the most numerous large animal on Earth. They are now a major agent in modifying the land surface, in replacing natural ecosystems with extensive agriculture and other "invasive species," and in significantly altering atmospheric composition and climatic trends. The impacts of humans have accelerated since the Industrial Revolution of the 1800s. The informal term "*Anthropocene*" for this extensive influence on the Earth system was introduced by Crutzen and Stoermer (2000) and suggested as a geologic unit by Crutzen (2002). The stratigraphic and geological-mapping importance includes the direct humanmade deposits ranging from pyramids to massive urban reworking of the landscape and the indirect impacts on sediment production (e.g., Zalasiewicz et al., 2011, 2012), as well as erosion in the form of excavation for mining or exploitation of resources. Therefore, many stratigraphers have proposed to formalize the Anthropocene concept as a stratigraphic unit with epoch/series rank that is distinct from the Holocene Epoch (e.g., Walters et al., 2016).

The deposition of significant humanmade stratigraphic deposits and other impacts of humans is not synchronous across the globe. This diachroneity presents the main challenge to designating a standardized base to the Anthropocene concept in the geologic record (e.g., Zalasiewicz et al., 2015). The initial proposal assigned the beginning of the Anthropocene to the unnatural rise of atmospheric carbon-dioxide levels during the early Industrial Revolution (Crutzen, 2002); however, this is a gradual change that cannot be easily recognized or assigned within stratigraphic sections. The global marker of radioactive isotopes from the atomic bomb usage in 1945, especially the massive testing in the 1950s, is a distinctive event of anomalous carbon-14 ratios; but this AD1945 level occurs after some of the main human-produced stratigraphic deposits in much of the world.

The Working Group on Anthropocene (Subcommission on Quaternary Stratigraphy) is working to resolve several questions: Should the geologic and stratigraphic significance of human impacts be a distinct stratigraphic unit, or is such a unit unnecessary (e.g., critiques by Walker et al., 2015; Klein, 2015; and others)? Should the Anthropocene have the status of a geologic epoch/series or of a stage within the Holocene Epoch? Can it be given a standardized definition, or should it be an informal term applied to all human-influenced deposits? These and other questions are actively debated (e.g., volume edited by Waters et al., 2014). Regardless whether it is formalized, the concept of the Anthropocene has both fascinated and enlightened the general public about our massive impacts on our Earth system.

Selected main stratigraphic scales and events

(1) Regional terrestrial stages

The major climatic oscillations and glacial advances and retreats are recorded in stacked successions of glaciogenic sediments, including diamictites, interbedded soil, and windblown loess layers, and other Quaternary terrestrial and shallow coastal deposits. These have been correlated through magnetostratigraphy, stable and radiogenic isotopic stratigraphy, and other dating methods (reviewed with correlation charts in Pillans and Gibbard, 2012). The Subcommission on Quaternary Stratigraphy produce a series of detailed Quaternary charts showing the status of the correlation of these regional divisions and other stratigraphic events (*http://quaternary.stratigraphy.org/*).

Holocene Time Scale

AGE (ka)	Age/Stage	Greenland NGRIP 18O (per-mille PDB) -39 -38 -37 -36 -35 -34	Cultural Stages	Europe	Egypt	Mediterranean	Mesopotamia	China
	Late Holocene	cooler ... warmer / Ice Rafting Events	AD	Industrial Age		Independ. States		Republic
				Renaissance	Islamic	Ottoman	Islamic rule	Yuan, Ming, Qing Dynasties
1		1.4 ka		Medieval		Byzantine		Sui, Tang, Song Dynasties
2				Dark Ages	Byzantine Period		Sasanian rule	
				Roman	Roman Period	Roman	Parthian rule	Qin (Chin), Han Dynasties
				Celtic	Ptolemaic	Etruscan	Hellenistic rule	
		2.8 ka	Iron Age	Iron Age	Late Period		Persian rule	
3					Third intermediate Period	Early Iron Age	Assyrian / Babylonian period	Shang, Zhou Dynasties
				Bronze Age	New Kingdom	Late Bronze Age		
4	4.2	4.2 ka			Middle Kingdom	Mid. Bronze Age		
			Bronze Age	Beaker	Old Kindom (pyramids)		Akkadian	Longshan
5						Early Bronze Age	Sumerian	Late Neolithic
				Baden	Early Dynastic			
6	Middle Holocene	5.9 ka	Chalcolithic (Copper Age)	Chalcolithic (Copper Age)	Protodynastic (Naqadan Culture)		Protoliterate (Uruk)	Dawenkou
						Neolithic (Impressed Ware pottery)	Late Chalcolithic (Ubaid)	
7			Neolithic	Neolithic	Bashendi pastorism		Early Chalcolithic (Halafian)	Peilingang, Yangshao, Hongshan, Daxi, Majiabang
8	8.2	8.2 ka				Neolithic (pre-pottery)	Samarra/Hassuna	
			Mesolithic-Neolithic transition				Pre-Pottery Neolithic B (animal domestication)	Early Neolithic (Dapenkeng)
9		9.4 ka		Mesolithic	Masara	Mesolithic (Epipaleolithic)	Pre-Pottery Neolithic A (early farming)	
10	Early Holocene	10.3 ka	Mesolithic					Ordosian
11		11.1 ka					Natufian (cereal collecting)	
	11.78				Qadan			
12	Late Pleistocene	Younger Dryas	Late Paleolithic	Gravettian Culture			Epipaleolithic (Kebaran, etc.)	

◀ **Figure 16.5 Holocene. Proposed stages, North Hemisphere climatic changes recorded in Greenland ice cores (Seierstad et al., 2014), and selected intervals in regional human civilizations.** The Holocene Epoch begins at the rapid warming recorded in Greenland ice cores at 11,700 ± 99 years ago (relative to AD2000; or "b2k"); the proposed Early–Middle stage boundary is the cooling excursion recorded in those ice-core records at 8200 years ago, and Middle–Late boundary at 4200 years ago (Walker et al., 2012). Greenland NGRIP ice-core curve is the 20-years average $\delta^{18}O$ from Seierstad et al. (2014). [Note: A colder climate in Greenland is associated with a greater "snow-out" of moisture before the storms reach the center of its ice cap (from where the ice cap was drill cored). Oxygen-18 is progressively removed during snowfall; therefore, the remaining moisture reaching central Greenland is more depleted in Oxygen-18 (more negative in value).] Ice-rafting events in North Atlantic are from Bond et al. (1997). Human civilization intervals in Eurasia (other than China) are generalized and modified from compilations by Sherratt (1991), who advises that *"Archeologists use a variety of terms to label the periods and areas which have been important in the past, in a way which is often confusing – both to themselves and to the layman."* [These columns and additional composites of global archeology intervals with documentation are available as a *"Human Culture"* datapack for the TSCreator visualization system at *www.tscreator.org*.]

(2) Biostratigraphy (marine; terrestrial)

The extensive micropaleontology studies of Cenozoic deposits drilled from the ocean basins, continental shelves, and interior basins have enabled the compilation of very detailed intercalibrated datums and zones for calcareous, siliceous, and organic-walled microfossils that are correlated to orbital–climate cycles and magnetic polarity chrons. A selection of the main marine reference scales are summarized in Fig. 15.3 in the Neogene chapter. However, despite the major global climatic changes, the Quaternary has very few reliable microfossil datums for correlation of its stage boundaries or within those stages.

Terrestrial deposits record the evolution and migratory exchanges of mammals across the distributed continents through the Quaternary. Of these mammals, the development, inter-regional migrations, and impacts of different hominid species have received particular attention (e.g., review in Catt and Maslin, 2012; among many others) (Fig. 16.2). The earliest record of migrations of hominids from their main habitat in African savannas is the ca. 1.8 Ma occurrence of *Homo erectus* at the Dmanisi site in the Caucasus of Georgia followed by its arrival in southeastern Asia. At ca. 800 to 600 ka, the *Homo heidelbergensis* and its

relatives arrived in Eurasia and *Homo neanderthalensis* were their descendants. *Homo sapiens* had one or more migration episodes out of Africa at ca. 50 ka (Fig. 16.3), which profoundly impacted the ecosystems of all the continents.

(3) Magnetostratigraphy, stable-isotope stratigraphy, and selected events

The Quaternary magnetic polarity pattern has a dual system of nomenclature (Fig. 16.3). The "C" chron terminology is associated with the Cenozoic marine magnetic anomaly nomenclature; but the main polarity episodes are also named after important workers in magnetic science (e.g., Gilbert, Gauss, Matuyama, Brunhes) and the shorter polarity events after geographic locations (Olduvai, Jaramillo, etc.). Superimposed on the polarity chrons are many excursions of the magnetic field during which the intensity of the main dipole field either nearly vanished or the Earth's geomagnetic field briefly inverted (reviewed in Singer, 2014). When these geomagnetic excursions can be unambiguously identified, they potentially enable high-resolution global correlation horizons.

The main method for high-resolution correlations in oceanic and continental ice-cap strata is the changes of oxygen-isotope ratios in seawater or ice composition. Marine isotope stages (MIS) are based on oscillations in the $\delta^{18}O$ in carbonate shells of benthic foraminifera. Even-numbered peaks (elevated $\delta^{18}O$ values) are cold or glacial intervals (Fig. 16.2). Frequent brief warming events (Dansgaard–Oeschger events) are superimposed on the last glacial cycle according to Greenland ice cores; and some of these warming events immediately follow horizons of ice-rafted debris in the North Atlantic sediments (Heinrich Events) caused by anomalous release of icebergs. Ice-rafting events also occur within the warm Holocene interglacial interval in the North Atlantic, but these are interpreted as cooling excursions (Bond et al., 1997) (Fig. 16.5).

The eruption of the Toba volcano in Sumatra at 74 ka, the largest volcanic episode in the Quaternary, deposited thick volcanic ash throughout southeast Asia and caused a brief, but major, cooling of global climate with probably severe impacts on regional hominids (e.g., Storey et al., 2012). At ca. 40 ka, the eruption of the Campanian Ignimbrite centered at near Naples, Italy, the largest volcanic episode in Europe in the past 200 kyr, deposited ashfalls from northern Africa to the Russian Plain and may have been one factor in the decline of the *Homo neanderthalensis* population (e.g., Fitzsimmons et al., 2013; Black et al., 2015).

Numerical age model

The GSSP levels for Pleistocene stages were selected to correspond to a particular Milankovitch cycle; therefore, the uncertainty in their assigned ages according to astronomical calculations is generally less than one-half precession cycle or less than 10,000 years. The Holocene GSSP and its potential stages are calibrated to detailed paleoclimate records from ice cores of Greenland and from cored lake sediments with uncertainties less than 100 years.

Revised ages compared to GTS2012

GTS2012 had used the estimated ages for stage boundaries according to the values published in the original GSSP documents. The Subcommission on Quaternary Stratigraphy has fine-tuned some of the ages for these stage boundaries, and those revised values are used here to main consistency.

Quaternary (Gelasian) base (**2.58 Ma** vs. 2.59 Ma in GTS2012): The base of the Quaternary is at the base of the marly later overlying sapropel MPRS 250. "*astrochronological age of sapropel MPRS 250 (mid-point), corresponding to precessional cycle 250 from the present, is 2.588 Ma* (Lourens et al., 1996), *which can be assumed as the age of the boundary*" (Rio et al., 1998). However, Gibbard and Head (2009) note that the GSSP is a partial-precession-cycle above that sapropel, therefore recommend a rounded astronomical age of 2.58 Ma for the Quaternary GSSP. This 2.58 Ma is used on charts by ICS, and is adopted here to avoid confusion in scales. However, the Matuyama/Gauss reversal boundary, which is 1 m lower or ca. "20 kyr older" according to Rio et al. (1998) in the GSSP section, is independently calibrated to oxygen-isotope records in ocean drilling cores and bracketed by recalibrated high-precision radioisotopic dating in other sections that both indicate an age of 2.61 Ma (Singer, 2014). Until this discrepancy is resolved, the recommended age of 2.58 Ma of Head and Pillans (2015), which is used on the current Quaternary charts of the Subcommission and of INQUA, is adopted here for consistency.

Calabrian base (**1.80 Ma** vs. 1.806 Ma in GTS2012): The base of the **Calabrian** Stage at its GSSP at Vrica section in Calabria of southern Italy is also at the base of a claystone that overlies a sapropel (MPRS 176), that has an astronomical-calibrated age of 1.806 Ma at its midpoint (Cita et al., 2012).

GSSPs of the Quaternary Stages, with location and primary correlation criteria

Stage	GSSP Location	Latitude, Longitude	Boundary Level	Correlation Events	Reference
Anthropocene Series	*informal term awaiting possible formal definition*			*Major human impact on Earth's surface*	
Upper Holocene	*Mawmluh Cave, NE India*			4.2 ka climate event	
Middle Holocene	*NorthGRIP ice core, central Greenland*			8.2 ka climate event	
Lower Holocene	NorthGRIP ice core, central Greenland	75.10°N 42.32°W	1492.45m depth in Borehole NGRIP2	End of the Younger Dryas cold spell, which is reflected in a shift in deuterium excess values	Journal of Quaternary Science **24**/1, 2009
Upper Pleistocene	Taranto, Italy			*Base of warm marine isotope stage 5e, before final glacial episode of Pleistocene*	
Middle Pleistocene	*Chiba,, Japan; Montalbano Jonico, Valle di Manche, Italy*			*Brunhes-Matuyama magnetic reversal (base of Chron 1n)*	
Calabrian	Vrica, Italy	39°02'18.61"N 17°08'05.79"E	base of the marine claystone overlying the sapropelic marker Bed 'e'	Top of Oldovai Subchron is about 8 m above GSSP	Episodes **8**/2 1985 Episodes **35**/3, 2012
Gelasian	Monte San Nicola, Sicily, Italy	37°08'48.8"N 14°12'12.6"E*	base of marly layer overlying sapropel MPRS 250 with an age of 2.588 Ma	corresponds to Gauss/Matuyama magnetic epoch boundary; precessional cycle 250 from the present, Marine Isotope Stage 103, with an age of ~2.58 Ma	Episodes **21**/2, 1998 Episodes **33**/3, 2010

* according to Google Earth

Figure 16.6 Ratified GSSPs and potential primary markers under consideration for defining the Quaternary stages *(status as of early 2016).* (Details of each GSSP are available at http://www.stratigraphy.org, https://engineering. purdue.edu/Stratigraphy/gssp/, and in the Episodes publications.)

Therefore, after adjusting for the offset of the GSSP within that precession cycle, Cita et al. (2012) recommend a rounded age of 1.80 Ma, which is adopted here.

"Middle Pleistocene" base (**773 ka** vs. 781 ka in GTS2012): There is enhanced cycle-stratigraphy and $^{40}Ar/^{39}Ar$ dating of the Matuyama–Brunhes polarity reversal, which is the primary correlation marker for the stage boundary (e.g., Singer et al., 2014).

Acknowledgments

The compilation of this Quaternary summary relied extensively on discussions with Phil Gibbard, Martin Head, Brad Pillans, Brad Singer, and Jan Zalasiewicz (in alphabetical order), and upon their detailed publications. Phil Gibbard and Ann Jennison provided valuable recommendations and edits on an early draft.

Selected publications and websites

Cited publications

Only select publications were cited in this review with an emphasis on aspects of post-2011 updates. Pre-2011 literature is well summarized in the syntheses by Pillans et al. (2012) and by Zalasiewicz et al. (2012) and in some of the publications cited below.

Black, B.A., Neely, R.R., Manga, M., 2015. Campanian Ignimbrite volcanism, climate, and the final decline of the Neanderthals. *Geology* **43**: 411–415. http://dx.doi.org/10.1130/G36514.1.

Bond, G., Showers, W., Cheseby, M., Lotti, R., Almasi, P., deMenocal, P., Priore, P., Cullen, H., Hajdas, I., Bonani, G., 1997. A pervasive millennial-scale cycle in North Atlantic Holocene and glacial climates. *Science* **278**(5341): 1257–1266. http://dx.doi.org/10.1126/science.278.5341.1257.

Capraro, L., Macrì, P., Scarponi, D., Rio, D., 2015. The lower to Middle Pleistocene Valle di Manche section (Calabria, Southern Italy): state of the art and current advances: the Quaternary System and its formal subdivision. In: Head, M.J., Gibbard, P.L., van Kolfschoten, T. (Eds.), *Special Volume, Quaternary International*, **383**, pp. 36–46.

Catt, J.A., Maslin, M.A., 2012. The prehistoric human time scale. In: Gradstein, F.M., Ogg, J.G., Schmitz, M., Ogg, G., coordinators (Eds.), *The Geologic Time Scale 2012*. Elsevier Publication, pp. 1011–1032 (A synthesis of hominin evolution, migration episodes, and Pleistocene through early Holocene divisions based on tool technologies).

Cita, M.B., Capraro, L., Ciaranfi, N., Di Stefano, E., Marino, M., Rio, D., Sprovieri, R., Vai, G.B., 2006. Calabrian and Ionian: a proposal for the definition of Mediterranean stages for the Lower and Middle Pleistocene. *Episodes* **29**(2): 107–114.

Cita, M.B., Gibbard, P.L., Head, M.J., The ICS Subcommission on Quaternary Stratigraphy, 2012. Formal ratification of the base Calabrian Stage GSSP (second stage of the Pleistocene Series, Quaternary System). *Episodes* **35**(3): 388–397.

Crutzen, P.J., Stoermer, E.F., 2000. The Anthropocene. *Global Change Newsletter* **41**: 17–18.

Crutzen, P.J., 2002. Geology of mankind. *Nature* **415**: 23.

Fitzsimmons, K.E., Hambach, U., Veres, D., Iovita, R., 2013. The Campanian Ignimbrite eruption: new data on volcanic ash dispersal and its potential impact on human evolution. *PLoS ONE* **8**: e65839. http://dx.doi.org/10.1371/journal.pone.0065839 13 pp.

Gibbard, P.L., Head, M.J., 2009. IUGS ratification of the Quaternary System/Period and the Pleistocene Series/Epoch with a base at 2.58 Ma. *Quaternaire* **20**: 411–412.

Gibbard, P.L., Head, M.J., Walker, M., Alloway, B., Beu, A.G., Coltorti, M., Hall, V.M., Liu, J., Knudsen, K.-L., Van Kolfschoten, T., Litt, T., Marks, L., McManus, J., Partridge, T.C., Piotrowski, J.A., Pillans, B., Rousseau, D.-D., Suc, J.-P., Tesakov, A.S., Turner, C., Zazo, C., 2010. Formal ratification of the Quaternary System/Period and the Pleistocene Series/Epoch with a base at 2.588 Ma. *Journal of Quaternary Science* **25**: 96–102.

Haug, G.H., Ganopolski, A., Sigman, D.M., Rosell-Mele, A., Swann, G.E.A., Tiedemann, R., Jaccard, S.L., Bollmann, J., Maslin, M.A., Leng, M.J., Eglinton, G., 2005. North Pacific seasonality and the glaciation of North America 2.7 million years ago. *Nature* **33**: 821–825.

Head, M.J., Gibbard, P.L., Salvador, A., 2008. The Quaternary: its character and definition. *Episodes* **31**(2): 234–238.

Head, M.J., Gibbard, P.L., 2015a. Formal subdivision of the Quaternary System/Period: past, present, and future. In: Head, M.J., Gibbard, P.L., van Kolfschoten, T. (Eds.), *The Quaternary System and Its Formal Subdivision* Special Volume, Quaternary International, **389**, pp. 4–35. http://dx.doi.org/10.1016/j.quaint.2015.06.039 (Extensively illustrated summary of ratified and candidate GSSP divisions, their dating and the global stratigraphic relationships).

Head, M.J., Gibbard, P.L., 2015b. Early-Middle Pleistocene transitions: linking terrestrial and marine realms. *Quaternary International* **389**: 7–46. http://dx.doi.org/10.1016/j.quaint.2015.09.042.

Head, M.J., Gibbard, P.L., van Kolfschoten, T., 2015. The Quaternary System and its formal subdivision. Special volume. *Quaternary International* **383** 208 pp.

Kazaoka, O., Suganuma, Y., Okada, M., Kameo, K., Head, M.J., Yoshida, T., Sugaya, M., Kameyama, S., Ogitsu, I., Nirei, H., Aida, N., Kumai, H., 2015. Stratigraphy of the Kazusa Group, Boso Peninsula: an expanded and highly-resolved marine sedimentary record from the Lower and Middle Pleistocene of Central Japan: the Quaternary System and its formal subdivision. In: Head, M.J., Gibbard, P.L., van Kolfschoten, T. (Eds.), *Special Volume, Quaternary International*, **383**, pp. 116–135.

Klein, G.D., 2015. The "Anthropocene": what is its geological utility? (Answer: it has none!). *Episodes* **38**: 218.

Lisiecki, L.E., Raymo, M.E., 2005. A Pliocenee Pleistocene stack of 57 globally distributed benthic $\delta^{18}O$ records. *Paleoceanography* **20**: PA1003. http://dx.doi.org/10.1029/2004PA001071.

Litt, T., Gibbard, P., 2008. A proposed Global Stratotype Section and Point (GSSP) for the base of the Upper (Late) Pleistocene Subseries (Quaternary System/Period). *Episodes* **31**(2): 260–263.

Lourens, L.J., Antonarakou, A., Hilgen, F.J., Van Hoof, A.A.M., Vergnaud-Grazzini, C., Zachariasse, W.J., 1996. Evaluation of the Plio-Pleistocene astronomical timescale. *Paleoceanography* **11**(4): 391–413.

Marino, M., Bertini, A., Ciaranfi, N., Aiello, G., Barra, D., Gallicchio, S., Girone, A., La Perna, R., Lirer, F., Maiorano, P., Petrosino, P., Toti, F., 2015. Paleoenvironmental and climatostratigraphic insights for Marine isotope stage 19 (Pleistocene) at the Montalbano Jonico succession, South Italy. In: Head, M.J., Gibbard, P.L., van Kolfschoten, T. (Eds.), *The Quaternary System and Its Formal Subdivision*. Special Volume, Quaternary International, **383**, pp. 104–115.

Negri, A., Amorosi, A., Antonioli, F., Bertini, A., Florindo, F., Lurcock, P.C., Marabini, S., Mastronuzzi, G., Regattieri, E., Rossi, V., Scarponi, D., Taviani, M., Zanchetta, G., Vai, G.B., 2015. A potential global boundary stratotype section and point (GSSP) for the Tarentian Stage, Upper Pleistocene, from the Taranto area (Italy): results and future perspectives. In: Head, M.J., Gibbard, P.L., van Kolfschoten, T. (Eds.), *The Quaternary System and its Formal Subdivision*. Special Volume, Quaternary International, **383**, pp. 145–157.

Ogg, J.G., Pillans, B., 2008. Establishing the Quaternary as a formal international Period/System. *Episodes* **31**(2): 230–233.

Pillans, B., Gibbard, P.L., 2012. The Quaternary Period. In: Gradstein, F.M., Ogg, J.G., Schmitz, M., Ogg, G., (Coordinators), *The Geologic Time Scale 2012*. Elsevier Publ., pp. 980–1010 (A comprehensive Quaternary overview with details and graphics on the stages, terrestrial and marine stratigraphy, and dating methods).

Rio, D., Sprovieri, R., Castradori, D., Di Stefano, E., 1998. The Gelasian Stage (Upper Pliocene): a new unit of the global standard chronostratigraphic scale. *Episodes* **21**: 82–87.

Scotese, C.R., 2014. *Atlas of Neogene Paleogeographic Maps (Mollweide Projection), Maps 1-7, PALEOMAP Atlas for ArcGIS, Volume 1, the Cenozoic*. PALEOMAP Project, Evanston, IL. https://www.academia.edu/11082185/Atlas_of_Neogene_Paleogeographic_Maps.

Seierstad, I.K., Abbott, P.M., Bigler, M., Blunier, T., Bourne, A., Brook, E., Buchardt, S.L., Buizert, C., Clausen, H.B., Cook, E., Dahl-Jensen, D., Davies, S., Guillevic, M., Johnsen, S.J., Pedersen, D.S., Popp, T.J., Rasmussen, S.O., Severinghaus, J., Svensson, A., Vinther, B.M., 2014. Consistently dated records from the Greenland GRIP, GISP2 and NGRIP ice cores for the past 104 ka reveal regional millennial-scale $\delta^{18}O$ gradients with possible Heinrich Event imprint. *Quaternary Science Reviews* **106**: 29–46. http://dx.doi.org/10.1016/j.quascirev.2014.10.032. 4th INTIMATE special issue http://www.iceandclimate.nbi.ku.dk/data. Data files are also available from.

Sherratt, A., 1991. *Ancient Times: An Archeological Map and Timescale for Europe, Western Asia and Egypt*. Chart available through the Department of Antiquities, Asmolean Museum, University of Oxford.

Singer, B.S., 2014. A Quaternary geomagnetic instability time scale. *Quaternary Geochronology* **21**: 29–52. http://dx.doi.org/10.1016/j.quageo.2013.10.003.

Storey, M., Roberts, R.G., Saidin, M., 2012. Astronomically calibrated $^{40}Ar/^{39}Ar$ age for the Toba supereruption and global synchronization of late Quaternary records. *Proceedings of the National Academy of Sciences of the United States of America (PNAS)* **109**: 18684–18688.

Walker, M., Johnsen, S., Rasmussen, S.O., Steffensen, J.P., Popp, T., Gibbard, P., Hoek, W., Lowe, J., Bjorck, S., Cwynar, L., Hughen, K., Kershaw, P., Kromer, B., Litt, T., Lowe, D.J., Nakagawa, T., Newnham, R., Schwander, J., 2009. Formal definition and dating of the GSSP (Global Stratotype Section and Point) for the base of the Holocene using the Greenland NGRIP ice core and selected auxiliary records. *Journal of Quaternary Science* **24**: 3–17.

Walker, M.J.C., Berkelhammer, M., Bj€orck, S., Cwynar, L.C., Fisher, D.A., Long, A.J., Lowe, J.J., Newnham, R.M., Rasmussen, S.O., Weiss, H., 2012. Formal subdivision of the Holocene Series/Epoch: a discussion paper by a Working Group of INTIMATE (Integration of ice-core marine and terrestrial records) and the Subcommission on Quaternary Stratigraphy (International Commission on Stratigraphy). *Journal of Quaternary Science* **27**: 649–659.

Walker, M.J.C., Gibbard, P.L., Lowe, J., 2015. Comment on "When did the Anthropocene begin? A mid-twentieth century boundary is stratigraphically optimal" by Jan Zalasiewicz et al. (2015). In: Head, M.J., Gibbard, P.L., van Kolfschoten, T. (Eds.), *The Quaternary System and its Formal Subdivision*. Special Volume, Quaternary International, **383**, pp. 204–207. http://dx.doi.org/10.1016/j.quaint.2015.04.007.

Waters, C.N., Zalasiewicz, J.A., Williams, M., Ellis, M.A., Snelling, A.M., 2014. A stratigraphic basis for the Anthropocene. *Geological Society Special Publication No. 395* 321.

Waters, C.N., Zalasiewicz, J., Summerhayes, C., Barnosky, A.D., Poirer, C., Galuszka, A., Cearreta, A., Edgeworth, M., Ellis, E.C., Ellis, M., Jeandel, C., Leinfelder, R., McNeill, J.R., Richter, D.deB., Steffen, W., Syvitski, J., Vidas, D., Wagreich, M., Williams, M., Zhisheng, A., Grinevald, J., Odada, E., Oreskes, N., Wolfe, A.P., 2016. The Anthropocene is functionally distinct from the Holocene. *Science* **351**: 137. http://dx.doi.org/10.1126/science.aad2622 and on-line main text: aad2622 (10 pp).

Zalasiewicz, J., Williams, M., Fortey, R., Smith, A., Barry, T., Coe, A., Bown, P., Rawson, P., Gale, A., Gibbard, P., Gregory, F., Hounslow, M., Kerr, A., Pearson, P., Knox, R., Powell, J., Waters, C., Marshall, J., Oates, M., Stone, P., 2011. Stratigraphy of the Anthropocene. *Philosophical Transactions of the Royal Society A* **369**: 1036–1055.

Zalasiewicz, J., Crutzen, P.J., Steffen, W., 2012. The Anthropocene. In: Gradstein, F.M., Ogg, J.G., Schmitz, M., Ogg, G., (Coordinators), *The Geologic Time Scale 2012.* Elsevier Publication, pp. 1033–1040 (Summary of importance and of debates on formalization as a geologic unit).

Zalasiewicz, J., Waters, C.N., Williams, M., Barnosky, A.D., Cearreta, A., Crutzen, P., Ellis, E., Ellis, M.A., Fairchild, I.J., Grinevald, J., Haff, P.K., Hajdas, I., Leinfelder, R., McNeill, J., Odada, E.O., Poirer, C., Richter, D., Steffen, W., Summerhayes, C., Syvitski, J.P.M., Vidas, D., Wagerich, M., Wing, S.L., Wolfe, A.P., Zhisheng, A., Oreskes, N., 2015. When did the Anthropocene begin? A mid-twentieth century boundary level is stratigraphically optimal. In: Head, M.J., Gibbard, P.L., van Kolfschoten, T. (Eds.), *The Quaternary System and its Formal Subdivision.* Special Volume, Quaternary International, **383**, pp. 196–203. http://dx.doi.org/10.1016/j.quaint.2014.11.0450.

Websites (selected)

Subcommission on Quaternary Stratigraphy (ICS)—*http://quaternary.stratigraphy.org/*—GSSPs and divisions, detailed regional and correlation charts, and many PDFs.

Working group on the "Anthropocene" (under Subcommission on Quaternary Stratigraphy)—*http://quaternary.stratigraphy.org/workinggroups/anthropocene/*—newsletters, articles, etc.

International Union for Quaternary Research (INQUA)—*www.inqua.org*—Subcommissions with reports/newsletters on paleoclimate, marine and terrestrial processes, humans and biosphere, and stratigraphy. Hosts *Quaternary International* journal.

Smithsonian Human Origins Program—*http://humanorigins.si.edu*—Hominid evolution and migration, stratigraphy, online exhibits, videos, newsletters, and other information.

Palaeos: The Neogene—*http://palaeos.com/cenozoic/neogene.html*—A well-presented suite of diverse topics for a general science audience that was originally compiled by M. Alan Kazlev in 1998–2002. Includes the Pleistocene and Holocene epochs that now constitute the Quaternary Period.

About Archeology—*http://archaeology.about.com*—Thousands of articles covering anthropology, ancient civilizations, art history, climate change, and other aspects.

Heilbrunn Timeline of Art History (hosted at The Metropolitan Museum of Art)—*http://www.metmuseum.org/toah/*—Holocene cultural intervals for the entire world with compact essays illustrated by typical arts.

APPENDIX

CMYK Color Code according to the Commission for the Geological Map of the World (CGMW), Paris, France

The CMYK color code is an additive model with percentages of Cyan, Magenta, Yellow and Black. For example: the CMYK color for Devonian (20/40/75/0) is a mixture of 20% Cyan, 40% Magenta, 75% Yellow and 0% Black. The CMYK values are the primary reference system for designating the official colors for these geological units.

Precambrian (0/75/30/0)

Proterozoic (0/80/35/0)

Neoproterozoic (0/30/70/0)
- Ediacaran (0/15/55/0)
- Cryogenian (0/20/60/0)
- Tonian (0/25/65/0)

Mesoproterozoic (0/30/55/0)
- Stenian (0/15/35/0)
- Ectasian (0/20/40/0)
- Calymmian (0/25/45/0)

Paleoproterozoic (0/75/30/0)
- Statherian (0/55/10/0)
- Orosirian (0/60/15/0)
- Rhyacian (0/65/20/0)
- Siderian (0/70/25/0)

Archean (0/100/0/0)
- Neoarchean (0/40/5/0) — (0/35/5/0)
- Mesoarchean (0/60/5/0) — (0/50/5/0)
- Paleoarchean (0/75/0/0) — (0/60/0/0)
- Eoarchean (10/100/0/0) — (5/90/0/0)

Hadean (30/100/0/0)

Phanerozoic (40/0/5/0) — Paleozoic (40/10/40/0)

Cambrian (50/20/65/0)

Terreneuvian (45/15/55/0)
- Fortunian (45/15/50/0)
- Stage 2 (35/15/45/0)

Series 2 (40/10/50/0)
- Stage 3 (35/10/45/0)
- Stage 4 (30/10/40/0)

Series 3 (35/5/45/0)
- Stage 5 (30/5/40/0)
- Drumian (25/5/35/0)
- Guzhangian (20/5/30/0)

Furongian (30/0/40/0)
- Paibian (20/0/30/0)
- Jiangshanian (15/0/25/0)
- Stage 10 (10/0/20/0)

Ordovician (100/0/60/0)

Lower (90/0/60/0)
- Tremadocian (80/0/50/0)
- Floian (75/0/45/0)

Middle (70/0/50/0)
- Dapingian (60/0/40/0)
- Darriwilian (55/0/35/0)

Upper (50/0/40/0)
- Sandbian (45/0/40/0)
- Katian (40/0/35/0)
- Hirnantian (35/0/30/0)

Silurian (30/0/25/0)

Llandovery (40/0/25/0)
- Rhuddanian (30/0/20/0)
- Aeronian (30/0/20/0)
- Telychian (25/0/15/0)

Wenlock (30/0/20/0)
- Sheinwoodian (25/0/20/0)
- Homerian (20/0/15/0)

Ludlow (25/0/15/0)
- Gorstian (20/0/10/0)
- Ludfordian (15/0/10/0)

Pridoli (10/0/10/0)
- (10/0/10/0)

Devonian (20/40/75/0)

Lower (10/30/65/0)
- Lochkovian (10/25/60/0)
- Pragian (10/20/55/0)
- Emsian (10/15/50/0)

Middle (5/20/55/0)
- Eifelian (5/15/50/0)
- Givetian (5/10/45/0)

Upper (5/10/35/0)
- Frasnian (5/5/30/0)
- Famennian (5/5/25/0)

Phanerozoic (40/0/5/0) — Mesozoic (60/0/10/0) / Paleozoic (40/10/40/0)

Jurassic (80/0/5/0)

Upper (30/0/0/0)
- Tithonian (15/0/0/0)
- Kimmeridgian (20/0/0/0)
- Oxfordian (25/0/0/0)

Middle (50/0/5/0)
- Callovian (25/0/5/0)
- Bathonian (30/0/5/0)
- Bajocian (35/0/5/0)
- Aalenian (40/0/5/0)

Lower (75/5/0/0)
- Toarcian (40/5/0/0)
- Pliensbachian (50/5/0/0)
- Sinemurian (60/5/0/0)
- Hettangian (70/5/0/0)

Triassic (50/75/75/0)

Upper (25/40/0/0)
- Rhaetian (10/25/0/0)
- Norian (15/30/0/0)
- Carnian (20/35/0/0)

Middle (30/30/0/0)
- Ladinian (20/45/0/0)
- Anisian (25/35/0/0)

Lower (40/75/0/0)
- Olenekian (30/65/0/0)
- Induan (35/70/0/0)

Permian (5/75/75/0)

Lopingian (0/35/30/0)
- Changhsingian (0/25/20/0)
- Wuchiapingian (0/30/25/0)

Guadalupian (0/55/50/0)
- Capitanian (0/40/35/0)
- Wordian (0/45/40/0)
- Roadian (0/50/45/0)

Cisuralian (5/65/60/0)
- Kungurian (10/45/40/0)
- Artinskian (10/50/45/0)
- Sakmarian (10/55/50/0)
- Asselian (10/60/55/0)

Carboniferous (60/15/30/0)

Pennsylvanian (40/10/20/0)
- Upper (25/10/20/0): Gzhelian (20/10/15/0), Kasimovian (25/10/15/0)
- Middle (35/10/20/0): Moscovian (30/10/20/0), Bashkirian (40/10/20/0)

Mississippian (60/25/55/0)
- Lower (45/15/55/0): Serpukhovian (25/15/55/0), Visean (35/15/55/0), Tournaisian (45/15/55/0)

Phanerozoic (40/0/5/0) — Cenozoic (5/0/90/0) / Mesozoic (60/0/10/0)

Quaternary (0/5/50/0)

Holocene (0/5/10/0)
- (0/5/10/0)

Pleistocene (0/5/30/0)
- Upper (0/5/15/0)
- "Ionian" (0/5/20/0)
- Calabrian (0/5/25/0)
- Gelasian (0/5/35/0)

Neogene (0/10/90/0)

Pliocene (0/0/40/0)
- Piacenzian (0/0/25/0)
- Zanclean (0/0/30/0)

Miocene (0/0/100/0)
- Messinian (0/0/55/0)
- Tortonian (0/0/60/0)
- Serravallian (0/0/65/0)
- Langhian (0/0/70/0)
- Burdigalian (0/0/75/0)
- Aquitanian (0/0/80/0)

Paleogene (0/40/60/0)

Oligocene (0/25/45/0)
- Chattian (0/10/30/0)
- Rupelian (0/15/35/0)

Eocene (0/30/50/0)
- Priabonian (0/20/30/0)
- Bartonian (0/25/35/0)
- Lutetian (0/30/40/0)
- Ypresian (0/35/45/0)

Paleocene (0/35/55/0)
- Thanetian (0/25/50/0)
- Selandian (0/25/50/0)
- Danian (0/30/55/0)

Cretaceous (50/0/75/0)

Upper (35/0/75/0)
- Maastrichtian (5/0/45/0)
- Campanian (10/0/50/0)
- Santonian (15/0/55/0)
- Coniacian (20/0/60/0)
- Turonian (25/0/65/0)
- Cenomanian (30/0/70/0)

Lower (45/0/70/0)
- Albian (20/0/40/0)
- Aptian (25/0/45/0)
- Barremian (30/0/50/0)
- Hauterivian (35/0/55/0)
- Valanginian (40/0/60/0)
- Berriasian (45/0/65/0)

Color composition by J.M. Pellé (BRGM, France)
This chart was designed by Gabi Ogg

RGB Color Code according to the Commission for the Geological Map of the World (CGMW), Paris, France

The RGB color code is an additive model of Red, Green and Blue. Each is indicated on a scale from 0 (no pigment) to 255 (saturation of this pigment). "Devonian (203/140/205)" indicates a mixture of 203 Red, 140 Green and 205 Blue.

The conversion from the reference CMYK values to the RGB codes utilizes Adobe® Illustrator® CS3's color function of "Emulate Adobe® Illustrator® 6.0" (menu Edit / Color Settings / Settings).

ATTENTION: For color conversions using a program other than Adobe® Illustrator®, it is necessary to conserve the reference CMYK, even if the resulting RGB values are slightly different.

Color composition by J.M. Pellé (BRGM, France)
This chart was designed by Gabi Ogg

Precambrian 247/67/112

Proterozoic 247/53/99

- Neoproterozoic 254/179/66
  - Ediacaran 254/217/106
  - Cryogenian 254/204/92
  - Tonian 254/191/78
- Mesoproterozoic 253/180/98
  - Stenian 254/217/154
  - Ectasian 253/204/138
  - Calymmian 253/192/122
- Paleoproterozoic 247/67/112
  - Statherian 248/117/167
  - Orosirian 247/104/152
  - Rhyacian 247/91/137
  - Siderian 247/79/124

Archean 240/4/127

- Neoarchean 249/155/193 / 250/167/200
- Mesoarchean 247/104/169 / 248/129/181
- Paleoarchean 244/68/159 / 246/104/178
- Eoarchean 218/3/127 / 230/29/140

Hadean 174/2/126

Phanerozoic 154/217/221 — Paleozoic 153/192/141

Devonian 203/140/55

- Upper 241/225/157
  - Famennian 242/237/197
  - Frasnian 242/237/173
- Middle 241/200/104
  - Givetian 241/225/133
  - Eifelian 241/213/118
- Lower 229/172/77
  - Emsian 229/208/117
  - Pragian 229/196/104
  - Lochkovian 229/183/90

Silurian 179/225/182

- Pridoli 230/245/225
- Ludlow 191/230/207
  - Ludfordian 217/240/223
  - Gorstian 204/236/221
- Wenlock 179/225/194
  - Homerian 204/235/209
  - Sheinwoodian 191/230/195
- Llandovery 153/215/179
  - Telychian 191/230/207
  - Aeronian 179/225/194
  - Rhuddanian 166/220/181

Ordovician 0/146/112

- Upper 127/202/147
  - Hirnantian 166/219/171
  - Katian 153/214/159
  - Sandbian 153/214/159
- Middle 77/180/126
  - Darriwilian 116/198/156
  - Dapingian 102/192/146
- Lower 26/157/111
  - Floian 65/176/135
  - Tremadocian 51/169/126

Cambrian 127/160/86

- Furongian 179/224/149
  - Stage 10 230/245/201
  - Jiangshanian 217/240/187
  - Paibian 204/235/174
- Series 3 166/207/134
  - Guzhangian 204/223/170
  - Drumian 191/217/157
  - Stage 5 179/212/146
- Series 2 153/192/120
  - Stage 4 179/202/142
  - Stage 3 166/197/131
- Terreneuvian 140/176/108
  - Stage 2 166/186/128
  - Fortunian 153/181/117

Mesozoic 103/197/202 — Paleozoic 153/192/141

Jurassic 52/178/201

- Upper 179/227/238
  - Tithonian 217/241/247
  - Kimmeridgian 204/236/244
  - Oxfordian 191/231/241
- Middle 128/207/216
  - Callovian 191/231/229
  - Bathonian 179/226/227
  - Bajocian 166/221/224
  - Aalenian 154/217/221
- Lower 66/174/208
  - Toarcian 153/206/227
  - Pliensbachian 128/197/221
  - Sinemurian 103/188/216
  - Hettangian 78/179/211

Triassic 129/43/146

- Upper 189/140/195
  - Rhaetian 227/185/219
  - Norian 214/170/211
  - Carnian 201/155/203
- Middle 177/104/177
  - Ladinian 201/131/191
  - Anisian 188/117/183
- Lower 152/57/153
  - Olenekian 176/81/165
  - Induan 164/70/159

Permian 240/64/40

- Lopingian 251/167/148
  - Changhsingian 252/192/178
  - Wuchiapingian 252/180/162
- Guadalupian 251/116/92
  - Capitanian 251/154/133
  - Wordian 251/141/118
  - Roadian 251/128/105
- Cisuralian 239/88/69
  - Kungurian 227/135/118
  - Artinskian 227/123/104
  - Sakmarian 227/111/92
  - Asselian 227/99/80

Carboniferous 103/165/153

- Pennsylvanian 153/194/181
  - Upper 191/208/186
    - Gzhelian 204/212/199
    - Kasimovian 191/208/197
  - Middle 166/199/183
    - Moscovian 179/203/185
  - Lower 140/190/180
    - Bashkirian 153/194/181
- Mississippian 103/143/102
  - Upper 179/190/108
    - Serpukhovian 191/194/107
  - Middle 153/168/108
  - Lower 128/171/108
    - Visean 166/185/108
    - Tournaisian 140/176/108

Cenozoic 242/249/29 — Mesozoic 103/197/202

Quaternary 249/249/127

- Holocene 254/242/224
- Pleistocene 255/242/174
  - "Ionian" 255/242/199
  - Calabrian 255/242/186
  - Gelasian 255/237/179

Neogene 255/230/25

- Pliocene 255/255/153
  - Piacenzian 255/255/191
  - Zanclean 255/255/179
- Miocene 255/255/0
  - Messinian 255/255/115
  - Tortonian 255/255/102
  - Serravallian 255/255/89
  - Langhian 255/255/77
  - Burdigalian 255/255/65
  - Aquitanian 255/255/51

Paleogene 253/154/82

- Oligocene 253/192/122
  - Chattian 254/230/170
  - Rupelian 254/217/154
- Eocene 253/180/108
  - Priabonian 253/205/161
  - Bartonian 253/192/145
  - Lutetian 252/180/130
  - Ypresian 252/167/115
- Paleocene 253/167/95
  - Thanetian 253/191/111
  - Selandian 254/191/101
  - Danian 253/180/98

Cretaceous 127/198/78

- Upper 166/216/74
  - Maastrichtian 242/250/140
  - Campanian 230/244/127
  - Santonian 217/239/116
  - Coniacian 204/233/104
  - Turonian 191/227/93
  - Cenomanian 179/222/83
- Lower 140/205/87
  - Albian 204/234/151
  - Aptian 191/228/138
  - Barremian 179/223/127
  - Hauterivian 166/217/117
  - Valanginian 153/211/106
  - Berriasian 140/205/96

Phanerozoic 154/217/221

AUTHOR BIOGRAPHIES

James G. Ogg (Professor at Purdue University, Indiana, United States, and visiting professor at China University of Geosciences, Wuhan, China) served as Secretary General of the International Commission on Stratigraphy (ICS) (2000–2008) and coordinated the ICS stratigraphy information service (2008–2012). His Mesozoic Stratigraphy Lab group works on aspects of climate cycles, magnetic polarity correlations, and integration of stratigraphic information. Their *TimeScale Creator* array of visualization tools for extensive databases in global and regional Earth history (www.tscreator.org) was used to generate many of the diagrams in this book.

Gabi M. Ogg applied micropaleontology to Jurassic–Cretaceous correlations before concentrating on public outreach in geosciences. In addition to co-authoring the *Concise Geologic TimeScale* (GTS2008) book, she was a coordinator of GTS2012 and produced most of its extensive array of graphics. She is the webmaster for the Geologic TimeScale Foundation (http://stratigraphy.science.purdue.edu) and for the *TimeScale Creator* visualization and database suites (www.tscreator.org), and has produced numerous posters and time scale cards for public audiences.

Felix M. Gradstein (Professor at Geology Museum, Oslo University, Norway, and visiting professor at University of Rio Grande do Sul, Sao Leopoldo, Brazil, the University of Nebraska, United States, and the University of Portsmouth, United Kingdom) was Chair of the International Commission on Stratigraphy from 2000 to 2008. Under his tenure, major progress was made with the formal definition of chronostratigraphic units from Precambrian through Quaternary. For his fundamental work with regard to the geologic timescale and stratigraphy, micropaleontology, and geochronology in general, the European Geosciences Union awarded him in 2010 the Jean Baptiste Lamarck Medal. He teaches applied biostratigraphy and paleoenvironment courses and is the current Chair of the Geologic TimeScale Foundation (http://stratigraphy.science.purdue.edu).

Felix Gradstein (left) and James Ogg (right) on Isle of Portland, Dorset, England.

INDEX

'*Note*: Page numbers followed by "f" indicate figures and "t" indicate tables.'

ADDRESSES/INSTITUTIONS

James G. Ogg

Department of Earth, Atmospheric and Planetary Sciences, Purdue University, 550 Stadium Mall Drive, West Lafayette, Indiana 47907-2051, USA. E-mail: jogg@purdue.edu
and
State Key Laboratory of Biogeology and Environmental Geology, School of Earth Sciences, China University of Geosciences, Wuhan 430074, China

Gabi M. Ogg

Geologic TimeScale Foundation, 1224 North Salisbury St., West Lafayette, Indiana 47906, USA. E-mail: gabiogg@hotmail.com

Felix M. Gradstein

Geology Museum, University of Oslo, N-0318 Oslo, Norway. E-mail: felix.gradstein@gmail.com
and
ITT Fossil, Unisinos, University of Rio Grande do Sul, Sao Leopoldo, Brazil

Printed in the United States
By Bookmasters